# Oil and Gas Engineering

# Oil and Gas Engineering

Edited by **Jane Urry**

SYRAWOOD
PUBLISHING HOUSE

New York

Published by Syrawood Publishing House,
750 Third Avenue, 9th Floor,
New York, NY 10017, USA
www.syrawoodpublishinghouse.com

**Oil and Gas Engineering**
Edited by Jane Urry

International Standard Book Number: 978-1-68286-102-8 (Hardback)

Printed in the United States of America.

# Contents

# Preface

Oil and gas engineering pertains to the effective extraction and management of oil and gas resources. This book covers individual branches associated with oil and gas engineering and explains their need and contribution in the context of a growing economy. Included herein are topics like techniques of oil refining, natural gas utilization, petrochemicals, etc. As this field is emerging at a rapid pace, the contents of this text will help the readers understand the modern concepts and applications of the subject. This book on oil and gas engineering is a collective contribution of a renowned group of international experts.

This book unites the global concepts and researches in an organized manner for a comprehensive understanding of the subject. It is a ripe text for all researchers, students, scientists or anyone else who is interested in acquiring a better knowledge of this dynamic field.

I extend my sincere thanks to the contributors for such eloquent research chapters. Finally, I thank my family for being a source of support and help.

**Editor**

# Piperazine/N-methylpiperazine/ N,N'-dimethylpiperazine as an Aqueous Solvent for Carbon Dioxide Capture

Stephanie A. Freeman, Xi Chen, Thu Nguyen, Humera Rafique, Qing Xu and Gary T. Rochelle*

*Department of Chemical Engineering, The University of Texas at Austin, 1 University Station C0400, Austin, TX 78712 - USA*
*e-mail: sfreeman@che.utexas.edu - xi.a.chen@power.alstom.com - tnguyen02_2000@yahoo.com - hrafique@bechtel.com*
*qing.xu06@gmail.com - gtr@che.utexas.edu*

* Corresponding author

**Résumé** — **Mélange pipérazine/N-méthylpipérazine/N,N'-diméthylpipérazine en solution aqueuse pour le captage du $CO_2$** — Un mélange pipérazine (PZ), N-méthylpipérazine (MPZ) et N,N'-diméthylpipérazine (DMPZ) en solution aqueuse est décrit pour le captage du $CO_2$ par absorption-régénération. Par rapport à la pipérazine concentrée, ce mélange permet un gain sur la solubilité des produits solides et sur la chaleur d'absorption. Aucune insolubilité n'a été observée à fort taux de charge en $CO_2$ contrairement aux solvants à base de pipérazine. Ce mélange a montré des performances équivalentes à la pipérazine concentrée en termes de capacité et de cinétique d'absorption du $CO_2$, qui sont plus que doublées par rapport à la traditionnelle MonoEthanolAmine (MEA) à 7 M (30 % massique). Un équilibre s'est établi entre les trois constituants qui augmente la stabilité thermique en comparaison des mélanges traditionnels de solvants. Le principal inconvénient de ce nouveau système est une volatilité plus importante des amines dans les conditions de l'absorbeur en comparaison avec le cas de la pipérazine concentrée et de la MEA.

*Abstract* — *Piperazine/N-methylpiperazine/N,N'-dimethylpiperazine as an Aqueous Solvent for Carbon Dioxide Capture* — *A blend of piperazine (PZ), N-methylpiperazine (MPZ) and N,N'-dimethylpiperazine (DMPZ) is described as a novel $CO_2$ capture solvent for aqueous absorption-stripping. This blend provides improved solid solubility and heat of absorption compared to concentrated PZ. No insolubility was observed for regions of high $CO_2$ loading, unlike PZ solvents. The blend performed like concentrated PZ in terms of $CO_2$ capacity and $CO_2$ absorption rate, both of which were more than double that of a traditional 7 molal (30 wt%) MonoEthanolAmine (MEA). Thermal equilibrium was established between the three constituent amines that increases the thermal stability compared to traditional blended solvents. The primary drawback of this novel solvent system is enhanced amine volatility at absorber conditions compared with both concentrated PZ and MEA.*

## INTRODUCTION

Amine-based absorption-stripping is the most promising solution for post-combustion carbon dioxide ($CO_2$) capture from coal-fired power plant flue gas. Advanced amine solvents such as concentrated, aqueous piperazine (PZ) are being proposed to replace traditional alkanol-amines such as MonoEthanolAmine (MEA) due to advantageous $CO_2$ absorption rates and overall energy performance [1, 2]. A new solvent concept proposed here is the use of amines that demonstrate noticeable levels of overall thermal degradation but degrade to useful products that can react with $CO_2$ and maintain alkalinity of the solution. One example of this solvent concept presented here is the use of a blend of PZ, N-methylpipera-zine (MPZ), and N,N'-dimethylpiperazine (DMPZ) [3].

Thermal degradation and oxidation are the two primary routes of amine loss during absorption-stripping along with volatilization of the amine in the absorber. Thermal degradation and oxidation of MEA and other alkanolamines have been investigated extensively previously and the impact of degradation was determined to be an important factor in solvent selection [4-12]. The thermal degradation of MEA generated products that have none or only limited $CO_2$ reactivity, effectively reducing the $CO_2$ capacity of the solvent. Other classes of amines have been found to generate $CO_2$-carrying molecules as products that mitigate the overall effect of thermal degradation. For example, concentrated PZ solutions has been found to produce high concentrations of N-formylpiperazine (FPZ), N-(2-aminoethyl)-piperazine (AEP) and ethylenediamine (EDA) during thermal degradation which can react with $CO_2$ directly (AEP and EDA), or be easily reacted to create PZ (FPZ and other amides) [13]. The same has been seen for N-methyldiethanolamine (MDEA)/PZ which degrades to diethanolamine (DEA) and substituted PZ molecules (1-methylpiperazine, N-(2-hydroxyethyl) piperazine and others) which can react and carry $CO_2$ during absorption-stripping [14, 15].

The blend of PZ, MPZ and DMPZ offers a solvent that takes advantage of the mixture of multiple amines with $CO_2$-carrying capacity that can interconvert to become degradation products of each other. The concept is to develop the solvent with a composition stable to thermal degradation that maximizes $CO_2$ capacity and absorption rate. In developing this solvent, the pertinent physical and thermodynamic properties of the blend that affect its expected performance in a $CO_2$ capture unit were investigated. Solid solubility, density and viscosity were evaluated over the expected range of $CO_2$ loading. $CO_2$ solubility, $CO_2$ capacity, heat of absorption, amine volatility and $CO_2$ absorption rate were studied over the

pertinent range of lean and rich conditions expected. Finally, the thermal stability and thermal equilibrium of the amine blend were evaluated to determine the degradation potential of the solvent. In each case, the properties of the blend are compared to 8 molal (m) PZ, 7 m MEA, or both solvents to determine the advantages and disadvantages of the solvent compared to current baseline solvents.

## 1 MATERIALS AND METHODS

### 1.1 Solution Preparation

Aqueous amine solutions were prepared as described previously with few modifications [13, 16, 17]. Anhydrous PZ (IUPAC: 1,4-diazacyclohexane, CAS 110-85-0, purity 99%, *Acros Organics N.V.*, Geel, Belgium), MPZ (IUPAC: 1-methyl-piperazine, CAS 109-01-3, purity > 99%, *Acros Organics*), and DMPZ (IUPAC: 1,4-dimethyl-piperazine, CAS 106-58-1, purity 98.5%, *Acros Organics*) were obtained commercially and used without further purification. Blends were prepared gravimetrically by combining all three amines with water at the desired concentrations and then heating to melt the PZ. Then, the hot, aqueous solution was transferred to a gas washing bottle where $CO_2$ (99.5%, *Matheson* Tri Gas, Basking Ridge, NJ, USA) was gravimetrically sparged to achieve the desired $CO_2$ concentration [16, 17]. Total alkalinity was verified through acid titration and $CO_2$ concentration was verified through Total Inorganic Carbon (TIC) analysis as described previously [13, 18]. $CO_2$ loading, $\alpha$, is reported as mol $CO_2$/mol alkalinity, which corresponds to mol $CO_2$/equivalent of amine nitrogen.

### 1.2 Solid-Liquid Equilibrium Measurements

Solid-liquid equilibrium of blended amine solutions was measured through observation of the conditions under which precipitation occurred. For the amine blend of interest, solutions containing a range of $CO_2$ loading were prepared by combining unloaded and a maximum rich loaded solution. The maximum rich loaded condition was the highest $CO_2$ loading found to remain soluble at controlled temperature at or below room temperature. The transition temperature between solid and liquid phases was observed by slowly heating a solution at a given $CO_2$ loading containing precipitation and observing the phase change, as previously described [16]. All solutions were observed for any precipitation at temperatures between 0°C and the melting point of the solid phase of the amine solution.

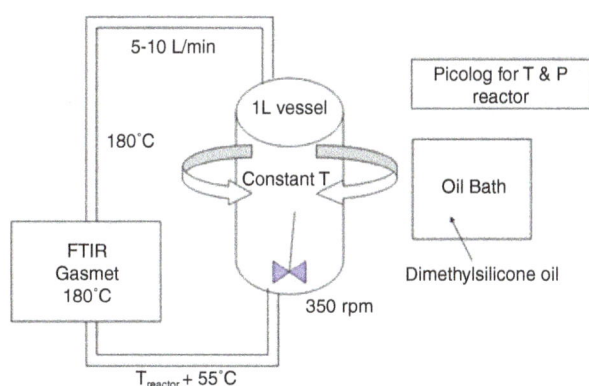

Figure 1

Schematic of equilibrium cell for amine volatility measurement.

## 1.3 Equilibrium Cell for Volatility

Amine volatility was measured in a stirred 1 L equilibrium cell coupled with a hot gas DX-4000 Fourier Transform Infrared Spectroscopy (FTIR) analyzer (*Gasmet Technologies Inc.*, La Prairie, Canada) as shown in Figure 1 and described previously [19]. The insulated glass reactor was agitated at 350 rpm using an overhead stirrer and temperature in the reactor was controlled by recirculating dimethylsilicone oil. Vapor from the headspace of the reactor was pumped at 5 to 10 L/minute through a heated *Teflon®* line to the FTIR. Both the sample line and analyzer were maintained at 180°C to prevent liquid condensation or adsorption of amine. The FTIR quantified amine, CO₂, and water concentration in the vapor phase. After the gas was analyzed in the FTIR, it was returned to the reactor through a second heated line maintained 55°C hotter than the reactor temperature. The 55°C difference was sufficient to ensure that the returned gas did not upset the liquid-vapor equilibrium inside the reactor and to prevent potential heat loss at the bottom of the reactor.

## 1.4 High Temperature VLE Apparatus for $CO_2$ Solubility

A high temperature vapor-liquid equilibrium apparatus was used to measure $CO_2$ solubility in aqueous amine solutions. A 500 mL stainless steel autoclave with an agitator and a heating jacket was used as the equilibrium cell. The method details have been described previously [20]. Total pressure of the $CO_2$ loaded blend at 100-160°C was measured with ± 2.4 kPa accuracy. $CO_2$ partial pressure was derived by subtracting partial

pressure of $N_2$ and water from the equilibrium total pressure. Data for the partial pressure of water were taken from the DIPPR database and was assumed to follow Raoult's Law [21]. The partial pressure of nitrogen at the experimental condition was calculated using a low temperature pressure measurement and the assumption that nitrogen behaves as an ideal gas. Liquid samples were collected before and after each experiment at room temperature, analyzed for amine concentration by acid titration and total $CO_2$ by TIC analysis, and corrected for the gas phase $CO_2$ at each condition to get $CO_2$ loading at high temperature.

## 1.5 Wetted Wall Column for Absorption Rates and $CO_2$ Solubility

The wetted wall column used in this study to characterize the mass transfer coefficient of amine solution ($k'_g$) is the same apparatus that had been described previously by Cullinane *et al.* and Dugas *et al.* [22, 23]. In a typical experiment, a gas mixture of nitrogen and $CO_2$ counter currently contacted the amine solution under study. The flux of $CO_2$ ($N_{CO_2}$) between gas and liquid phases for any given condition was calculated from the difference between inlet and outlet $CO_2$ concentrations quantified using an IR analyzer. By varying the inlet $CO_2$ partial pressure, both absorption and desorption were observed. The equilibrium $CO_2$ partial pressure of the amine solution, $P^*_{CO_2,l}$, was calculated using absorption and desorption data to estimate the pressure where zero flux occurs.

The overall mass transfer coefficient ($K_G$) was calculated as the ratio of the $N_{CO_2}$ and the gas phase partial pressure driving force between the inlet $CO_2$ partial pressure, $P_{CO_2,g}$ and $P^*_{CO_2,l}$, as shown in Equation (1) :

$$K_G = \frac{N_{CO_2}}{P_{CO_2,\,g} - P^*_{CO_2,\,l}} \tag{1}$$

The series resistance correlation for gas absorption into liquid is shown in Equation (2). The liquid film mass transfer coefficient based on a partial pressure driving force, $k_g'$, correlates with the experimental conditions and geometry of the apparatus as established above (2):

$$\frac{1}{k'_g} = \frac{1}{K_G} - \frac{1}{k_g} \tag{2}$$

The values for $k_g$ were estimated from a previous correlation [1]. Data for the kinetic absorption rate and $CO_2$ solubility were obtained simultaneously with this technique.

## 1.6 Thermal Cylinders for Thermal Degradation

Thermal degradation of amine solutions was assessed using thermal cylinders as described previously [4, 13, 18]. Cylinders consisted of (1/2)-inch OD 316 stainless steel tubing with two stainless steel *Swagelok®* fittings and end caps (*Swagelok Company*, Solon, OH, USA). Cylinders were filled with 10 mL of the amine solution of interest, sealed according to *Swagelok®* specifications, and placed in forced convection ovens maintained at the experimental temperature. Individual cylinders were removed from the oven periodically and the liquid contents were analyzed for total alkalinity and $CO_2$ concentration, and analyzed using cation ion chromatography to quantify the parent amines and cationic degradation products.

## 1.7 Cation Ion Chromatography (IC)

Cation Ion Chromatography (cation IC) was used to quantify PZ, MPZ, DMPZ and other cation degradation products in solution. A *Dionex* ICS-2500 Ion Chromatography System with AS autosampler was used as described previously (*Dionex Corporation*, Sunnyvale, CA, USA) [4, 13, 18]. Separation occurred in an IonPac GC17 guard (4 × 50 mm) and CS17 analytical column (4 × 250 mm) with varying concentrations of methanesulfonic acid in analytical grade water and conductivity detector cell.

## 2 RESULTS AND DISCUSSION

### 2.1 Physical Properties of PZ/MPZ/DMPZ Blends

#### 2.1.1 Solid Solubility

The solid solubility of 3.75 m PZ/3.75 m MPZ/0.5 m DMPZ with varying concentrations of $CO_2$ was observed. At room temperature (25°C), a maximum $CO_2$ loading of 0.39 mol $CO_2$/mol alkalinity without any solid precipitation was obtained for 3.75 m PZ/3.75 m MPZ/0.5 m DMPZ. At 20°C, a higher maximum loading of 0.43 mol $CO_2$/mol alkalinity was obtained without solids precipitation. Neither MPZ nor DMPZ were found to precipitate in aqueous solution in the absence of PZ. PZ apparently precipitates as PZ·6H$_2$O in the blended solvent.

The transition temperature of this blend at 0 to 0.39 mol $CO_2$/mol alkalinity was observed in order to quantify the solubility envelope where an aqueous amine solution is known to exist. The transition temperature over this range of $CO_2$ loading for 3.75 m PZ/3.75 m MPZ/0.5 m DMPZ is shown in Figure 2. The transition temperatures of 8 m PZ are included in the

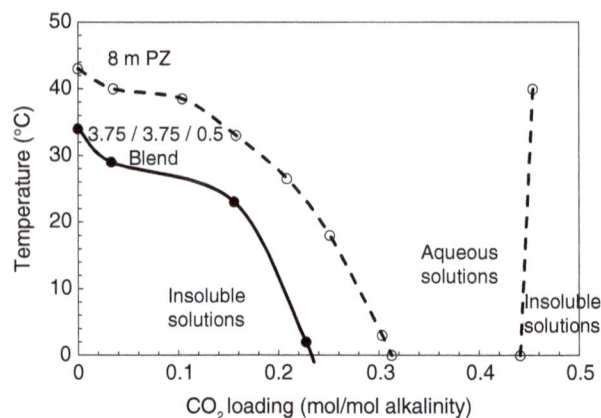

Figure 2

Comparison of solid-liquid transition temperatures of 8 m PZ [18] and 3.75 m PZ/3.75 m MPZ/0.5 m DMPZ.

figure for comparison purposes [18]. For the blend, the temperature at which any precipitation in the solution completely melted or the liquid precipitated into a solid was obtained and the transition point recorded.

The blended system has a larger solubility envelope and this solvent is predicted to have fewer solubility issues than a concentrated PZ system. In $CO_2$-loaded, concentrated PZ, solid PZ hexahydrate (PZ·6H$_2$O) is formed where the loading is less than 0.22 mol $CO_2$/mol alkalinity and the temperature is less than 25°C [13]. The blend contains less PZ and PZ·6H$_2$O, the expected solid formed in the blend in the same region of Figure 2, precipitates at lower temperature and $CO_2$ loading. This behavior expands the solubility envelope and usefulness of the blend.

At $CO_2$ loading greater than 0.43 mol $CO_2$/mol alkalinity, 8 m PZ was found to form hydrated protonated PZ carbamate (H$^+$PZCOO$^-$·H$_2$O) when precipitation occurred [16]. No solids precipitated in 3.75 m PZ/3.75 m MPZ/0.5 m DMPZ at any of the high $CO_2$ concentrations tested. It appears that at least 4 m PZ is required to form significant concentrations of H$^+$PZCOO$^-$, so this species is not present to precipitate in its hydrated form in the blended solvent.

### 2.1.2 Density and Viscosity of 5 M PZ/2 M MPZ/1 M DMPZ

The density and viscosity of 5 m PZ/2 m MPZ/1 m DMPZ were measured from 0 to 0.39 mol $CO_2$/mol alkalinity and 10 to 80°C (283.15 to 353.15 K). Data were obtained for all solutions where solid precipitation and $CO_2$ evolution were absent. The raw data are shown in Table 1 with $CO_2$ concentration reported in units of $CO_2$ loading and mol/kg. The density and viscosity data are compared to that of 8 m PZ at 40 and 60°C in Figure 3 and Figure 4, respectively. The data demonstrate the expected trends of increased density and viscosity with increasing $CO_2$ loading and decreasing temperature. In comparison with 8 m PZ, the blend demonstrates a lower density and higher viscosity at a given $CO_2$ loading at 40 and 60°C. The higher viscosity of the blend,

approximately 25% higher across the board, is a slight disadvantage of this solvent as it is expected to decrease mass transfer and reduce heat transfer coefficients.

The viscosity of 5 m PZ/2 m MPZ/1 m DMPZ is correlated well in terms of the viscosity of water obtained from DIPPR, $\mu_{water}$, $CO_2$ concentration in mol/kg, $C_{CO_2}$ and temperature in Kelvin, $T$ [21]. The details of the regression and parameters are shown in Equations (3, 4) and Table 2. The regression was able to fit all data

TABLE 1

Density and viscosity of 5 m PZ/2 m MPZ/1 m DMPZ

| $CO_2$ loading (mol/mol alk) | $CO_2$ (mol/kg) | $T$ (°C) | Density ($\rho$) (g/cm$^3$) | Viscosity ($\mu$) (mPa.s) |
|---|---|---|---|---|
| 0.2 | 1.661 | 10 | 1.0937 | 42.320 ± 0.042 |
| 0.23 | 1.900 | 10 | 1.1031 | 42.940 ± 0.070 |
| 0.29 | 2.411 | 10 | 1.1221 | 44.240 ± 0.070 |
| 0.33 | 2.680 | 10 | 1.1358 | 44.440 ± 0.070 |
| 0.37 | 2.940 | 10 | 1.1490 | 44.390 ± 0.088 |
| 0.39 | 3.105 | 10 | 1.1553 | 44.290 ± 0.120 |
| 0.1 | 0.894 | 25 | 1.0553 | 18.030 ± 0.082 |
| 0.15 | 1.286 | 25 | 1.0707 | 19.290 ± 0.110 |
| 0.2 | 1.661 | 25 | 1.0856 | 20.540 ± 0.178 |
| 0.23 | 1.900 | 25 | 1.0955 | 21.310 ± 0.173 |
| 0.29 | 2.411 | 25 | 1.1147 | 22.790 ± 0.242 |
| 0.33 | 2.680 | 25 | 1.1287 | 23.380 ± 0.199 |
| 0.37 | 2.940 | 25 | 1.1421 | 23.870 ± 0.200 |
| 0.39 | 3.105 | 25 | 1.1481 | 23.960 ± 0.212 |
| 0 | 0 | 40 | 1.0117 | 7.420 ± 0.095 |
| 0.05 | 0.478 | 40 | 1.0290 | 8.261 ± 0.122 |
| 0.1 | 0.894 | 40 | 1.0457 | 9.154 ± 0.148 |
| 0.15 | 1.286 | 40 | 1.0618 | 9.993 ± 0.157 |
| 0.2 | 1.661 | 40 | 1.0772 | 10.900 ± 0.226 |
| 0.23 | 1.900 | 40 | 1.0874 | 11.390 ± 0.213 |
| 0.29 | 2.411 | 40 | 1.1073 | 12.570 ± 0.206 |
| 0.33 | 2.680 | 40 | 1.1214 | 13.290 ± 0.311 |
| 0.37 | 2.940 | 40 | 1.1347 | 13.680 ± 0.319 |
| 0.39 | 3.105 | 40 | 1.1408 | 13.960 ± 0.337 |
| 0 | 0 | 60 | 0.9957 | 3.666 ± 0.125 |

(continued)

TABLE 1 *(continued)*

| CO$_2$ loading (mol/mol alk) | CO$_2$ (mol/kg) | $T$ (°C) | Density ($\rho$) (g/cm$^3$) | Viscosity ($\mu$) (mPa.s) |
|---|---|---|---|---|
| 0.05 | 0.478 | 60 | 1.0139 | 4.157 ± 0.124 |
| 0.1 | 0.894 | 60 | 1.0317 | 4.698 ± 0.180 |
| 0.15 | 1.286 | 60 | 1.0486 | 5.310 ± 0.226 |
| 0.2 | 1.661 | 60 | 1.0644 | 5.900 ± 0.243 |
| 0.23 | 1.900 | 60 | 1.0750 | 6.354 ± 0.246 |
| 0.29 | 2.411 | 60 | 1.0951 | 7.045 ± 0.293 |
| 0.33 | 2.680 | 60 | 1.1097 | 7.698 ± 0.369 |
| 0 | 0 | 80 | 0.9786 | 2.035 ± 0.051 |
| 0.05 | 0.478 | 80 | 0.9978 | 2.289 ± 0.104 |
| 0.1 | 0.894 | 80 | 1.0165 | 2.628 ± 0.123 |
| 0.15 | 1.286 | 80 | 1.0337 | 3.039 ± 0.142 |
| 0.20 | 1.661 | 80 | 1.0511 | 3.277 ± 0.160 |
| 0.23 | 1.900 | 80 | 1.0614 | 3.696 ± 0.103 |
| 0.29 | 2.411 | 80 | - | 4.163 ± 0.192 |
| 0.33 | 2.680 | 80 | - | 5.350 ± 0.196 |

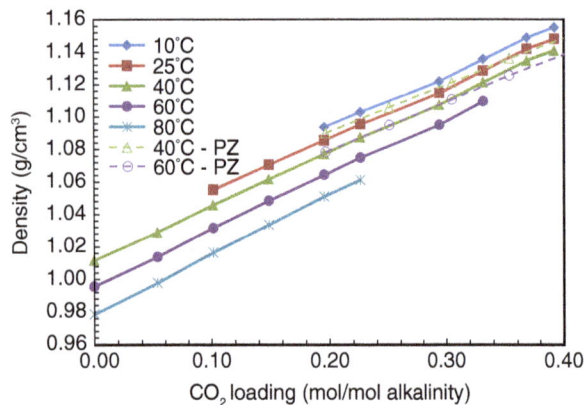

Figure 3

A comparison of the density of 5 m PZ/2 m MPZ/1 m DMPZ at 10 (◆), 25 (■), 40 (▲), 60 (●), and 80°C (*) and 8 m PZ at 40 (△) and 60°C (○).

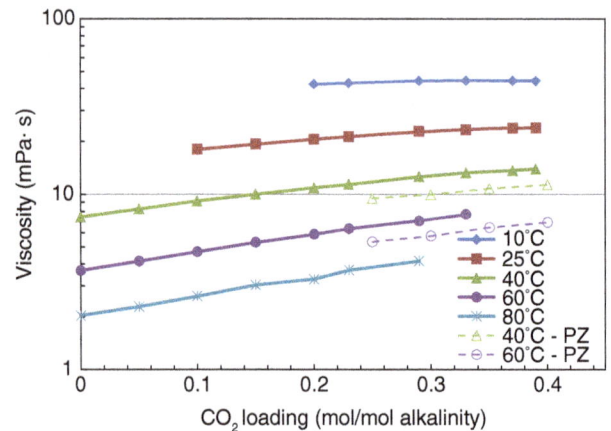

Figure 4

A comparison of the viscosity of 5 m PZ/2 m MPZ/1 m DMPZ at 10 (◆), 25 (■), 40 (▲), 60 (●), and 80 °C (*) and 8 m PZ at 40 (△) and 60°C (○).

with an average error of only 2.0% and a maximum deviation of 5.1%:

$$\ln\left(\frac{\mu}{\mu_{\text{water}}}\right) = \phi_1 + \phi_2 \cdot T + \frac{\phi_3}{T} \quad (3)$$

$$\phi_i = a_i + b_i \times C_{CO_2} \quad (4)$$

## 2.2 Thermodynamic Properties of PZ/MPZ/DMPZ

### 2.2.1 CO$_2$ Solubility

The CO$_2$ partial pressure, $P_{CO_2}$, of 3.75 m PZ/3.75 m MPZ/0.5 m DMPZ was measured from 40 to 160°C for a range of CO$_2$ loading. Data were obtained from the

TABLE 2

Value of parameters in Equation (4)

| Parameter | $i = 1$ | $i = 2$ | $i = 3$ |
|---|---|---|---|
| $a$ | −29.12 | 0.0379 | 6 167.9 |
| $b$ | 6.151 | −0.00746 | −1 131.04 |

equilibrium cell at 40°C, the wetted wall column for 40 to 100°C, and the high temperature VLE apparatus from 100 to 160°C. The total pressure, $P_{total}$, was obtained from the high temperature measurements and the $P_{CO_2}$ was estimated as described above. The $CO_2$ solubility values measured directly from the equilibrium cell and wetted wall column are compared in Table 3. The estimated $P_{CO_2}$ and $P_{total}$ measurements are shown in Table 4.

The $CO_2$ solubility measured from 40 to 160°C is shown in Figure 5 as the $P_{CO_2}$ plotted against the $CO_2$ loading. Data from three separate experimental apparatuses, the wetted wall column, high temperature VLE apparatus and equilibrium cell, are shown and demonstrate strong agreement that provides confidence in the data. An empirical model for the prediction of $P_{CO_2}$ was developed using the data from 40 to 160°C (excluding the single point from the equilibrium cell). The regressed form of the model is shown in Equation (5) where the natural log of $P_{CO_2}$ in Pa is predicted as a function of $CO_2$ loading, $\alpha$, and temperature in K, $T$.

The uncertainty of the regression is demonstrated by the error for each of the three regressed coefficients:

$$\ln P_{CO_2} = (34.4 \pm 0.2) + \frac{(-10597 \pm 81)}{T} + (7627 \pm 177) \cdot \frac{\alpha}{T} \tag{5}$$

### 2.2.2 CO₂ Capacity

The intrinsic $CO_2$ capacity of 3.75 m PZ/3.75 m MPZ/0.5 m DMPZ was estimated assuming lean loading (0.23 mol $CO_2$/mol alkalinity) and rich loading (0.33 mol $CO_2$/mol alkalinity) that correspond to $P_{CO_2}$ values at 40°C of 0.5 and 5 kPa, respectively. These two pressure values are estimates of the $P_{CO_2}$ that would provide an appropriate driving force at the bottom or top of the absorber [1, 24]. With the appropriate unit conversions, the capacity of the blend is 0.88 mol $CO_2$/kg (amine + water). This compares to an intrinsic $CO_2$ capacity of 0.79 and 0.47 mol $CO_2$/kg (amine + water) for 8 m PZ and 7 m MEA, respectively [25, 26].

### 2.2.3 Estimation of Heat of Absorption

The heat of $CO_2$ absorption ($\Delta H_{abs}$) in 3.75 m PZ/ 3.75 m MPZ/0.5 m DMPZ was estimated from the derivative of Equation (5) according to the Gibbs-Helmholtz Equation (*Eq. 6*) and is given by Equation (7). Over the expected $CO_2$ loading range, 0.23 to 0.33 mol $CO_2$/mol

TABLE 3

$CO_2$ partial pressure and mass transfer coefficient for 3.75 m PZ/3.75 m MPZ/0.5 m DMPZ

| $T$ (°C) | $CO_2$ loading (mol/mol alk) | $P_{CO_2}$ (kPa) | $k'_g$ (mol/s.Pa.m²) | Source[1] |
|---|---|---|---|---|
| 40 | 0.21 | 0.3 | $2.5 \times 10^{-6}$ | WWC |
| 40 | 0.23 | 0.3 | – | EC |
| 40 | 0.25 | 0.8 | $1.2 \times 10^{-6}$ | WWC |
| 40 | 0.29 | 2.1 | $9.2 \times 10^{-7}$ | WWC |
| 40 | 0.32 | 4.5 | $5.6 \times 10^{-7}$ | WWC |
| 60 | 0.21 | 1.7 | $2.6 \times 10^{-6}$ | WWC |
| 60 | 0.25 | 3.8 | $1.5 \times 10^{-6}$ | WWC |
| 60 | 0.29 | 9.9 | $8.7 \times 10^{-7}$ | WWC |
| 60 | 0.32 | 19.0 | $5.1 \times 10^{-7}$ | WWC |
| 80 | 0.21 | 8.0 | $2.1 \times 10^{-6}$ | WWC |
| 80 | 0.25 | 16.8 | $1.1 \times 10^{-6}$ | WWC |
| 100 | 0.21 | 29.2 | $1.3 \times 10^{-6}$ | WWC |

(1) WWC = wetted wall column; EC = equilibrium cell.

TABLE 4

Total pressure and estimated $P_{CO_2}$ of 3.75 m PZ/3.75 m MPZ/0.5 m DMPZ from high temperature VLE apparatus

| $T$ (°C) | $CO_2$ loading (mol/mol alk) | $P_{total}$ (kPa) | Estimated $P_{CO_2}$ (kPa) |
|---|---|---|---|
| 100 | 0.24 | 129 | 40.8 |
| 100 | 0.32 | 360 | 271 |
| 110 | 0.32 | 573 | 447 |
| 120 | 0.24 | 334 | 160 |
| 120 | 0.31 | 865 | 690 |
| 130 | 0.24 | 506 | 270 |
| 130 | 0.30 | 1 257 | 1 020 |
| 140 | 0.23 | 771 | 456 |
| 140 | 0.29 | 1 790 | 1 470 |
| 150 | 0.23 | 1 142 | 730 |
| 150 | 0.28 | 2 437 | 2 020 |
| 160 | 0.22 | 1 647 | 1 110 |
| 160 | 0.27 | 3 269 | 2 730 |

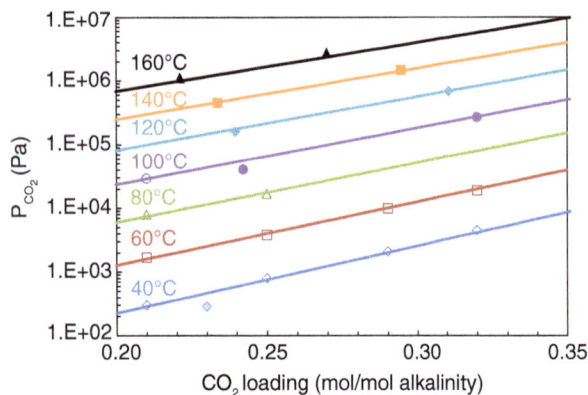

Figure 5

Solubility of $CO_2$ in 3.75 m PZ/3.75 m MPZ/0.5 m DMPZ from 40 to 160°C. Data: wetted wall column (open points), high temperature VLE apparatus (closed points), equilibrium cell (shaded point); lines: Equation (5).

alkalinity, the $\Delta H_{abs}$ predicted with Equation (7) varies from 73.5 to 67.2 kJ/mol:

$$\Delta H_{abs} = -R \frac{\partial (\ln P_{CO_2})}{\partial \left(\frac{1}{T}\right)} \qquad (6)$$

$$\Delta H_{abs} = -R \cdot (-10597 + 7627 \cdot \alpha) \qquad (7)$$

### 2.2.4 Amine Volatility

The volatility of amines in 3.75 m PZ/3.75 m MPZ/0.5 m DMPZ was measured at 40°C and a nominal lean loading of 0.23 mol $CO_2$/mol alkalinity, the conditions expected at the top of the absorber. The conditions that exist at the top of the absorber are the most important for volatility comparisons in order to assess potential fugitive losses to the atmosphere. For this solvent, this condition resulted in a $P_{CO_2}$ of 287 Pa and partial pressure of water, $P_{H_2O}$, or 5 875 Pa. The measured amine volatility of each of the three amine components in the blend is presented in Table 5 as the partial pressure of each amine ($P_{PZ}$, $P_{MPZ}$ and $P_{DMPZ}$). The amine volatility for 8 m PZ and 7 m MEA are also shown for comparison purposes [19]. The $CO_2$ loading for these two baseline systems were also chosen to mimic the operating conditions expected at the top of the absorber.

Based on the results, it is clear that methylation increases volatility. The volatility of MPZ and DMPZ are significantly higher than that of PZ in the blend and by itself, suggesting a significant impact of methylation on the behavior of the amino function. Even though DMPZ is present only in a small concentration, 0.5 m, its volatility is the highest and 65 times greater than PZ. The fact that this molecule has two methyl groups renders it fairly nonpolar in the polar environment consisting of water and electrolytes. Therefore, DMPZ has

TABLE 5

Amine volatility for 3.75 m PZ/3.75 m MPZ/0.5 m DMPZ, 8 m PZ and 7 m MEA at 40°C

| Solvent | 3.75 m PZ/3.75 m MPZ/ 0.5 m DMPZ | 8 m PZ | 7 m MEA |
|---|---|---|---|
| $CO_2$ loading (mol/mol alk) | 0.23 | 0.29 | 0.47 |
| $P_{PZ}$ (Pa) | 0.95 | 0.8 | - |
| $P_{MPZ}$ (Pa) | 16.4 | - | - |
| $P_{DMPZ}$ (Pa) | 64.1 | - | - |
| $P_{MEA}$ (Pa) | - | - | 2.7 |

unfavorable interactions with the surrounding species and is highly volatile. MPZ also suffers the effects of a nonpolar methyl group but to a lesser extent than DMPZ. This observation highlights the fact that degradation products can be more volatile than the parent amine itself, even at small concentrations, an important fact in solvent selection.

## 2.2.5 CO₂ Absorption Rate

The liquid film mass transfer coefficient ($k_g'$) for 3.75 m PZ/3.75 m MPZ/0.5 m DMPZ was measured from 40 to 100°C at a range of loadings (*Fig. 6* and *Tab. 3*). The use of $P_{CO_2}$ at 40°C as the abscissa allows direct comparison of rate data at different temperature as well as for different amine solvents on the same basis without influences from differences in $CO_2$ loading. The range of $P_{CO_2}$ from 0.5 to 5 kPa, as used previously for capacity calculations,

represents a reasonable driving force between gas and liquid at the top and bottom of an absorber expected during the treatment of flue gas from coal-fired power plant. At 0.5 kPa and 40°C, the expected conditions at the lean end of the absorber, the $k_g'$ of the blend is $2.0 \times 10^{-6}$ mol/s.Pa.m² compared to $2.1 \times 10^{-6}$ and $7.6 \times 10^{-7}$ mol/s.Pa.m² for PZ and MEA, respectively.

The temperature range expected for the absorber is 40 to 60°C. In this temperature range, the rate for the blend stays relatively constant, approximately equal to that for 8 m PZ and over 2.5 times faster than 7 m MEA. Temperature above 60°C leads to a decrease in the mass transfer rate of the blend, as has been observed for other amine systems [24-26]. The blend performed as well as concentrated PZ and represents another solvent that can outperform the baseline MEA solvent.

## 2.3 Behavior of the Blend at High Temperature

### 2.3.1 Thermal Degradation of the PZ-Based Blends

The thermal behavior of PZ/MPZ/DMPZ was studied at 150°C. This temperature represents the expected temperature of the stripper and reboiler section of an absorber-stripper system utilizing a thermally stable solvent in order to maximize high temperature energy savings.

Six PZ-based solvent blends were investigated for their thermal stability. All blends are compared to 8 m PZ as the baseline case of a highly stable solvent. In the blends tested, the total alkalinity of the solvent was maintained at 8 m, while PZ was replaced with methylated PZ molecules (MPZ or DMPZ). Blend compositions are indicated as the concentration of PZ/MPZ/DMPZ, all in units of m. For example, one two-component blend of 4 m PZ/4 m MPZ, or 4/4/0, was investigated. Five three-component blends (5/2/1, 5/2.5/0.5, 5/1.5/1.5, 3.9/3.9/0.2, and 3.75/3.75/0.5) were also investigated.

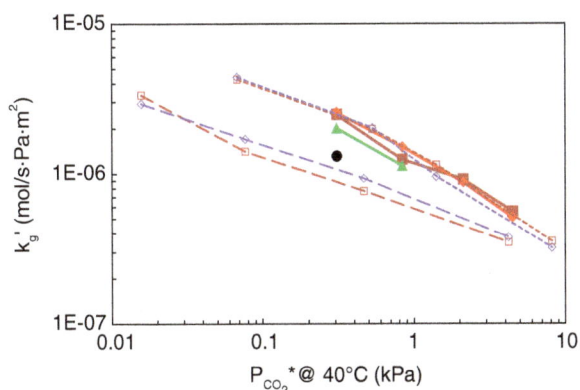

Figure 6

Mass transfer coefficient of 3.75 m PZ/3.75 m MPZ/0.5 m DMPZ (solid points) at temperatures 40 (■), 60 (♦), 80 (▲), and 100°C (●), compared with 8 m PZ (short dash with open points) and 7 m MEA (long dash with open points) at 40 and 60°C.

The overall rate of thermal degradation was assessed using a first order rate constant of amine loss, $k_1$, as described in detail previously [13, 27]. For comparing the blends, the rate was determined based on the total amine loss as a sum of PZ, MPZ and DMPZ rather than the rate of loss of the individual amine species. This was done to allow an overall comparison of the blends on a similar basis. The $k_1$ values for thermal degradation at 150°C of the 8 m PZ and the blends are compared in Table 6. The $k_1$ values of the blends vary from $8.6 \times 10^{-9}$ to $7.5 \times 10^{-8}\,s^{-1}$ and are substantially less than the comparable value for 7 m MEA with 0.4 mol $CO_2$/mol alkalinity degraded at 150°C of $8.3 \times 10^{-7}\,s^{-1}$ [4].

Blending of PZ with MPZ and DMPZ also reduces solvent stability as shown with the high $k_1$ values for all blends studied. The changes in the rate constant are small for most of the blends with the exception of the 5/1.5/1.5 solvent, so the slight increase in degradation may be balanced by the improvement in solid solubility.

### 2.3.2 Thermal Equilibrium of PZ, MPZ and DMPZ

A thermal equilibrium was found to establish itself in the presence of methylated PZ. In the degradation of 8 m MPZ, PZ and DMPZ were generated as products through a $S_N2$-type substitution or arm-switching reaction. Observations of this system indicate it tends toward an equilibrium set of concentrations as the species interconvert to produce a solvent that is thermally stable at its equilibrium concentrations. The thermal equilibrium established between PZ, MPZ, and DMPZ was studied in the thermal degradation experiments performed at 150°C described in the previous section. At this temperature, some degradation was expected to occur, but the interconversion of PZ, MPZ and DMPZ was assumed to occur more rapidly than degradation processes based on the slow degradation rates discussed in the previous section.

The expected equilibrium reaction of MPZ is as follows, where the brackets indicate the concentration of the given species in units of mmol/kg (Eq. 8). The equilibrium constant, $K_{eq}$, of Equation (8) can then be calculated (Eq. 9):

$$2[MPZ] \leftrightarrow [PZ] + [DMPZ] \qquad (8)$$

$$K_{eq} = \frac{[PZ]\,[DMPZ]}{[MPZ]^2} \qquad (9)$$

The goal of this $K_{eq}$ analysis is to determine the concentrations of the three amine species in equilibrium at 150°C. These concentrations would be what were expected to be stable in a system with a reboiler temperature of 150°C. The $K_{eq}$ values for the thermal equilibrium of PZ, MPZ, and DMPZ are shown in Figure 7. For each blend, the $K_{eq}$ was calculated at each time point according to Equation (9) and plotted against degradation time.

All of the experiments, including those with limited data, were found to tend toward a $K_{eq}$ value of 0.27-0.30, indicating the equilibrium concentrations of amine expected at 150°C. The experiments containing 3.9/3.9/0.2 and 3.75/3.75/0.5 demonstrate this tendency toward a $K_{eq}$ of 0.29. Experiments with initial amine concentrations far from equilibrium, such as the 5/1.5/1.5 blend, demonstrated a higher thermal degradation rate as the amines reacted toward equilibrium, as shown in Table 6.

Degradation of PZ alone will not yield these results as MPZ and DMPZ are only minor products [13]. However, when a blend begins with at least some N-methylated PZ (i.e., MPZ or DMPZ), these methyl groups can easily

TABLE 6

First-order rate constant for amine thermal degradation at 150°C with 0.3 mol $CO_2$/mol alkalinity calculated based on total amine loss

| PZ/MPZ/DMPZ concentration (m) | $k_1 \times 10^{-9}$ at 150°C $(s^{-1})$ |
|---|---|
| 0/8/0 | 8.6 |
| 3.75/3.75/0.5 | 10.3 |
| 3.9/3.9/0.2 | 8.4 |
| 4/4/0 | 13.9 |
| 5/1.5/1.5 | 75.3 |
| 5/2/1 | 13.7 |
| 5/2.5/0.5 | 14.4 |
| 8 m PZ | 6.1 |
| 7 m MEA | 828 |

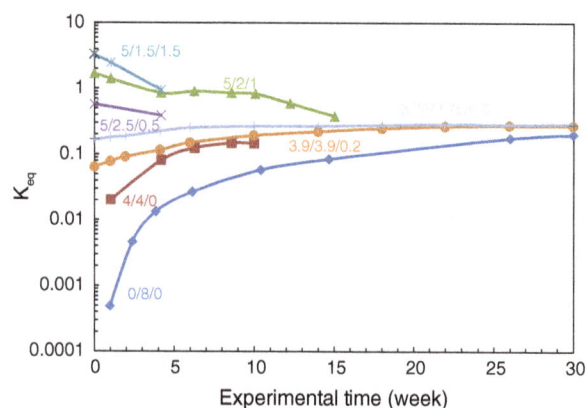

Figure 7

$K_{eq}$ for PZ + MPZ + DMPZ solutions thermally degraded at 150°C ($\alpha = 0.3$). Labels indicate the concentrations of PZ, MPZ, and DMPZ in solution in molal (m).

TABLE 7
Comparison of $CO_2$ capture solvents

| Solvent | 7 m MEA | 8 m PZ | 3.75 m PZ/3.75 m MPZ/ 0.5 m DMPZ |
|---|---|---|---|
| Solubility range at 20°C (mol $CO_2$/mol alkalinity) | 0 to max | 0.25-0.45 | 0.18 to max |
| Operating loading range (mol $CO_2$/mol alkalinity) | 0.45-0.55 | 0.31-0.39 | 0.23-0.32 |
| $CO_2$ capacity (mol/kg amine + water) | 0.47 | 0.79 | 0.88 |
| $k_g'_{avg}$ (40°C) (mol/s.Pa.m$^2$) | $4.3 \times 10^{-7}$ | $8.5 \times 10^{-7}$ | $8.5 \times 10^{-7}$ |
| $\Delta H_{abs}$ (kJ/mol $CO_2$) | 72 ± 4 | 66 ± 2 | 70 ± 1 |
| $P_{amine}$ at lean ldg and 40°C (Pa) | 2.7 | 0.8 | 81.5 |
| Thermal degradation, $k_1 \times 10^{-9}$ at 150°C (s$^{-1}$) | 828 | 6.1 | 10.3 |

undergo a straightforward nucleophilic attack and disproportionation, resulting in switching of the methyl groups between molecules. Overall, the data suggest that any system starting with MPZ will tend toward a $K_{eq}$ of 0.28 to 0.30 at 150°C.

## 2.4 Overall Comparison of Blends

A comparison of the important parameters for $CO_2$ capture solvents is presented in Table 7 for 7 m MEA, 8 m PZ, and 3.75 m PZ/3.75 m MPZ/0.5 m DMPZ. The $CO_2$ loading range when the solvent is completely soluble at 20°C is shown based on solubility data. The improved solubility of both MEA and the blend is a strong advantage over 8 m PZ. The $CO_2$ capacity was calculated for each system as described in Section 2.2.2 based on $CO_2$ solubility at $P_{CO_2}$ of 0.5 and 5 kPa. The blend and 8 m PZ have approximately double the $CO_2$ capacity as MEA, an important advantage. The average $k'_g$, $k'_{g\,avg}$, based on a log-mean flux and log-mean driving force across a typical absorber design, as described in detail previously, is compared at 40°C [26]. Both PZ and the blend have $k_g'_{avg}$ values double that of MEA.

Empirical regressions for the $\Delta H_{abs}$ are shown for each solvent based on $CO_2$ solubility regressions developed for MEA and PZ previously or for the blend as described in this manuscript [20]. The $\Delta H_{abs}$ was calculated from the empirical regressions at a midpoint in the expected loading range for each amine system. The midpoint was determined to be at a $P_{CO_2}$ of 1.5 kPa, which corresponded to $CO_2$ loadings of 0.49, 0.35, and 0.28 mol $CO_2$/mol alkalinity for the MEA, PZ and blended systems, respectively. Error in the calculation of this value was assessed based on the confidence of

each regressed coefficient. The larger $\Delta H_{abs}$ for the blend over PZ is a crucial advantage that will reduce overall energy performance.

Volatility was reported as the overall amine volatility of each solvent at 40°C at the expected lean loading as this should correspond to the top of the absorber in a real system. For the blend, this is a sum of the partial pressure of PZ, MPZ and DMPZ at this condition. The volatility of the blend is significantly higher than the two baseline amines. Implementation of this amine system will require an adequately designed water wash system to avoid amine losses from the absorber.

The thermal stability of each solvent is expressed as the $k_1$ for thermal degradation at 150°C. Data for MEA, PZ and the blend were obtained at 0.4, 0.3 and 0.3 mol $CO_2$/mol alkalinity, respectively. The reported value for the blend is based on total amine loss from PZ, MPZ and DMPZ. Both PZ and the blend are significantly more thermally stable than MEA as demonstrated by the large $k_1$ value of MEA.

## CONCLUSIONS

A blend of PZ, MPZ and DMPZ is an example of a new solvent design based on the thermal equilibrium of multiple amines useful for $CO_2$ capture. Blends of this nature reach a thermal equilibrium at reboiler temperature where the overall thermal degradation rate is low while multiple species exist that are valuable to the $CO_2$ capture process. 3.75 m PZ/3.75 m MPZ/0.5 m DMPZ has been fully evaluated for this application.

In comparison with 8 m PZ, this blend was found to have advantageous solid solubility and $\Delta H_{abs}$, with approximately the same behavior in terms of $CO_2$

capacity, $CO_2$ absorption rate and thermal degradation as shown in Table 7. Disadvantages of the blend include mildly increased viscosity and substantially greater amine volatility as DMPZ. Both of these properties are important to the optimized design of a $CO_2$ capture system and would need to be addressed. A carefully designed water wash system would be needed to address the amine volatility. As with PZ, the blend was found to have significantly better $CO_2$ capacity, $CO_2$ absorption rates and thermal stability compared to 7 m MEA.

## ACKNOWLEDGMENTS

The Luminant Carbon Management Program provided support for this work.

## REFERENCES

1 Dugas R.E. (2009) Carbon Dioxide Absorption, Desorption, and Diffusion in Aqueous Piperazine and Monoethanolamine, *PhD Thesis*, The University of Texas, Austin.

2 Rochelle G.T., Chen E., Freeman S.A., Van Wagener D.H., Xu Q., Voice A.K. (2011) Aqueous piperazine as the new standard for $CO_2$ capture technology, *Chem. Eng. J.* **171**, 3, 725-733.

3 Rochelle G.T., Freeman S.A., Chen Xi, Nguyen Thu, Voice A., Rafique H. Acidic gas removal by aqueous amine solvents, U.S. Patent Application 2011/0171093 (PCT Patent Application W02011088008).

4 Davis J.D. (2009) Thermal Degradation of Aqueous Amines Used for Carbon Dioxide Capture, *PhD Thesis*, The University of Texas, Austin.

5 Davis J.D., Rochelle G.T. (2009) Thermal degradation of monoethanolamine at stripper conditions, *Energy Procedia* **1**, 327-333.

6 Meisen A., Kennard M.L. (1982) DEA degradation mechanism, *Hydrocarbon Process.* **61**, 105-108.

7 Rochelle G.T. (2009) Amine Scrubbing for $CO_2$ Capture, *Science* **325**, 1652-1654.

8 Sexton A.J. (2008) Amine Oxidation in $CO_2$ Capture Processes, *PhD Thesis*, The University of Texas, Austin.

9 Sexton A.J., Rochelle G.T. (2011) Reaction Products from the Oxidative Degradation of Monoethanolamine, *Ind. Eng. Chem. Res.* **50**, 1, 667-673.

10 Strazisar B.R., Anderson R.R., White C.M. (2003) Degradation Pathways for Monoethanolamine in a $CO_2$ Capture Facility, *Energy Fuels* **17**, 4, 1034-1039.

11 Chakma A., Meisen A. (1987) Degradation of aqueous DEA solutions in a heat transfer tube, *Can. J. Chem. Eng.* **65**, 2, 264-273.

12 Chakma A., Meisen A. (1997) Methyl-diethanolamine degradation - Mechanism and kinetics, *Can. J. Chem. Eng.* **75**, 5, 861-871.

13 Freeman S.A. (2011) Thermal Degradation and Oxidation of Aqueous Piperazine for Carbon Dioxide Capture, *PhD Thesis*, The University of Texas, Austin.

14 Closmann F., Nguyen T., Rochelle G.T. (2009) MDEA/Piperazine as a solvent for $CO_2$ capture, *Energy Procedia* **1**, 1351-1357.

15 Closmann F., Rochelle G.T. (2011) Degradation of aqueous methyldiethanolamine by temperature and oxygen cycling, *Energy Procedia* **4**, 23-28.

16 Freeman S.A., Dugas R.E., Van Wagener D.H., Nguyen T., Rochelle G.T. (2010) Carbon dioxide capture with concentrated, aqueous piperazine, *Int. J. Greenhouse Gas Control* **4**, 2, 119-124.

17 Hilliard M.D. (2008) A Predictive Thermodynamic Model for an Aqueous Blend of Potassium Carbonate, Piperazine, and Monoethanolamine for Carbon Dioxide Capture from Flue Gas, *PhD Thesis*, The University of Texas, Austin.

18 Freeman S.A., Davis J.D., Rochelle G.T. (2010) Degradation of aqueous piperazine in carbon dioxide capture, *Int. J. Greenhouse Gas Control* **4**, 5, 756-761.

19 Nguyen T., Hilliard M.D., Rochelle G.T. (2010) Amine volatility in $CO_2$ capture, *Int J. Greenhouse Gas Control* **4**, 5, 707-715.

20 Xu Q., Rochelle G.T. (2011) Pressure and $CO_2$ Solubility at High Temperature in Aqueous Amines, *Energy Procedia* **4**, 117-124.

21 DIPPR (2010) *DIPPR 801 Database of Physical and Thermodynamic Properties of Pure Chemicals*, Brigham Young University.

22 Cullinane J.T., Rochelle G.T. (2006) Kinetics of carbon dioxide absorption into aqueous potassium carbonate and piperazine, *Ind. Eng. Chem. Res.* **45**, 8, 2531-2545.

23 Dugas R.E., Rochelle G.T. (2009) Absorption and desorption rates of carbon dioxide with monoethanolamine and piperazine, *Energy Procedia* **1**, 1163-1169.

24 Dugas R.E., Rochelle G.T. (2011) $CO_2$ Absorption Rate into Concentrated Aqueous Monoethanolamine and Piperazine, *J. Chem. Eng. Data* **56**, 2187-2195.

25 Chen X., Closmann F., Rochelle G.T. (2011) Accurate screening of amines by the Wetted Wall Column, *Energy Procedia* **4**, 101-108.

26 Chen X., Rochelle G.T. (2011) Aqueous Piperazine Derivatives for $CO_2$ Capture: Accurate Screening by a Wetted Wall Column, *Chem. Eng. Res. Des.* **89**, 9, 1693-1710.

27 Freeman S.A., Rochelle G.T. (2011) Thermal degradation of piperazine and its structural analogs, *Energy Procedia* **4**, 43-50.

# 2

# Investigation of Primary Recovery in Low-Permeability Oil Formations: A Look at the Cardium Formation, Alberta (Canada)

S.M. Ghaderi*, C.R. Clarkson and D. Kaviani

Department of Chemical and Petroleum Engineering, University of Calgary, 2500 University Drive N.W., Calgary, Alberta - Canada
e-mail: smghader@ucalgary.ca - clarksoc@ucalgary.ca - dkaviani@ucalgary.ca

* Corresponding author

**Résumé** — **Recherches sur la récupération primaire dans les formations de pétrole de faible perméabilité : étude de la formation de Cardium, Alberta (Canada)** — Les formations compactes de pétrole (perméabilité < 1 mD) dans l'Ouest canadien sont récemment apparues comme étant une ressource sûre d'approvisionnement en pétrole léger en raison de l'utilisation de puits horizontaux multifracturés. La formation de Cardium, qui contient 25 % du total de pétrole léger découvert dans l'Alberta (selon l'*Alberta Energy Resources Conservation Board*), comprend des zones de production conventionnelles et non conventionnelles (de faible perméabilité ou compacts). Les zones de type conventionnel ont été développées dès 1957. En revanche, le développement des zones de type non conventionnel représente un événement récent à cause des propriétés considérablement moins bonnes des réservoirs, ce qui augmente le risque lié au placement de capitaux. Cela implique donc la nécessité d'une étude globale et critique de la zone avant de prévoir toute stratégie de développement.

Cet article présente des résultats à propos de la performance des portions de faible perméabilité de la formation de Cardium où de nouveaux puits horizontaux ont été forés et stimulés en multiples étapes afin de favoriser des fractures hydrauliques transversales. Le développement de la formation compacte de Cardium au moyen d'une récupération primaire est envisagé. Les données de production de ces puits ont d'abord été recréées à l'aide d'un simulateur d'écoulement. Le modèle calibré a été utilisé pour prédire les performances en production à partir d'analyses de sensibilité par rapport à différents paramètres dont des facteurs de conception comme l'espacement des puits, les propriétés des fractures et les contraintes opératoires.

*Abstract* — *Investigation of Primary Recovery in Low-Permeability Oil Formations: A Look at the Cardium Formation, Alberta (Canada)* — *Tight oil formations (permeability < 1 mD) in Western Canada have recently emerged as a reliable resource of light oil supply owing to the use of multi-fractured horizontal wells. The Cardium formation, which contains 25% of Alberta's total discovered light oil (according to Alberta Energy Resources Conservation Board), consists of conventional and unconventional (low-permeability or tight) play areas. The conventional play areas have been developed since 1957. Contrarily, the development of unconventional play is a recent event, due to considerably poorer reservoir properties which increases the risk associated with capital investment. This in turn implies the need for a comprehensive and critical study of the area before planning any development strategy.*

*This paper presents performance results from the low permeability portions of the Cardium formation where new horizontal wells have been drilled and stimulated in multiple stages to promote transverse hydraulic fractures. Development of the tight Cardium formation using primary recovery is considered. The production data of these wells was first matched using a black oil simulator. The calibrated model presented was used for performance perditions based on sensitivity studies and investigations that encompassed design factors such as well spacing, fracture properties and operational constraints.*

## NOMENCLATURE

| | |
|---|---|
| $F_{cd}$ | Fracture conductivity (dimensionless) |
| $k$ | Matrix permeability, $L^2$ (mD) |
| $k_f$ | Permeability of hydraulic fracture, $L^2$ (mD) |
| $k'_f$ | Permeability of hydraulic fracture in simulation model, $L^2$ (mD) |
| $n$ | Dimensionless well length |
| $p_b$ | Bubble point pressure (m/Lt$^2$) (psi) |
| $p_{wf}$ | Well flowing pressure (m/Lt$^2$) (psi) |
| $w_f$ | Fracture width $L$ (ft) |
| $x_f$ | Fracture half length $L$ (ft) |

### Greek symbols

| | |
|---|---|
| $\Delta_f$ | Equivalent fracture width in the simulation model, $L$ (ft) |

## SI CONVERSION FACTORS

acre × 4.046873.E−03 = m$^2$

bbl × 1.589873.E−01 = m$^2$

cp × 1.0*.E−03 = Pa.s

ft × 3.048*.E−01 = M

lbm × 4.53592.E−01 = kg

mD × 9.869223.E−16 = m$^2$

psi × 6.894757.E+03 = Pa

* Conversion factors are exact.

## INTRODUCTION

Economic production of hydrocarbons from reservoir rocks requires reservoir-specific solutions. Until recently, oil and gas exploitation was restricted to relatively high permeability and porosity reservoirs. In such reservoirs, wellbore contact through vertical wells was sufficient to obtain economic rates and recovery. Recent advances in drilling and completion technology have enabled commercial production from reservoirs with poorer properties. The low-permeability area of the Pembina Cardium field in Western Canada, referred to as a "Halo Oil" play by Clarkson and Pedersen (2011), is an example.

The Pembina Cardium field is the principal conventional oil pool in Canada covering an area of over 3 000 km$^2$ with more than 6 100 wells (approximately 4 400 producers and 1 700 injectors which are mainly vertical). Original oil in place is in excess of 7 780 MMbbl with recovery of less than 17% to date. The field is located in a stratigraphic trap of northwest-southeast oriented shoreface sands with the eastern up-dip margin being defined by shale out of the sands and the western downdip margin by decreasing reservoir quality (Krause et al., 1987). Horizontal wells have been drilled in both the Cardium sands and conglomerates within the Pembina field with limited success (Adegbesan et al., 1996).

In 2008, the first horizontal well with seven hydraulic fracture stages was completed in the unconventional (low-permeability) portion of the Cardium. The production rates from this well were so promising (average production rates of approximately 123 bbl/day without any water production) that another 12 wells were drilled in 2009 (Viau and Nielsen, 2010). Since then, more and more multi-fractured horizontal wells are being completed or planned in the area by different companies.

Because the unconventional portion of the Pembina is at early stages of production and development, more detailed studies are required for better management and exploitation of the resources. In this study, our purpose is to determine the major parameters that will affect the recovery of oil from these reservoirs under primary production scheme when fractured horizontal wells are being used. Oil production rates from the wells are usually high at the beginning, nevertheless limited permeability of the bulk of reservoir impede favorable rates over the long term. Therefore short-term and long-term production performance will be considered separately in this work. As part of the study, extensive numerical simulations were performed using the ECLIPSE 100[TM] simulator[1] (*Schlumberger*, 2010). The Design of Experiment (DOE) approach was employed to define the simulation

---

[1] ECLIPSE 100, ECLIPSE and PVTi are trademarks of *Schlumberger*.

scenarios and the parameters involved. In each simulation run, several parameters can be varied simultaneously to capture the effect of all main parameters and their likelihood of interaction. The oil recovery from simulation runs will be the main response in the statistical interpretation of the results. Simulation strategies and results will first be discussed.

## 1 BASE PROPERTIES USED IN SIMULATION MODELS

The properties required for generating simulation models were obtained from one of the active operators in the area. To avoid issues related to reservoir heterogeneity, all reservoir properties are considered constant at their average values. Some of these properties are summarized in Table 1 (modified from Clarkson and Pedersen, 2010). The data in this table have been obtained from an area of Cardium where the pay zone varies between 16 to 26 ft and the reservoir rock consists of muddy fine-grained sandstones with low permeability (≈ 0.3 mD) but relatively suitable porosity (≈ 0.12). It is also known that the reservoir is not supported by any initial gas cap (note the initial reservoir pressure and oil bubble point pressure in Tab. 1, 2) or bottom water support and therefore the main mechanism of primary production would be through solution gas drive (Clarkson and Pedersen, 2010).

Table 2 provides the fluid data used in the simulation models which are based on the real data obtained from one of the wells in the study area. The oil is relatively light with °API equal to 37. Figure 1 depicts the results obtained from the simulation of the PVT data for oil and gas using the PVTi$^{TM}$ module of Eclipse (*Schlumberger*, 2010). Figures 1a, b show the results of the differential liberation test and the viscosity measurements (*Fig. 1c, d*) obtained from constant composition expansion test. Pedersen's correlation (Pedersen *et al.*, 1984) was used to match the viscosity data. For the range of pressure applicable to this study (200-2 000 psia), the match of the PVT data is acceptable. The results of Table 2 are similar to what was used by Clarkson and Pedersen (2010) using standard correlations.

One missing element in our simulation models was a relative permeability data set. To obtain the relative permeability or at least a rough estimate of its shape, we tried to history match the production data of the wells in the study area and used the relative permeability data as the matching parameter. We received about one year production data for three separate wells in the region from an operator. For the well selected for analysis, the pressure is somewhat less than the virgin pressure (approximately 2 500 psi), suggesting that some

TABLE 1

General properties of the simulation models

| Simulation model parameter | Parameter value |
|---|---|
| Thickness (ft) | 16.4 |
| Porosity (%) | 12 |
| Absolute permeability (mD) | 0.28 |
| Initial reservoir pressure (psia) | 2 017 |
| Initial oil saturation (%) | 86 |
| Initial water saturation (%) | 14 |
| Reservoir temperature (°F) | 115 |
| Model area (acres) | 640 |
| Number of blocks ($NX \times NY \times NZ$) | $105 \times 105 \times 1$ |
| Dimension of blocks (ft × ft × ft) | $50 \times 50 \times 16.4$ |

TABLE 2

Fluid properties used in the simulations

| Oil | |
|---|---|
| $Pb$ (psia) | 1 602 |
| $Rs$ at $Pb$ (scf/stb) | 430 |
| Viscosity at $Pb$ (cp) | 1.43 |
| $Bo$ at $Pb$ (res. bbl/stb) | 1.22 |
| Density at STP (lbm/ft$^3$) | 51.8 |
| Gas | |
| Ave. Viscosity (cp) | 0.011 |
| Density at STP (lbm/ft$^3$) | 0.069 |
| Water | |
| Viscosity (cp) | 0.58 |

depletion due to offsetting vertical wells may have occurred. Nonetheless, we analyzed the well individually, realizing that a field simulation may be justified.

Well completion data were obtained from the operator. The length of the horizontal well is 3 865 ft and its wellbore diameter is 7.2 inches. It was frac'd in 10 stages with an average fracture spacing of 429 ft (assuming 1 fracture per stage). We estimate the fracture half-length to be equal to 181 ft and fracture conductivity close to 300 md-ft. The well is completed open-hole (Clarkson and Pedersen, 2010).

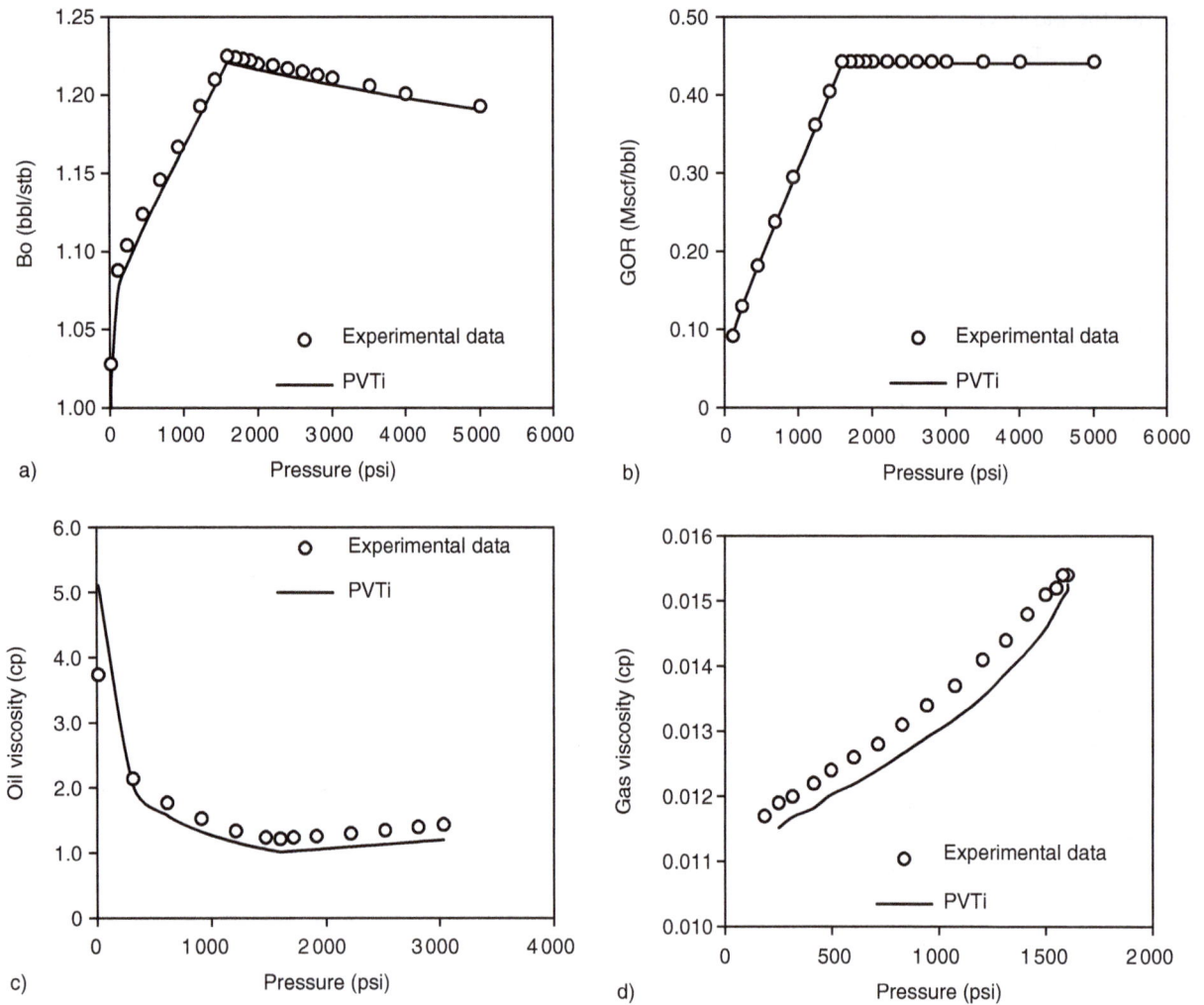

Figure 1

Comparison between the experimental and simulated PVT behavior of the fluid used in this study: a) oil formation volume factor *vs* pressure, b) oil solution gas *vs* pressure, c) oil viscosity *vs* pressure, d) gas viscosity *vs*. pressure.

Figure 2a displays a close-up view of the simulated transverse fractures planes which are perpendicular to the well trajectory. To precisely capture the physics of the fluid flow in the models, the Local Grid Refinement (LGR) feature is used to explicitly construct the hydraulic fractures along the well. However, exact duplication of the fractures' width in a simulation model will require extremely refined grids which results in computationally very expensive models (Shaoul *et al.*, 2007). Therefore, a sensitivity study was performed to determine the optimal degree of refinement for the fractures blocks as well as the parent block dimensions. Also, transmissibility of the fractures blocks should be adjusted such that following relationship holds:

$$k_f \times w_f = k'_f \times \Delta_f \quad (1)$$

where $k_f \times w_f$ is the product of fracture permeability and width and $k'_f \times \Delta_f$ is the corresponding product in the simulation model. The hydraulic fractures in our study have a constant width equal to 2.0 ft which are obtained by dividing the parent grid block into 25 parts in the direction perpendicular to the fracture orientation.

Figure 2b shows the historic oil production rates from the subject well for almost one year. It should be noted that in Figure 2 time zero corresponds to the time when the well has reached the "pump-off state", signified by a stabilized fluid level in the annulus. As the well bottom-hole pressure (BHP) was estimated from periodic fluid

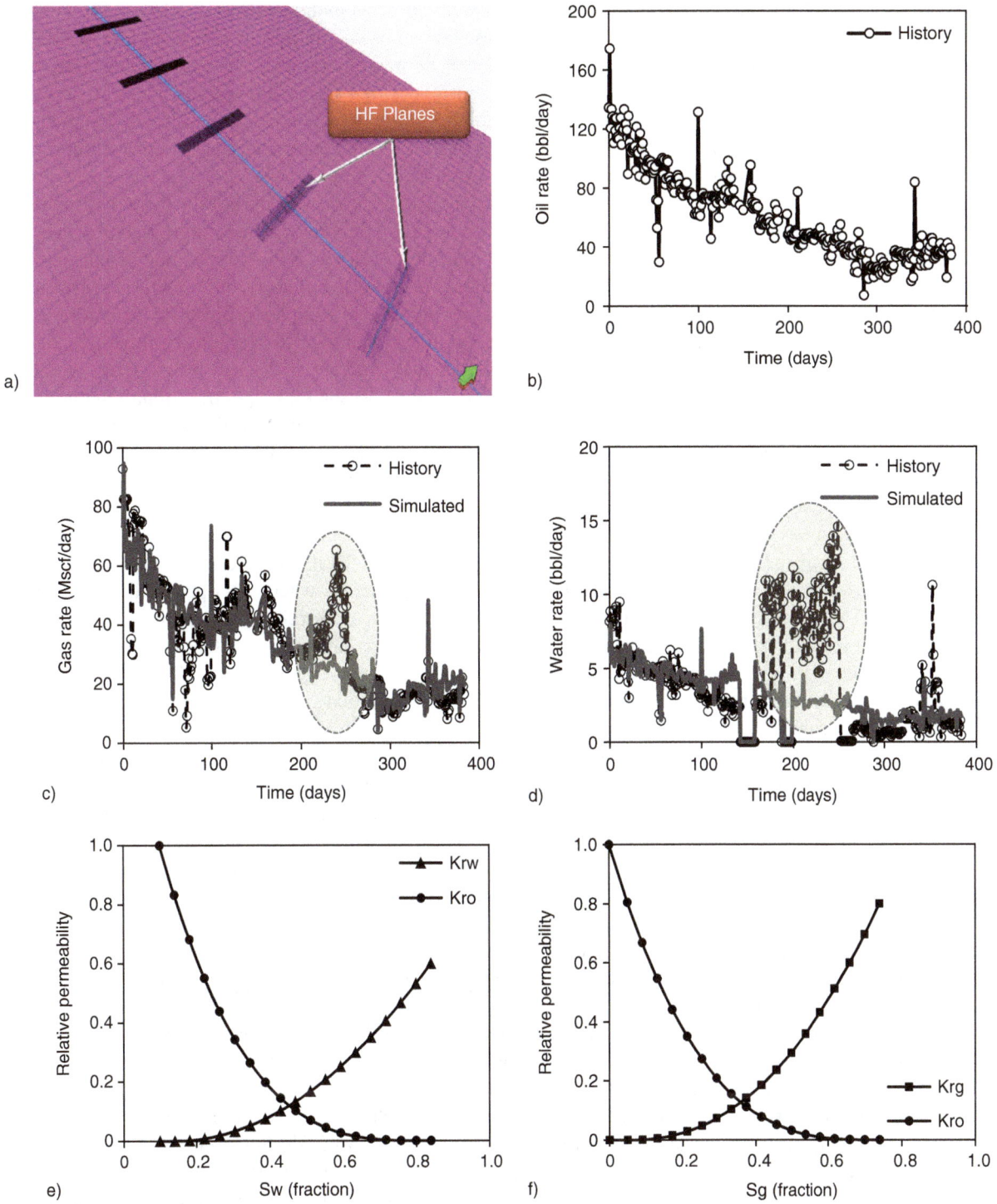

Figure 2

a) A close-up view of the simulation model showing the fractures' plane crossing the well lateral; b) historical oil production data for the subject well; c-d) comparison of gas and water production data from well history and simulation, respectively; e-f) two-phase relative permeability of water-oil and gas-oil obtained from history matching of production data, respectively.

shots combined with casing pressure, a stabilized fluid level in the annulus indicates that the well is producing under constant BHP. For the period of production displayed in Figure 2b this BHP is close to 250 psia (the pressure data is not shown).

For history-matching, oil production rate and flowing pressure were honored in the simulator and gas and water production rates were predicted. Figure 2c shows the gas production history from the well and the simulation results from the history-matching process, while Figure 2d shows the water production rate match. Although the overall match in both graphs looks acceptable and the trends are honored, there is a period (between 185 to 240 days) in which the simulation results are quite different from the historical data. The anomalous well production during this period is due to an unknown operational change in the well and is ignored in the simulation model.

Figures 2e, f show the two phase water-oil and gas-oil relative permeability curves which were obtained as a result of history-matching the production data. The curves were generated with the Corey correlation for initial simulations. These were then adjusted for matching historical water and gas rate measurements. It is important to note that because the changes in the gas rate follow the changes in the oil rate (compare *Fig. 2b, c*) and also the gas rate has stabilized at the end of the matching period, the reservoir is interpreted to be above the bubble point pressure (if the pressure falls below the bubble point pressure we will experience a sharp increase in gas rate and consequently gas-oil ratio). As also noted by Clarkson and Pedersen (2011), the GOR is relatively flat through the production history. Therefore, because only oil and water are flowing in the reservoir, the production data are not ideal for deriving the gas-oil relative permeability curve. Nonetheless, the best match was obtained by using this set of permeability data and we continue using it in the rest of this study.

From the history-matching process, we have obtained base values for our simulation runs. In the next section, we will define the modeling strategy for our sensitivity study and the parameters considered.

## 2 MODELING STRATEGY

To fully understand the requirements for successful exploitation of low-permeability reservoirs, one should perform extensive simulation runs. Before any simulation, the objective function or response (*e.g.* cumulative oil production) should be defined and all impactful factors (those which are believed to affect the response,

which might be as many as tens) should be determined. These factors usually require two to three levels (values) to cover all possible range of variations under different circumstances. An important question is that, in each simulation run, which parameters should be used in the sensitivities and which level(s) (high, medium and low) should be employed.

Design of Experiment (DOE) is a formal structured technique for studying any situation that involves a response that varies as a function of one or more independent variables (Mathews, 2005). DOE is specifically designed to address complex problems where more than one variable may affect the response and two or more variables may interact with each other. DOE replaces inferior methods such as the traditional method of studying the effect of One Variable at a Time (OVAT). Compared to DOE, the OVAT method is an inefficient use of resources and is incapable of detecting the presence of or quantifying interactions between variables (Mathews, 2005).

Application of DOE and Response Surface (RS) in reservoir engineering where an optimization problem, history match or uncertainty analysis is of interest has recently received considerable attention (such as works done by Zabalza-Mezghani *et al.*, 2004; Feraille and Busby, 2009; Feraille and Marrel, 2012). DOE can considerably reduce the number of runs during a risk assessment problem which is an imperative concern for potentially expensive fluid flow simulations (Scheidt *et al.*, 2007). Therefore, we use the DOE technique for determining the variables and their associated levels for each simulation run and the sequence in which these runs should be performed.

## 3 SIMULATION RESPONSE AND VARIABLES

The main objective function that we are interested in is the final oil recovery factor (recovery efficiency) of the reservoir. However, production from multi-fractured horizontal wells starts with relatively high rates followed by a rapid decline. Therefore, short-term recovery factors (recovery at the end of 5 years) and long-term recovery factors (recovery at the end of 20 years) will be considered separately as the responses. Seven factors which seem to affect the recovery in tight oil formations the most were chosen. These factors and their different levels are summarized in Table 3. Five parameters are assigned three levels and the two remaining ones get two levels. In the following, we provide a brief description of the response, selected factors and their levels and the rationale for these choices.

TABLE 3
Factors considered for simulation runs

| Factor | Unit | Low | Medium | High |
|---|---|---|---|---|
| A: number of wells | Number/section | 2 | 4 | 6 |
| B: length of well | Ft | 1 500 | 3 000 | 4 500 |
| C: density of HF | Number of HF/(500 ft well) | 2 | - | 3 |
| D: half-length of HF | Ft | 125 | 225 | 325 |
| E: conductivity of HF | md-ft | 250 | 750 | 1 250 |
| F: well operating BHP | Psi | 300 | 600 | 1 200 |
| G: completion method | N/A | Open-hole ("1") | - | Cased-hole ("2") |

## Recovery Factor (Response)

The ultimate objective of any reservoir engineering design is to obtain the highest possible recovery from the reservoir, subject to economic viability of the design. However, we intentionally neglect the economic factors, as this is the focus of the second part of this study, which will be looking for the optimal design.

## Number of Wells per Section (Factor A)

The number of wells and the spacing between them are important factors in management of recovery. As the number of wells increases, the cumulative drainage area of the wells increases as does the recovery efficiency. To simplify the problem, in all conducted simulations, the wells are aligned in the same direction and are evenly spaced in the reservoir section.

## Length of Well Lateral (Factor B)

As the well length increases, more volume of the reservoir would participate in production leading to increased recovery. The minimum and maximum well lengths are 1 500 ft and 4 500 ft, respectively.

## Density of Hydraulic Fractures (Factor C)

The main goal of generating hydraulic fractures from the horizontal well is to increase the contacted area with the reservoir. It is reasonable to increase the number of fractures per well (to a point) in favor of obtaining higher recoveries. In this study, the hydraulic fractures will be evenly spaced along the well. Factor C controls the

number of fractures along the well which can be calculated from Equation (2).

$$\# \, HF/Well = \begin{cases} n+1 & ; \; C = 2 \\ 2n+1 & ; \; C = 3 \end{cases} \tag{2}$$

where $n$ is the well length (B)/500 ft. For instance, if a well is 3 000 ft in length, the number of fractures is 7 and 13 for $C$ equals to two and three, respectively.

## Half-Length of Hydraulic Fractures (Factor D)

In the case of transverse hydraulic factures, increasing the fracture half-length also increases the contacted area for each horizontal well. In simulation runs, one of these values may be encountered: 125 ft (short), 225 ft (moderate) or 325 ft (long).

## Conductivity of Hydraulic Fractures (Factor E)

This factor is important in determining the dimensionless fracture conductivity parameter as given in Equation (3):

$$F_{cd} = \frac{k_f \times w_f}{k \times x_f} \tag{3}$$

where $k_f \times w_f$ is the conductivity of the hydraulic fracture, $k$ is the matrix permeability and $x_f$ is the fracture half-length. Considering a matrix permeability equal to 0.28 mD the range of possible $F_{cd}$ in this study will be between 2.7 (for lowest fracture conductivity and highest fracture half-length) to 35.7 (for highest fracture conductivity and lowest fracture half-length). This range covers a spectrum of finite to almost infinite conductivity; *i.e.* $F_{cd} > 20$ (Economides *et al.*, 2002).

## Well Operating BHP (Factor F)

BHP values closer to the oil bubble point pressure hinder the formation of two phase flow of oil and gas and keep oil effective permeability at higher values but restrict the overall flow rates. BHP values closer to atmospheric pressure on the other hand have a reverse effect. Thus, this factor may also have a two-sided effect on the recovery factors and needs to be considered.

## Completion Strategy (Factor G)

In an "open-hole" completion, the whole length of the horizontal well + the hydraulic fractures contribute to the fluid flow while in the "cased-hole" method, the production only occurs through the hydraulic fractures to the well. This restricted flow path may adversely influence the productivity index of the well.

If OVAT technique is used to design the simulation runs, a total of 972 ($= 3^5 \times 2^2$) runs will be required. However, if an "optimal design" scheme through DOE is used, far fewer runs will be required to obtain the same level of information. By an optimal design, we mean a design that is "best" with respect to some criterion.

There are several popular design optimality criteria like *A*, *D* and *I* just to name a few (Montgomery, 2007). In this study, we used the "D-optimal" design which is the most widely used and is a suitable choice for deterministic computer simulation models. However, it should be mentioned that "optimal design" is only one of the available algorithms in DOE and other methods such as "factorial design", "latin hypercube" or more sophisticated techniques have been used in dealing with demanding reservoir simulation runs (*e.g.*, see Manceau *et al.*, 2001; Feraille and Busby, 2009; Maschio *et al.*, 2010; Ghomian *et al.*, 2010).

## 4 RESULTS AND DISCUSSION

The collection of simulation runs and the order in which these runs performed are tabulated in Table 4. As can be seen in this table, a total of 79 runs were carried out which is only 9% of the total runs required if the OVAT method were used. In each scenario, several factors may change simultaneously and acquire different levels. The design of the runs was performed using the Expert-Design® software (Stat-Ease, 2010). As can be seen in this table, no special pattern in the chosen factor levels

from one run to the next can be recognized, thus the random nature of selections is honored.

Despite the significant difference in the time-span of the two model responses, as can be seen in Table 3, for the majority of the simulation runs, the recovery factor at the end of 20th year (long-term recovery) is not considerably higher than that at the end of the 5th year. Therefore, it is reasonable to analyze these two responses separately. According to these results, the short-term recovery factor ranges between 0.9% (Run 64) and 14.9% (Run 19) and between 2% and 18.7% for the long-term recovery for the same set of runs.

A half-normal probability plot is used to determine whether the entire set of factors and interaction effects are by coincidence (and, thus, shows no effect significantly different from zero) or whether some factors and/or interactions have significant effect. This is a plot of the absolute value of the effect estimates against their normal cumulative probabilities. Any factors or interactions whose observed effects are due to chance are expected to be randomly distributed around zero. These effects will tend to fall along a straight line called "line of no effect". The straight line on this plot always passes through the origin and should also pass close to the fiftieth percentile data value. The effects that might be significant have average values different from zero and are located a substantial distance away from the straight line that represents no effect (Montgomery, 2007).

Figure 3a displays the half-normal probability plot of the factors that might affect the short-term recovery factors in simulation runs. As this figure suggests, with the model properties used in this paper (*Tab. 1*), only 5 factors out of 7 have a significant effect on the recovery. The order of their significance is:
- the number of wells (factor A);
- the length of the well (factor B);
- operating BHP of the wells (factor F);
- half-length of hydraulic fracture[2] (factor D);
- density of HF (factor C).

Interestingly, the conductivity of the fractures and the completion strategy in comparison with the other five factors have only a minor effect on the objective function (they are statistically insignificant). In other words, although increasing fracture conductivity and using the open-hole completion scheme may increase the recovery factor (which is the case for our simulation runs), investment in the other five factors possibly will yield better results. Figure 3b is similar graph but for the long-term recovery factors. The same factors as in the short-term recovery cases have emerged as significant, except the

---

[2] The term "Hydraulic Fracture(s)" will be henceforth abbreviated as HF.

TABLE 4

Simulation runs chosen based on the "d-optimal" design

| Run No. | Factor | | | | | | | Response | |
|---|---|---|---|---|---|---|---|---|---|
| | A | B | C | D | E | F | G | Rec. (%) short-term | Rec. (%) long-term |
| 1 | 4 | 4 500 | 3 | 225 | 250 | 1200 | 1 | 5.2 | 7.8 |
| 2 | 6 | 4 500 | 2 | 225 | 1 250 | 600 | 2 | 12.2 | 17.0 |
| 3 | 4 | 4 500 | 2 | 225 | 1 250 | 1 200 | 2 | 4.4 | 7.4 |
| 4 | 2 | 3 000 | 2 | 325 | 250 | 600 | 2 | 3.2 | 6.4 |
| 5 | 6 | 1 500 | 2 | 225 | 750 | 600 | 2 | 5.0 | 8.1 |
| 6 | 2 | 1 500 | 2 | 325 | 250 | 1200 | 1 | 1.1 | 2.3 |
| 7 | 6 | 3 000 | 2 | 325 | 250 | 300 | 1 | 10.5 | 14.3 |
| 8 | 2 | 1 500 | 3 | 225 | 1 250 | 1 200 | 1 | 1.2 | 2.3 |
| 9 | 4 | 1 500 | 3 | 225 | 750 | 600 | 1 | 4.0 | 6.8 |
| 10 | 2 | 1 500 | 3 | 225 | 750 | 300 | 1 | 2.3 | 4.5 |
| 11 | 2 | 1 500 | 3 | 225 | 250 | 1 200 | 2 | 1.1 | 2.2 |
| 12 | 6 | 1 500 | 3 | 225 | 1 250 | 600 | 1 | 5.6 | 7.9 |
| 13 | 4 | 3 000 | 2 | 125 | 250 | 1 200 | 1 | 2.9 | 5.2 |
| 14 | 2 | 3 000 | 2 | 225 | 750 | 600 | 1 | 3.3 | 6.2 |
| 15 | 4 | 1 500 | 2 | 325 | 1 250 | 300 | 1 | 5.1 | 8.1 |
| 16 | 2 | 3 000 | 3 | 125 | 1 250 | 600 | 1 | 2.9 | 5.8 |
| 17 | 6 | 4 500 | 2 | 225 | 750 | 300 | 1 | 14.1 | 18.5 |
| 18 | 4 | 1 500 | 2 | 225 | 250 | 300 | 2 | 3.6 | 7.2 |
| 19[†] | 6 | 4 500 | 3 | 225 | 250 | 300 | 2 | 14.9 | 18.7 |
| 20 | 4 | 1 500 | 2 | 125 | 750 | 300 | 2 | 3.1 | 6.7 |
| 21 | 6 | 4 500 | 3 | 125 | 250 | 600 | 2 | 11.2 | 16.2 |
| 22 | 2 | 4 500 | 2 | 225 | 250 | 1 200 | 2 | 2.0 | 4.1 |
| 23 | 6 | 1 500 | 3 | 325 | 1 250 | 600 | 2 | 6.3 | 8.5 |
| 24 | 2 | 3 000 | 3 | 325 | 250 | 600 | 1 | 4.0 | 6.8 |
| 25 | 6 | 4 500 | 3 | 225 | 750 | 1 200 | 2 | 7.4 | 9.1 |
| 26 | 2 | 3 000 | 3 | 325 | 750 | 300 | 2 | 5.0 | 8.2 |
| 27 | 2 | 1 500 | 2 | 225 | 1 250 | 600 | 2 | 1.9 | 3.9 |
| 28 | 6 | 3 000 | 3 | 325 | 1 250 | 1 200 | 1 | 5.9 | 7.2 |
| 29 | 4 | 3 000 | 3 | 325 | 750 | 1 200 | 2 | 4.6 | 6.4 |
| 30 | 6 | 4 500 | 3 | 125 | 1250 | 1 200 | 2 | 6.1 | 8.7 |
| 31 | 4 | 3 000 | 3 | 225 | 250 | 600 | 2 | 6.7 | 10.7 |
| 32 | 4 | 4 500 | 2 | 225 | 250 | 600 | 1 | 8.5 | 14.1 |
| 33 | 2 | 4 500 | 3 | 125 | 750 | 300 | 2 | 4.6 | 8.7 |
| 34 | 4 | 4 500 | 3 | 325 | 1 250 | 600 | 1 | 12.1 | 15.4 |
| 35 | 4 | 4 500 | 3 | 125 | 250 | 1 200 | 2 | 4.2 | 7.1 |
| 36 | 2 | 1 500 | 2 | 325 | 250 | 300 | 2 | 2.1 | 4.5 |
| 37 | 2 | 3 000 | 2 | 125 | 1 250 | 1 200 | 2 | 1.3 | 2.8 |
| 38 | 2 | 4 500 | 2 | 325 | 750 | 300 | 1 | 6.0 | 10.0 |

† Highest short and long-term recovery.
‡ Lowest short and long-term recovery.

TABLE 4  *(continued)*

| Run No. | Factor | | | | | | | Response | |
|---|---|---|---|---|---|---|---|---|---|
| | A | B | C | D | E | F | G | Rec. (%) short-term | Rec. (%) long-term |
| 39 | 6 | 1 500 | 3 | 225 | 1 250 | 300 | 2 | 6.4 | 9.0 |
| 40 | 6 | 3 000 | 3 | 225 | 750 | 600 | 2 | 9.9 | 12.8 |
| 41 | 6 | 1 500 | 2 | 325 | 250 | 600 | 2 | 4.9 | 8.1 |
| 42 | 4 | 3 000 | 2 | 225 | 750 | 300 | 2 | 6.9 | 12.1 |
| 43 | 4 | 3 000 | 3 | 125 | 750 | 600 | 2 | 5.7 | 10.2 |
| 44 | 4 | 3 000 | 3 | 325 | 1 250 | 300 | 2 | 9.7 | 13.2 |
| 45 | 6 | 3 000 | 2 | 125 | 250 | 600 | 1 | 7.7 | 12.1 |
| 46 | 6 | 1 500 | 2 | 125 | 750 | 300 | 1 | 5.1 | 8.5 |
| 47 | 4 | 1 500 | 2 | 325 | 750 | 600 | 1 | 4.4 | 7.2 |
| 48 | 4 | 4 500 | 2 | 125 | 250 | 300 | 1 | 8.4 | 15.2 |
| 49 | 4 | 1 500 | 3 | 225 | 250 | 300 | 1 | 4.3 | 7.4 |
| 50 | 6 | 1 500 | 2 | 125 | 1 250 | 1 200 | 2 | 2.0 | 3.9 |
| 51 | 2 | 4 500 | 2 | 225 | 1 250 | 300 | 1 | 5.1 | 9.3 |
| 52 | 4 | 4 500 | 2 | 125 | 750 | 1 200 | 1 | 4.2 | 7.1 |
| 53 | 6 | 3 000 | 2 | 325 | 750 | 1 200 | 1 | 5.3 | 7.1 |
| 54 | 2 | 1 500 | 2 | 125 | 250 | 300 | 1 | 1.8 | 3.9 |
| 55 | 6 | 1 500 | 3 | 225 | 750 | 1 200 | 1 | 3.0 | 4.4 |
| 56 | 2 | 3 000 | 2 | 125 | 1 250 | 300 | 1 | 3.1 | 6.4 |
| 57 | 6 | 4 500 | 3 | 125 | 750 | 600 | 1 | 11.5 | 16.2 |
| 58 | 6 | 1 500 | 3 | 125 | 250 | 1 200 | 1 | 2.5 | 4.1 |
| 59 | 6 | 4 500 | 2 | 125 | 750 | 300 | 2 | 10.9 | 18.0 |
| 60 | 6 | 3 000 | 2 | 225 | 250 | 1 200 | 1 | 4.6 | 6.7 |
| 61 | 2 | 4 500 | 2 | 325 | 1 250 | 1 200 | 2 | 2.8 | 5.0 |
| 62 | 6 | 4 500 | 3 | 325 | 250 | 1 200 | 2 | 7.5 | 9.2 |
| 63 | 4 | 1 500 | 2 | 325 | 1 250 | 1 200 | 2 | 2.3 | 3.9 |
| 64[‡] | 2 | 1 500 | 3 | 125 | 750 | 1 200 | 2 | 0.9 | 2.0 |
| 65 | 4 | 1 500 | 2 | 125 | 1 250 | 600 | 2 | 2.8 | 6.0 |
| 66 | 2 | 3 000 | 3 | 225 | 1 250 | 300 | 2 | 4.1 | 7.3 |
| 67 | 2 | 4 500 | 2 | 125 | 250 | 600 | 1 | 3.8 | 7.6 |
| 68 | 6 | 3 000 | 3 | 125 | 750 | 1 200 | 1 | 4.4 | 6.5 |
| 69 | 6 | 3 000 | 3 | 125 | 250 | 300 | 2 | 8.9 | 13.5 |
| 70 | 2 | 1 500 | 2 | 225 | 750 | 1 200 | 2 | 1.0 | 2.1 |
| 71 | 6 | 3 000 | 2 | 325 | 1 250 | 300 | 2 | 11.0 | 15.0 |
| 72 | 4 | 3 000 | 2 | 225 | 1 250 | 600 | 1 | 6.5 | 10.8 |
| 73 | 4 | 1 500 | 3 | 125 | 1 250 | 1 200 | 1 | 1.8 | 3.4 |
| 74 | 6 | 4 500 | 2 | 325 | 1 250 | 600 | 1 | 14.1 | 16.9 |
| 75 | 2 | 4 500 | 3 | 325 | 750 | 600 | 2 | 6.2 | 9.7 |
| 76 | 4 | 4 500 | 3 | 325 | 750 | 300 | 2 | 13.3 | 17.4 |
| 77 | 2 | 4 500 | 3 | 125 | 250 | 1 200 | 1 | 2.3 | 4.2 |
| 78 | 4 | 1 500 | 3 | 325 | 250 | 600 | 2 | 4.5 | 7.0 |
| 79 | 2 | 4 500 | 3 | 325 | 1 250 | 300 | 1 | 7.0 | 10.9 |

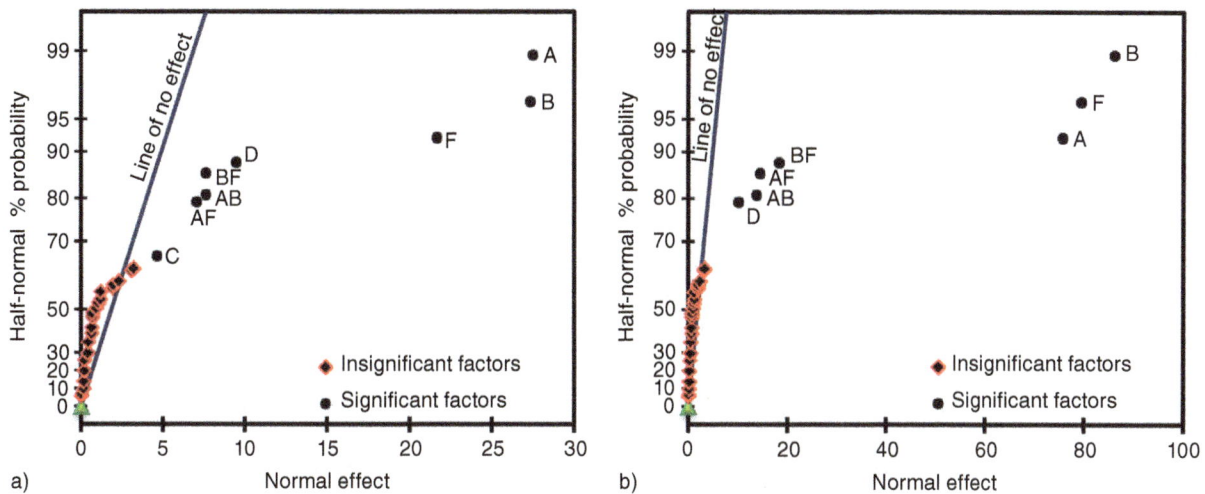

Figure 3

Half-normal probability plot of observed effect of factors and their interactions on a) short-term recovery factors and b) long-term recovery factor for reservoir model with absolute permeability equal to 0.28 mD.

density of HF (factor C). We will discuss the reasons for this difference later in the text.

In some situations, the difference in response between the levels of one factor is not the same at all levels of the other factors. When this occurs, there is interaction between the factors. For example in Figure 3, we see that interaction exists between A and B, denoted as AB, and this interaction is contributing to the recovery. However, factor D has no interaction with other factors. To clarify the interaction effect, consider Figure 4a in which recovery is plotted *versus* fracture half-length (factor D) at three different levels of factor A. The three depicted curves are approximately parallel to each other indicating that regardless of the level of A, increasing/decreasing the level of D from one level to the next one always causes the same amount of increase/decrease in recovery (similar slope for the curves). Since D has no interaction with any other factor, a similar plot will be obtained if the alternative factors (B, F, etc.) are used instead of A. The general trend is similar; if the level of factor D increases the recovery factor will increase and *vice versa*. Conversely, we see in Figure 4b that the curves describing the recovery factor *versus* the number of wells at different levels of B are non-linear; the amount of increase/decrease in the recovery factor depends on the changes in levels of both factors. For Figures 4c, d, similar comments can be made. Another possible conclusion from these figures is that increasing both the number and the length of the wells (*Fig. 4a*), increasing the number of wells and decreasing the BHP, increasing the well length

and decreasing the BHP causes an increase in the achievable recovery factor. Therefore, more wells with longer laterals which are operating at lower BHP when stimulated with longer hydraulic fractures increases both short and long-term recovery.

It is worth looking at the effect of factors which were recognized as having an insignificant impact on primary recovery, namely fracture density (factor C) for the long-term recovery efficiency, and fracture conductivity (factor E) and completion strategy (factor G) for both long-term and short-term efficiencies. Increasing the number of fractures per well, although improving the contacted area with the reservoir, causes the spacing between the fractures to reduce. Under constant BHP operation ($p_{wf} < p_b$) this may accelerate the pressure depletion between the fractures and the formation of the two phase flow regimes in this region. This can adversely reduce the oil effective permeability. Nevertheless, the amount of reduction in the oil effective permeability is a function of the relative permeability characteristics and especially the oil relative permeability end points and critical gas saturation. If the critical gas saturation is so high that it takes a long time to reach that gas saturation between the fractures, then increasing the number of fractures (neglecting economic factors) would be favorable, otherwise there is not much benefit expected. However, since the low-permeability reservoirs should be produced under very low-BHP conditions, this critical value of the gas saturation between the fractures will pass soon after the start of the production and

For: B = 1 500; C = 2; E = 250; F = 300; G = 1

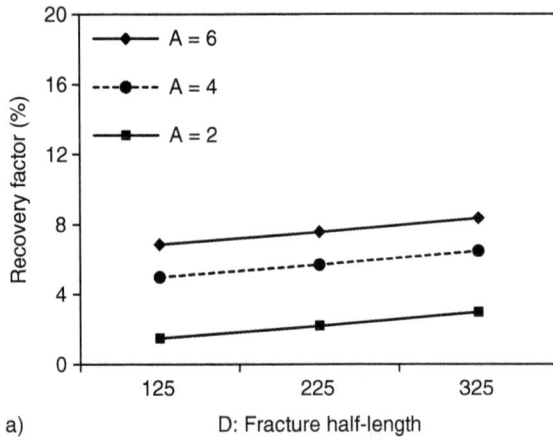

For: C = 2; D = 125; E = 250; F = 300; G = 1

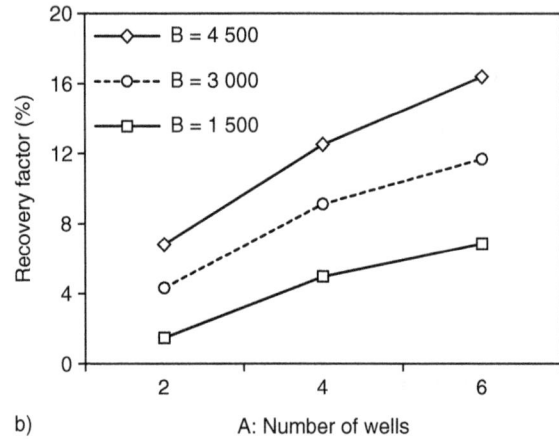

For: B = 1 500; C = 3; D = 325; E = 250; G = 1

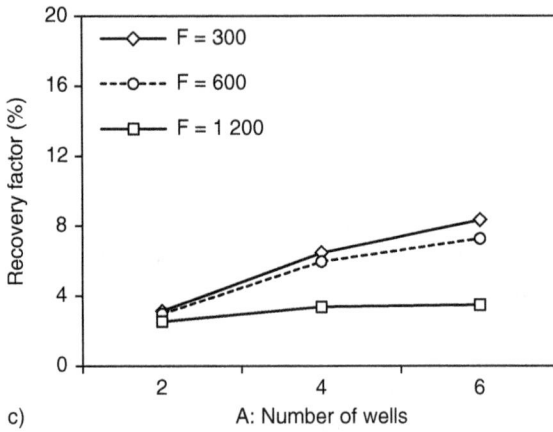

For: A = 4; C = 2; D = 125; E = 250; G = 1

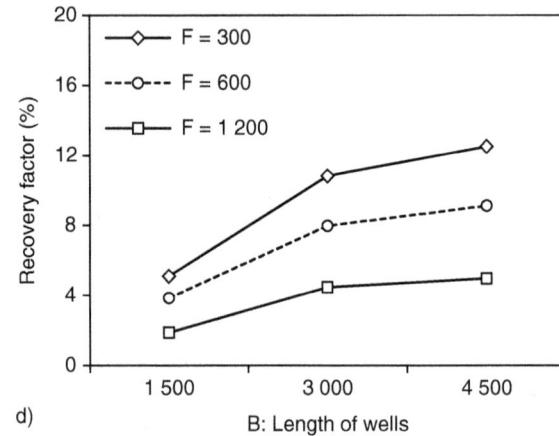

a)  D: Fracture half-length

b)  A: Number of wells

c)  A: Number of wells

d)  B: Length of wells

Figure 4

Effect interaction plot between different factors: a) AD, no interaction case, b) AB, c) AF, d) BF.

higher mobility of the gas with respect to oil hinders the sustainability of high oil production rates.

Figure 5a, b show the gas saturation distribution in the reservoir (top-view) after one year from start of production for two simulation runs which differ only in the number of fractures per well. The spacing between the fractures in Figure 5a is 500 ft and in Figure 5b is 250 ft. For the run with closer hydraulic-fracture spacing, the amount of the evolved gas saturation between fractures and also the extension of two phase oil-gas region is greater, both of which augment two phase flow interference. This in turn causes lower mobility to oil and hence reduces recovery efficiency. Figure 5c demonstrates that oil production rates from the model with larger number of fractures will produce oil at higher

rates early on (short-term effect) but at later times the two-phase flow interference becomes important and productivity drops off (long-term effect). The recovery curves (*Fig. 5d*) are consistent with the oil production rates and the results from two half-normal probability plots. According to this figure, there is a statistically significant difference between the short-term recovery factors but this difference gradually diminishes over time. Therefore, there exists an optimum number for the fracture density which primarily depends on the reservoir characteristics and secondly on the economic concerns.

Our simulation results confirm that similar comments made for fracture spacing can be made for the effect of fracture conductivity, as this factor was also recognized as insignificant for recovery efficiency. Figure 6a

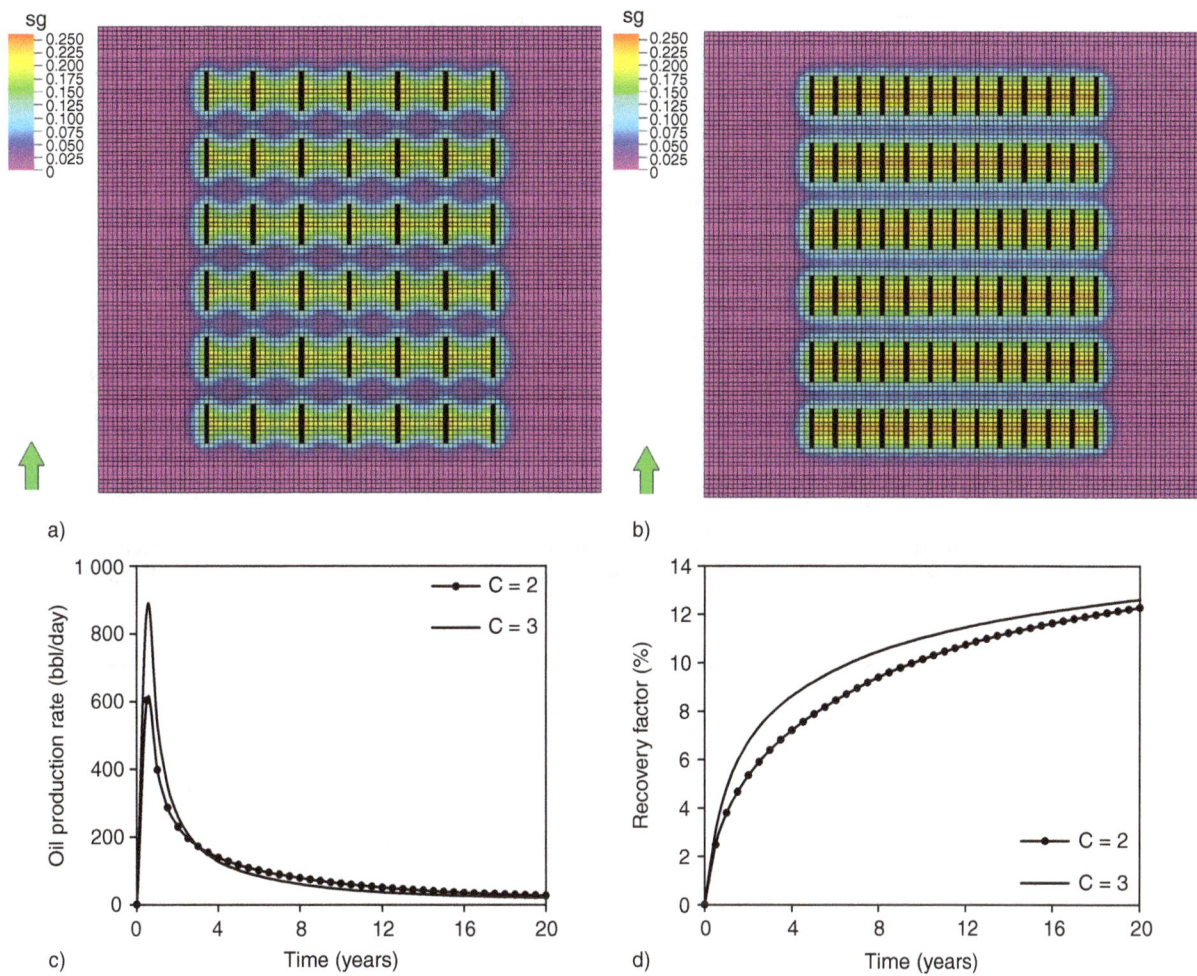

Figure 5

Gas saturation in two models with a) C = 2 and b) C = 3 one year after start of production; c,d) field oil production rates and the oil recovery factors for the models. Other parameters are as follow: A = 6; B = 3, 000 ft; D = 225 ft; E = 750 md-ft; F = 300 psia.

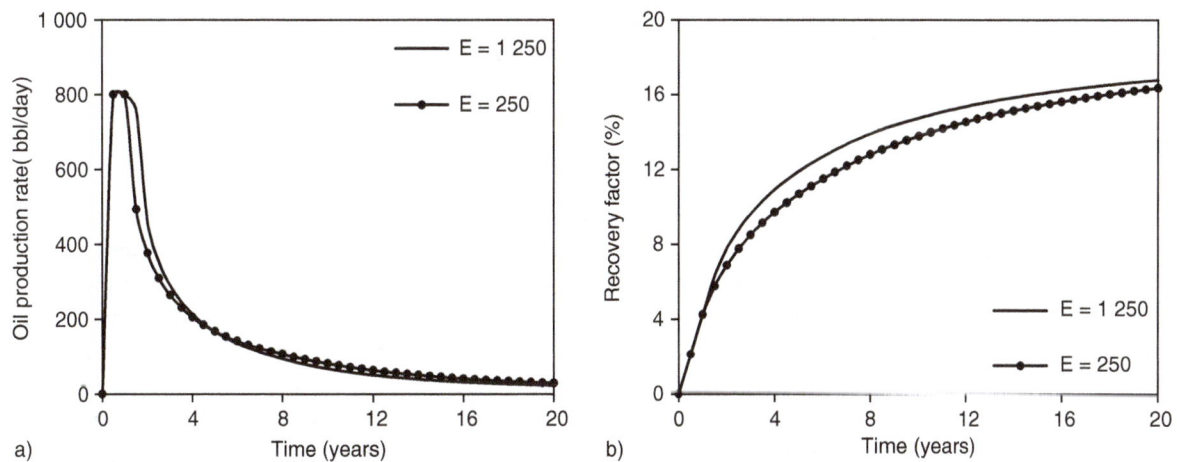

Figure 6

Field oil production rates a) and the oil recovery factors b) for the models with different levels of E factor; other parameters are as follow: A = 4; B = 4 500 ft; C = 2; D = 325 ft; F = 300 psia; G = 1.

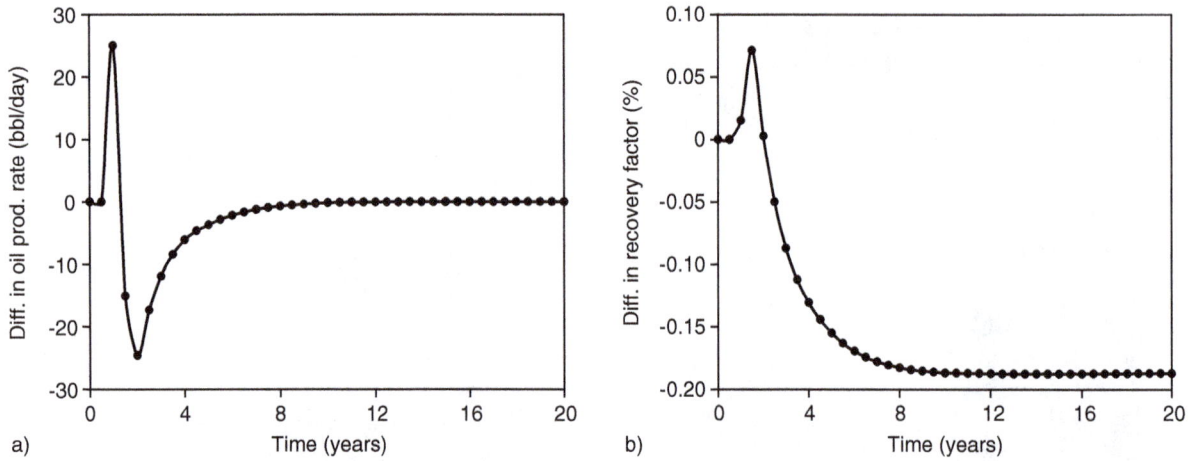

Figure 7

Difference in a) field oil production rates and b) the oil recovery factors for the models with different levels of G factor (graph results = "open-hole" results – "cased-hole" results); other parameters are as follow: A = 6; B = 3000 ft; C = 3; D = 325 ft; E = 750 md-ft; F = 300 psia.

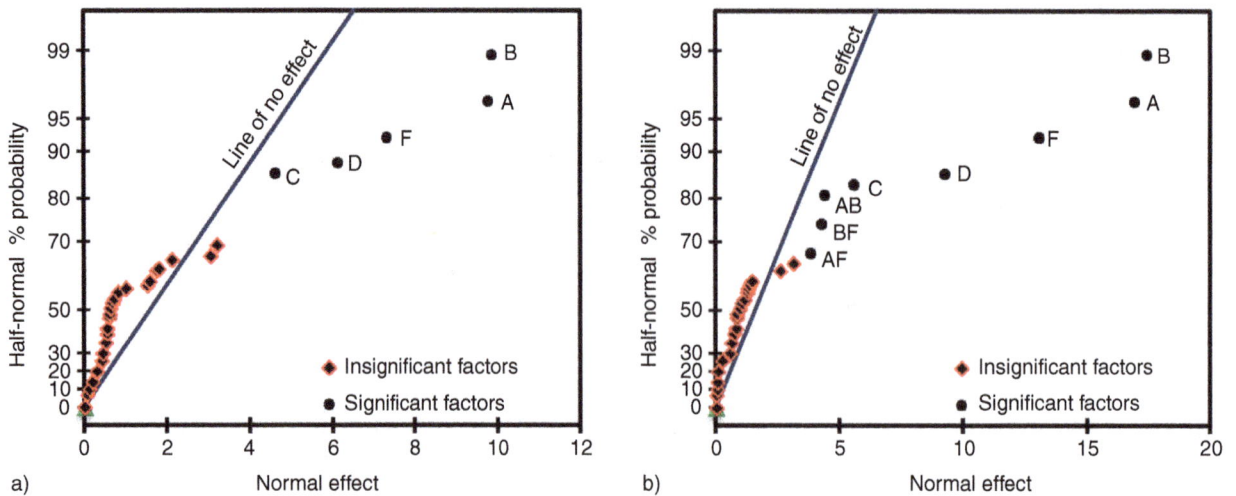

Figure 8

Half-normal probability plot of observed effect of factors and their interactions on a) short-term recovery factors and b) long-term recovery factor for reservoir model with absolute permeability equal to 0.028 mD.

compares the oil production rate from two models which differ only in fracture conductivity. Figure 6b demonstrates the corresponding recovery factors for the models. Embedding more conductive fractures will result in higher production rates earlier in the life of the project which might shorten the project pay-back period.

Finally, the difference in results for the open-hole and cased-hole methods (factor G) is depicted in Figure 7 and confirms the outcome of the previous analysis which demonstrated that there is little difference between the two completion methods. The largest contribution to

production occurs through the highly conductive fracture conduits and the contribution from other perforations along the horizontal well is of subtle importance. It should be also noted that a minor difference in the recovery factors between two models in Figure 7b may arise from the numerical errors.

Since the conventional methods for measurement of permeability sometimes fail when applied to tighter formations, there is risk in the estimation of this property. Therefore, we repeated the analysis above but with absolute permeability reduced one order of magnitude (absolute permeability = 0.028 mD). The half-normal probability plot of the factor effects is given in Figure 8. It is interesting to note that once more the same significant factors appear on both graphs but now factor C is significant for short and long-term recovery efficiency (compare *Fig. 8* and *Fig. 3*). Also comparing the two figures, we see that factor *C* either does not appear as a significant factor or it sits very close (in comparison with other factors) to the "line of no effect". There is a good chance that if the reservoir properties change, particularly the relative permeability data, fracture density may be totally removed from the list of significant factors and hence it should be treated carefully and optimized for the formation under consideration. This result signifies the importance of accurate reservoir characterization for tight formations.

## CONCLUSION

In this work, a systematic method for establishing the most important factors affecting primary recovery (short- and long-term) using multi-fractured horizontal wells is provided. The combination of Design of Experiment (DOE) and numerical simulation used in this work is superior to the One Variable at a Time approach (OVAT) used in many studies because a) fewer runs are required and b) the interaction between factors can be established and quantified.

For reservoir and fluid properties typical of the low-permeability Pembina Cardium, the most important factors controlling primary recovery in order of significance are: the number of wells per section; the length of the well; operating BHP pressure of the wells. Fracture conductivity has a lesser effect on recovery factor but will impact early time production rates. Higher fracture density may accelerate oil production but also accelerates gas saturation build-up between fractures, which reduces oil mobility and impairs long term oil production. Further, whether or not fracture density has a significant impact on recovery depends on the combination of reservoir properties, particularly relative and absolute permeability, underpinning the need for careful reservoir characterization. The difference in long-term recovery factor due to completion method (open-hole *versus* cased-hole) appears to be insignificant.

Although we performed our sensitivity runs at two different matrix permeability levels, we recognize that there is also considerable uncertainty in relative permeability characteristics – we will perform our analysis using different relative permeability scenarios in the future.

In future work, reservoir heterogeneity and changes in relative permeability will be considered. Economic criteria will also be incorporated into the analysis. Finally, a similar approach will be used to establish critical factors controlling $CO_2$-EOR in this tight oil play; $CO_2$-EOR in this play is worthy of investigation because of not only the potential for increased of oil recovery but also the possibility for $CO_2$ sequestration.

## ACKNOWLEDGMENTS

A preliminary version of this paper was published as SPE 148995 and one-time copyright permission has been received through license number 3035390338406 from the society of petroleum engineers. This research was made possible by funding provided by ISEEE-NRCan (project title "Co-optimization of $CO_2$ Sequestration and Oil Recovery in Tight Oil Formations"). Chris Clarkson would like to acknowledge Encana for support of his Chair position in Unconventional Gas at the University of Calgary, Department of Geosciences. He would also like to acknowledge the sponsors of Tight Oil Consortium (TOC), hosted at the University of Calgary. The authors would like to acknowledge ARC Resources for contributions of the Cardium well production.

## REFERENCES

Adegbesan K.O., Costello J.P., Elsborg C.C., King D.A., Mills M., King R.W. (1996) Key Performance Drivers for Horizontal Wells Under Waterflood Operations in the Layered Pembina Cardium Reservoir, *J. Can. Pet. Technol.* **35**, 8, 25-35, *SPE paper* 96-08-02.

Clarkson C.R., Pedersen P.K. (2010) Tight Oil Production Analysis: Adaptation of Existing Rate-Transient Analysis Techniques, *Canadian Unconventional Resources & International Petroleum Conference*, Calgary, Alberta, Canada, 19-21 Oct., *Paper CSUG/SPE* 137352, DOI: 10.2118/137352-MS.

Clarkson C.R., Pedersen P.K. (2011) Production Analysis of Western Canadian Unconventional Light Oil Plays, *Canadian Unconventional Resources Conference*, Calgary, Alberta, Canada, 15-17 Nov., *Paper CSUG/SPE* 149005.

Economides M.J., Oligney R.E., Valkó P. (2002) *Unified Fracture Design: Bridging the Gap between Theory and Practice*, Orsa Press, Alvin, Texas.

Feraille M., Busby D. (2009) Uncertainty Management on a Reservoir Workflow, International Petroleum Technology Conference, Doha, Qatar, 7-9 Dec.

Feraille M., Marrel A. (2012) Prediction under Uncertainty on a Mature Field, *Oil Gas Sci., Technol. – Rev. IFP Energies nouvelles* **67**, 2, 193-206.

Ghomian Y., Sepehrnoori K., Pope G.A. (2010) Efficient Investigation of Uncertainties in Flood Design Parameters for Coupled $CO_2$ Sequestration and Enhanced Oil Recovery, *SPE International Conference on $CO_2$ Capture, Storage and Utilization*, New Orleans, Louisiana, USA, 10-12 Nov., *SPE Paper* 139738.

Krause F.F., Collins H.N., Nelson D.A., Machemer S.D., French P.R. (1987) Multiscale Anatomy of a Reservoir: Geological Characterization of Pembina-Cardium Pool, West-central Alberta, Canada, *AAPG Bull.* **71**, 1233-1260.

Manceau E., Mezghani M., Zabalza-Mezghani I., Roggero F. (2001) Combination Of Experimental Design and Joint Modeling Methods for Quantifying the Risk Associated with Deterministic and Stochastic Uncertainties - An Integrated Test Study, *Annual Technical Conference and Exhibition*, New Orleans, Louisiana, New-Orleans, USA, 30 Sept.-3 Oct., *Paper SPE* 71620-MS.

Maschio C., Viegas de Carvalho C.P., Schiozer D.J. (2010) A New Methodology to Reduce Uncertainties in Reservoir Simulation Models Using Observed Data and Sampling Techniques, *J. Petrol. Sci. Eng.* **72**, 1, 110-119.

Mathews P. (2005) *Design of Experiments with MINITAB*, ASQ Quality Press, Milwaukee, Wisconsin.

Montgomery D.C. (2007) *Design and Analysis of Experiments*, seventh ed., John Wiley & Sons Inc.

Pedersen K.S., Fredenslund A., Christensen P.L., Thomassen P. (1984) Viscosity of Crude Oils, *Chem. Eng. Sci.* **39**, 6, 1011-1016.

*Schlumberger* (2010) Eclipse (version 2010.1) Reference Manual.

*Schlumberger* (2010) PVTi (version 2010.1) Reference Manual.

Shaoul J.R., Behr A., Mtchedlishvili G. (2007) Developing a Tool for 3D Reservoir Simulation of Hydraulically Fractured Wells, *SPE Reserv. Eval. Eng.* **10**, 1, 50-59.

Stat-Ease Inc. (2010) Design-Expert (version 8.0.6) Reference Manual.

Scheidt C., Zabalza-Mezghani I., Feraille M., Collombier D. (2007) Toward a Reliable Quantification of Uncertainty on Production Forecasts: Adaptive Experimental Designs, *Oil Gas Sci. Technol.* **62**, 2, 207-224.

Viau C., Nielsen K. (2010) Unconventional Oil Production in the Cardium Formation, East Pembina Area, Alberta, Canada, *CSPG Core Conference*, Calgary, Alberta, Canada, 16-17 Sept.

Zabalza-Mezghani I., Manceau E., Feraille M., Jourdan A. (2004) Uncertainty Management: From Geological Scenarios to Production Scheme Optimization, *J. Petrol. Sci. Eng.* **44**, 1-2, 11-25.

# 3

# Pilot Plant Studies for $CO_2$ Capture from Waste Incinerator Flue Gas Using MEA Based Solvent

Ismaël Aouini[1], Alain Ledoux[1]*, Lionel Estel[1] and Soazic Mary[2]

[1] Normandie Université, INSA de Rouen, LSPC EA4704, Avenue de l'Université, BP 8, 76801 Saint-Étienne-du-Rouvray Cedex - France
[2] Veolia Environnement Recherche et Innovation, 10 rue Jacques Daguerre, 92500 Rueil-Malmaison - France
e-mail: alain.ledoux@insa_rouen.fr

* Corresponding author

**Résumé — Étude du captage du $CO_2$ dans des gaz de combustion d'un incinérateur de déchets à l'aide d'un pilote utilisant un solvant à base de MEA** — L'étude évalue la faisabilité du captage du dioxyde de carbone ($CO_2$) dans des gaz de combustion d'un incinérateur de déchets spéciaux. Un pilote à l'échelle laboratoire a été mis en œuvre pour étudier le procédé d'absorption/désorption du $CO_2$ par un solvant à base de MonoEthanolAmine (MEA). L'étude a été conduite en laboratoire et sur un site industriel. L'installation expérimentale est instrumentée pour obtenir des bilans de matière avec précision. Une étude paramétrique en laboratoire a permis de mesurer le coefficient de transfert global $K_G a_w$ pour plusieurs régimes de fonctionnement du pilote. Des expériences de longue durée, en laboratoire et sur site industriel, ont analysé la résistance chimique de la MEA au gaz de combustion d'un incinérateur. Ces expériences ont également permis d'étudier l'absorption du dioxyde d'azote $NO_2$ et du dioxyde de soufre $SO_2$ dans le solvant de captage. Elles ont abouti à une estimation de la cinétique d'accumulation de sels stables dans un solvant à base de MEA confronté à des gaz de combustion d'un incinérateur.

*Abstract — Pilot Plant Studies for $CO_2$ Capture from Waste Incinerator Flue Gas Using MEA Based Solvent — Experimental study of carbon dioxide ($CO_2$) capture from waste incinerator flue gas is presented. A specific pilot plant has been achieved based on absorption/desorption process using MonoEthanolAmine (MEA) solvent. Several experiments have been carried out at laboratory and industrial site. The pilot is fully instrumented to establish precise balances. Laboratory experiments allow to measure overall mass transfer coefficient $K_G a_w$ for several pilot operating conditions. Long laboratory and industrial runs provide an estimation of MEA chemical resistance against waste incinerator flue gas. The experiments also allowed the analysis of $NO_2$ and $SO_2$ absorption through the solvent as well as the accumulation of Heat Stable Salts (HSS) for a full scale $CO_2$ capture unit fed by a waste incinerator flue gas.*

## NOMENCLATURE

| | |
|---|---|
| $K_Ga_W$ | Overall mass transfer coefficient (mol/Pa.s.m$^3$) |
| $P_{CO_2}^*$ | $CO_2$ partial pressure in equilibrium with the solvent (Pa) |
| $P_{CO_2}$ | $CO_2$ partial pressure (Pa) |
| $V_{pck}$ | Absorption packing volume (m$^3$) |
| $\Phi_{CO_2}$ | Absorption rate (mol/s) |
| $M_{ln}$ | Logarithmic average |
| MEA | MonoEthanolAmine |
| DEA | DiEthanolAmine |
| MDEA | MethylDiEthanolAmine |
| $CO_2$ | Carbon Dioxide |

## INTRODUCTION

The amine process is one of the most mature technologies on account of the industrial use for natural gas treatment in the oil industry (solvent are based on MonoEthanolAmine (MEA) and MethylDiEthanolAmine (MDEA)). The most used amine is MEA developed by *Fluor Daniel, MHI* (*Mitsubishi Heavy Industries*). Other actors also develop this process in order to adjust it to $CO_2$ capture from flue gases (*e.g. Alstom* (with *Dow Chemical*), *Prosernat* process (IFPEN)).

*Veolia Environnement* has a variety of $CO_2$ emissions with coal combustion plants and waste incinerators. Waste incinerators flue gases have specific composition compared to those from fossil fuel-based. Those specificities are composition and instabilities of flue gas in time related to the large variety of waste (*Tab. 1* for gas specifications and interval of values). Industrial waste incinerator are often close to waste producers. These

industrial areas lead to the creation of circular economy. Indeed, $CO_2$ captured by waste incinerator can be easily reused by other industrial. This is a particular illustration of the concept of industrial symbiosis.

Therefore, the aim of the research program was to validate the technical feasibility of $CO_2$ capture using MEA on hazardous waste incineration flue gas. Assuming this feasibility, the conventional methods of $CO_2$ capture with amines could be transposed to all combustion plants of the group without major modification of the combustion unit.

As a first step, a laboratory pilot scale was created with a synthetic combustion gas supply. Then, the pilot laboratory was moved to an industrial plant to validate the operation with real flue gas. In order to observe the behavior of the process related to the use of waste incinerator flue gas, methods have been developed for absorbing pollutants ($SO_x$, $NO_x$) and production of stable salts, in terms of mass transfer and degradation of the solvent.

First, the MEA process and the incinerator flue gas characteristics are presented. Second, a description of the pilot laboratory is given and finally the results of the main experimental trials are discussed.

## 1 $CO_2$ CAPTURE BY MEA BASED TECHNOLOGY FROM INDUSTRIAL INCINERATOR FLUE GAS

### 1.1 Industrial Incinerator Flue Gas Characteristics

The MonoEthanolAmine (MEA) based technology is widely used for commercial $CO_2$ production in small scale (< 30 ton$CO_2$/h) (Chapel *et al.*, 1999). Nevertheless, several technology gaps were clearly identified (Rochelle *et al.*, 2001; Steeneveldt *et al.*, 2006; Johnsen *et al.*, 2009) to use this technology in carbon

TABLE 1

Typical flue gas composition of actual coal-fired power plant and the target industrial incinerator unit

| | Coal fired power plant (Artanto *et al.*, 2012) | | Target industrial incinerator unit |
|---|---|---|---|
| Temperature | 160-180°C | | 70°C |
| Pressure | 1 atm | | 1 atm |
| H$_2$O | 20-23 (vol.% wet) | | 30 (vol.% wet) |
| CO$_2$ | 10-11 (vol.% wet) | 12.5-14.5 (vol.% dry)* | 5-11 (vol.% dry) |
| O$_2$ | 4-5 (vol.% wet) | 5.0-6.5 (vol.% dry)* | 8-12 (vol.% dry) |
| SO$_2$ | 120-200 (wet ppm volume) | 260-457 (dry mg/m$^3$)* | 6.2 (dry mg/m$^3$) |
| NO$_x$ | 150-250 (wet ppm volume) ($\approx$ 99% NO, balance NO$_2$ and N$_2$O) | 237-410 (dry mg/m$^3$)* | 91 (dry mg/m$^3$) ($\approx$ 80% NO, balance NO$_2$) |

* Converted data from literature.

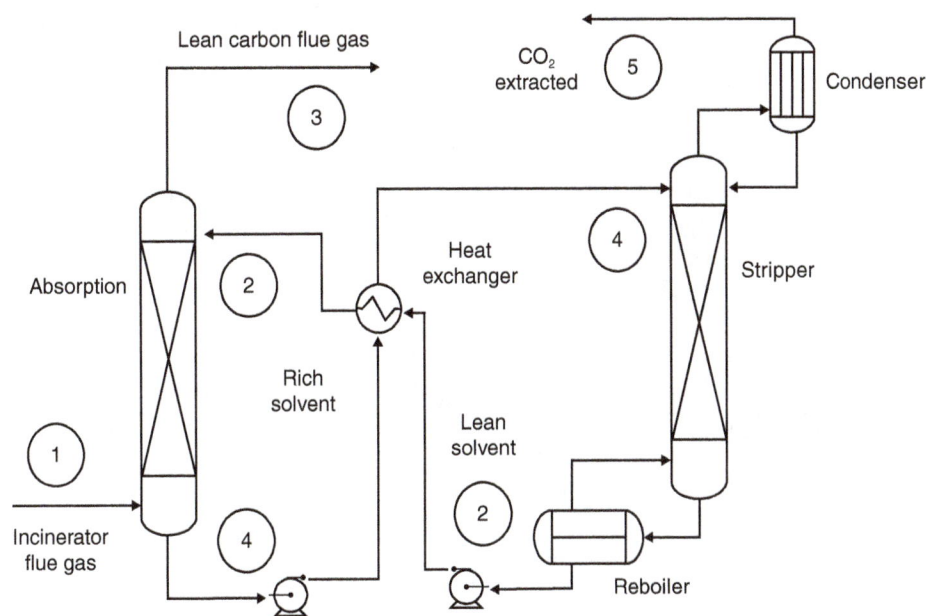

Figure 1

Schematic flow sheet of absorption/desorption process for $CO_2$ capture process.

capture and storage from power plant flue gas. Therefore, high amounts of R&D were launched including several pilot plant studies (Wang *et al.*, 2011) and mainly applied to carbon capture from coal fired power plant.

Results and conclusions of these studies are useful to check the sustainability of carbon dioxide capture from incinerator flue gas without providing an exhaustive answer. Table 1 shows typical compositions of an actual coal-fired power plant flue involved in pilot experiments and the associate flue gas composition of the target incinerator.

Table 1 shows that coal fired flue gas has a higher partial pressure of $CO_2$ and a lower partial pressure of $O_2$ compared to the incinerator flue gas. Moreover, coal fired flue gas has a low composition fluctuation. At the opposite, the incinerator inlet raw material composition varies depending on the waste to be treated and induces a constant fluctuation on outlet flue gas composition.

Coal power plant raw flue gas contains a high fraction of pollutants ($SO_2$, $NO_x$) and needs a gas pretreatment prior to its introduction in MEA capture unit. Incinerator already includes efficient gas purification units and provides flue gas with low level of pollutants ($SO_2$, $NO_x$).

To sum up, $CO_2$ capture from incinerator flue gas is in the commercial scale of MEA based technology but no industrial references could confirm the sustainability of this new application.

## 1.2 CO$_2$ Capture by MEA Based Technology

A schematic view of the $CO_2$ capture process is shown in Figure 1. The bottom of the absorption column is connected to the inlet flue gas. $CO_2$ transfer from gas to solvent along this column which is facilitated by a packing and a counter current flow. The lean carbon flue gas (3) leaves the top of the column to the atmosphere. The $CO_2$ rich solvent is then pumped from the bottom of the absorption column to the stripper (4). As the solvent progresses towards the stripping column, its temperature increases thanks to the heat provided by the reboiler at the bottom. $CO_2$ and steam are released and flow up to the condenser. Decreasing flue gas temperature leads to water condensation. The purified $CO_2$ leaves the condenser (5). The lean solvent is then conducted from the boiler to the absorption column for a new capture cycle. The crossover heat exchanger allows an energy saving by warming up the rich solvent and cooling the lean solvent.

## 1.3 Potential Interactions with the Solvent and the Incinerator Flue Gas

In the case of coal or gas power plant flue gas, comprehensive studies around the stability of 30 wt% MEA solvent have been published (Wang *et al.*, 2011). Although differences over the composition of these flue gases, the following degradation mechanisms of MEA can be transposed in our study.

First, thermal degradation of the solvent can occur through carbamate polymerization. This reaction takes place mainly in the stripper section of the process because a high level of temperature is needed. Several studies conclude that thermal degradation kinetics could be limited if temperature does not exceed 110°C (Bello and Idem, 2005; Davis and Rochelle, 2009; Davis, 2009; Lepaumier et al., 2009a; Rochelle, 2012).

Oxygen fraction of the flue gas creates a second way of degradation: oxidative degradation. Indeed, MEA has strong interaction with oxygen (Supap et al., 2001, 2006; Bello and Idem, 2005; Chi and Rochelle, 2002; Goff and Rochelle, 2004, 2006; Sexton and Rochelle, 2009; Bedell, 2009, 2011; Bedell et al., 2011; Lepaumier et al., 2009b, c). Working closer to the absorption conditions, Sexton and Rochelle (2009) have shown that ammonia is the major by-product in the gas phase and is a good indicator of MEA oxidative degradation. They found that organic acids are strongly present in the liquid phase. Moreover, Bello and Idem (2005) have identified from 10 to 30 oxidative mechanisms and products. Several suggested pathways exhibit the sensitivity of the chemistry system. It has been demonstrated that metal present in setup serves as catalyst for oxidize the reaction. Bedell (2009) proposed a pathway using radical mechanisms for the oxidative degradation. As summarized in Figure 2, the metal cation reacts with oxygen to generate radicals which attack the MEA molecules. The metal cation is then regenerated by the water for a new cycle of reaction.

The most active metallic ions in this kind of degradation are iron, chrome and nickel. Based on the strong interaction between metal and solvent degradation, the major source of metal comes from the corrosion of the facilities. The corrosion is a very significant problem in alkanolamines processes (Veawab and Aroonwilas, 2002) and can impact strongly process facilities

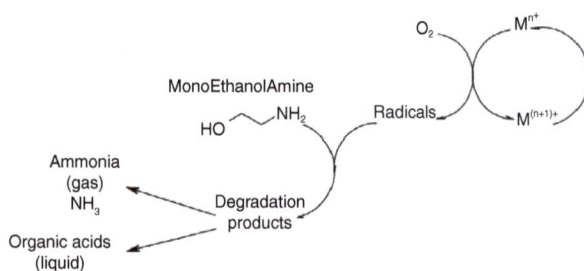

(Dupart et al., 1993a, b). The ability of the solvent in corrosion activity depends strongly on the gas stream composition. For incinerator flue gas, it is expected that, in presence of oxygen, MEA degradation generates acid molecules. Veawab and Aroonwilas (2002) proved that acids increase the corrosiveness of alkanolamine solvents, which increases the metal concentration in the solvent. Furthermore, some pollutants like chloride are well known as corrosion agents. Therefore, corrosion and degradation runaway is expected if an appropriate solution is not implemented. Several additives can be used to limit oxidation. For example, Sexton and Rochelle (2009) propose vanadium and copper based additives.

The incinerator flue gas contains acidic pollutants traces from waste. After dissolution and hydrolysis these pollutants will lead to strong acids which shall be added to the quantity of strong acids produce by MEA oxidative degradation. When the acid is weak like the $CO_2$, the equilibrium is reversed with a temperature increase. Conversely the reaction with strong acids is not reversible and leads to Heat Stable Salts (HSS). Accordingly, the process effectiveness decreases each time that strong acids are added to the solvent. For the incinerator flue gas, no experimental data quantify this phenomenon:

$$A H + HO \diagup NH_2 \rightleftharpoons HO \diagup NH_3^+ + A^-$$

$SO_2$ is the main acidic pollutants in incinerator flue gas. Kohl and Nielsen (1997) show clearly that the sulfur dioxide has a high reactivity with alkaline solution. Starting from $DeSO_x$ alkaline processes feedback, we assume the following mechanisms for $SO_2$ absorption in MEA solvent:

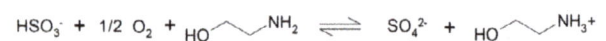

$$SO_2 + H_2O + HO \diagup NH_2 \rightleftharpoons HSO_3^- + HO \diagup NH_3^+$$

$$HSO_3^- + 1/2 O_2 + HO \diagup NH_2 \rightleftharpoons SO_4^{2-} + HO \diagup NH_3^+$$

In the case of $NO_x$ compounds absorption in alkaline solution is complex. First several gas phase reactions occur like described by Carta (1984):

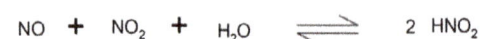

$$NO + 1/2 O_2 \longrightarrow NO_2$$

$$2 NO_2 \rightleftharpoons N_2O_4$$

$$NO + NO_2 \rightleftharpoons N_2O_3$$

$$NO + NO_2 + H_2O \rightleftharpoons 2 HNO_2$$

$$3 NO_2 + H_2O \rightleftharpoons 2 HNO_3 + NO$$

Figure 2

Catalytic way of oxidative degradation of MEA.

Once the $NO_x$ species (NO, $NO_2$, $N_2O_4$, $HNO_2$, $HNO_3$) have been absorbed in the liquid phase, the following mechanisms for the reaction with MEA are assumed (Decanini et al., 2000; Hüpen and Kenig, 2005):

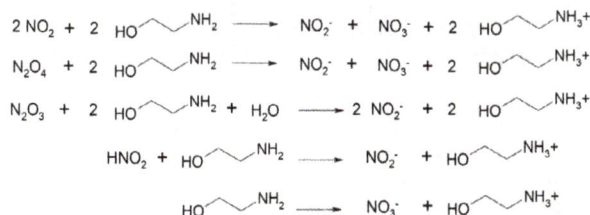

## 1.4 Study Goals

The study objective is to validate the sustainability of $CO_2$ capture from incinerator flue gas with a laboratory pilot. A target unit was chosen and a specific pilot was designed and built to analyze:

- the MEA solvent chemical stability in front of high oxygen fraction and incinerator flue gas pollutants,
- the process performance stability for $CO_2$ capture and the energy consumption relative to the flue gas fluctuations and the solvent degradation.

First, laboratory experiments allowed checking pilot performance for $CO_2$ capture and energy consumption. At the same time, we perform measurements on solvent stability against flue gas without the variation on inlet flue gas compositions.

Second, the pilot was installed in the target incinerator unit to validate the sustainability of the process.

## 2 EXPERIMENTAL SETUP

### 2.1 Previous Studies

During the last decade, numerous research and development projects have studied the $CO_2$ capture with a MEA technology using pilots (Wang et al., 2011). Table 2 presents the experimental units described on literature and on which our pilot design was based. Although the setups have a wide range of $CO_2$ production, from 1 kgCO2/h to 100 tonCO2/day, they share the following characteristics:

- dense instrumentation for the temperature, flow and composition measurements with high number of gas and liquid sampling points;
- multipurpose absorption and stripper equipment to perform parametric studies with a wide range of operating conditions.

All published results mainly focus on the process performances for $CO_2$ capture energy consumption by varying $L/G$ ratio and strippers operations conditions for different solvent formulation (Mimura et al., 1995; Gabrielsen, 2007; Chen, 2007; Idem et al., 2009; Knudsen et al., 2009; Tatsumi et al., 2011; Artanto et al., 2012; Notz et al., 2012). Notz et al. (2012) have provided the most complete published data for the MEA based solvent with detailed measurements of 47 pilot runs in various operating conditions.

Nevertheless, pilot results on the MEA chemical stability are rare (Mimura et al., 1995; Kittel et al., 2009; Knudsen et al., 2009) and a lack of pilot data is observed for the MEA degradation, the material corrosion and the foaming phenomena. To fill this gap, the pilot was built with the following specifications:

- setup specifications equivalent to the industrial processes for the temperature, the pressure, the column hydrodynamic and the mass transfer coefficients;
- temperature measurement of liquid and gas input and output streams and temperature profile for each column;
- flow rate measurement of liquid and gas input and output streams for each column;
- composition measurement of liquid and gas input and output streams and composition profile for each column;
- energy balance measurement tool for each heat exchanger and stripping column;
- gas generator section which controls temperature, flow rate and gas stream composition;
- no metal as raw material;
- gas analysis tools for concentration measurements of $CO_2$, $O_2$, pollutants, ammonia and MEA volatility;
- liquid analysis tools for MEA concentration, dissolved pollutants and organic acids;
- water balance measurement tool for the overall process;
- automatic control of pilot setup for long term experiments and thorough data system acquisition.

### 2.2 Pilot description

A classic flow diagram of alkanolamine process underlies the development of our unit (Fig. 1) (Aouini et al., 2011). The temperature and the pressure are identical to industrial process specifications as well as the column hydrodynamic and mass transfer conditions.

Absorption column gas feeding is operated by a mixing unit starting from pure gas cylinders. The controlled parameters are flow rates, temperatures and compositions. The chosen gas composition is achieved by using mass flow meters. First, the nitrogen and

TABLE 2

Pilot for $CO_2$ capture from flue gas and using MEA based technology

| Research group | | Flue gas source | $CO_2$ production | Columns dimensions | References |
|---|---|---|---|---|---|
| Institute of Thermodynamics and Thermal Process Engineering | University of Stuttgart, Germany | Natural gas burner | 10 kgCO₂/h | Absorption: $H$ = 4.2 m; Ø = 0.125 m Stripper: $H$ = 2.55 m; Ø = 0.125 m | Mangalapally and Hasse (2010), Notz et al. (2012) |
| SINTEF-NTNU | NTNU, Trondheim, Norway | Gas reconstituted | 10 kgCO₂/h | Absorption: $H$ = 4.3 m; Ø = 0.15 m Stripper: $H$ = 3.9 m; Ø = 0.1 m | Tobiesen et al. (2007), Gabrielsen (2007) |
| International Test Centre for $CO_2$ Capture (ITC) | University of Regina, Canada | Natural gas burner | 1 tonCO₂/day | Absorption: $H$ = 10 m; Ø = 0.33 m Stripper: $H$ = 10 m; Ø = 0.33 m | Idem et al. (2006), Idem et al. (2009), Kittel et al. (2009) |
| Projet SOLVit (SINTEF, NTNU, Aker clean carbon) | SINTEF, Trondheim, Norway | Not mentioned | 1 tonCO₂/day | Absorption: $H$: 19 m; Ø = 0.2 m Stripper: $H$ = 13.6 m; Ø = 0.162 m | Mejdell et al. (2011) |
| Luminant Carbon Management Program | University of Texas at Austin, USA | Gas reconstituted | 4 tonCO₂/day | Absorption: $H$ = 13.3 m; Ø = 0.43 m Stripper: $H$ = 13.3 m; Ø = 0.43 m | Dugas (2006), Chen (2007) |
| Nanko pilot KEPCO and MHI | Osaka, Japan | Natural gas turbine | 2 tonCO₂/day | Not mentioned | Mimura et al. (1995), Tatsumi et al. (2011) |
| International Test Centre for $CO_2$ Captage (ITC) | Boundary Dam, Canada | Coal fired power plant | 4 tonCO₂/day | Absorption: Ø = 0.46 m Stripper: Ø = 0.40 m | Idem et al. (2006) |
| Projet CASTOR and projet CESAR | Esbjervaerket, Danemark | Coal fired power plant | 24 tonCO₂/day | Absorption: H = 17 m; Ø = 1.1 m Stripper: $H$ = 10 m; Ø = 1.1 m | Knudsen et al. (2009), Kittel et al. (2009) |
| IFPEN and ENEL | Brindisi, Italy | Not mentioned | 54 tonCO₂/day | Not mentioned | Lemaire et al. (2011) |

oxygen mixture goes through a humidifier column where the gas flow is saturated with water. The desired temperature is reached by controlling the hot water flow. Due to their high solubility in water, the $CO_2$ and the pollutants are added afterwards. Electric tracing holds the temperature of the gas flow and avoids condensation between the humidifier and the absorption columns.

The absorption column is made of glass. It is about 1.5 m high and 0.06 m in diameter and is filled with ceramic packing (Tab. 3 for packing specifications). The packing section is divided in three parts including a temperature measurement, one liquid and two gas samplings. The gas flows from the bottom to the top and is mixed counter-currently with the MEA solvent. Temperature, pressure, flow rate and composition of gas and solvent are measured at the bottom and at the top of the

TABLE 3

Packing specifications

| | Unit | Beads |
|---|---|---|
| Size | mm | 6.4 |
| Number of elements per volume unit | Nb elements/m³ | 3 715 000 |
| porosity $\varepsilon$ | % vol. | 36.8 |
| density $\rho$ | kg/m³ | 1 500 |
| Specific area | m²/m³ | 480 |

column. To increase its temperature, the rich solvent goes through a heat exchanger controlled by a thermostatic bath.

TABLE 4

Experimental protocol

| Name | Number of runs | Experimental duration | Gas inlet composition | $L/G$ | Temperature & pressure | Results |
|---|---|---|---|---|---|---|
| 1 | 15 | $\approx$ 5 hours | $CO_2$: 5, 7, 9, 11, 13% vol. dry; $H_2O$ (saturation); $N_2$ (complementary) | From 6 to 9.5 | 50°C; 1 atm | Mass balance, $K_g a_w$ |
| 2 | 1 | $\approx$ 4 days | $CO_2$: 8 vol. dry; $H_2O$ (saturation); $N_2$ (complementary) | 6 | 50°C; 1 atm | Mass balance; $K_g a_w$, MEA degradation, HSS |
| 3 | 2 | $\approx$ 4 days | $CO_2$: 8 vol. dry; $O_2$: 12 vol. dry; $SO_2$: 40 mg/m$^3$ dry; $NO_2$: 25 mg/m$^3$ dry; $H_2O$ (saturation); $N_2$ (complementary) | 6 | 50°C; 1 atm | Mass balance, $K_g a_w$, MEA degradation, HSS |
| 4 | 5 | From 12 to 90 days | Real incinerator flue gas | From 2.5 to 9.5 | 50°C; 1 atm | MEA degradation, HSS |

The stripping column is featuring the absorption column except geometric dimensions (2 m high and 0.12 m in diameter). Heat is provided to the reboiler through the glass wall by an electrical resistance. Energy balance is done. Two liquid samplings give the solvent composition at the inlet and the outlet of the reboiler. An energy balance is done on the condenser upper the column and a gas sampling gives the purity of the outlet flow. The overall desorption pressure is controlled by proportional relief valve. On the reboiler outlet, the lean solvent goes through a second heat exchanger also controlled by a thermostatic bath. An energy balance is done by using temperature and flow rate measurements. For each column, a liquid distributor is placed at the top of the packing. Hydrodynamic validations have been conducted (Aouini et al., 2011).

The pilot is controlled by a *National Instrument* controller using an in-house Labview$^{TM}$ program. It allows getting a real-time acquisition of temperatures, pressures, flow rates, energy balances and gas compositions. Those profiles are reported on the screen and stored in a data acquisition system.

## 2.3 Analytical Protocols

The pilot gas samples are sent to an IR-analyser which gives the concentrations of:
- $CO_2$ (0 to 20 ± 0.5% vol. dry);
- $O_2$ (0 to 25 ± 0.5% vol. dry);
- $SO_2$ (0 to 250 ± 2 mg/m$^3$ dry);
- $NO_x$ (0 to 250 ± 2 mg/m$^3$ dry).

For liquid samples, a *Dionex* CS-11 cationic column allows the analysis of MEA concentration.

The $CO_2$ load of the solvent is quantified by gas chromatography by using the protocol of Jou et al. (1995). Finally, a *Dionex* AS-23 anionic column allows the analysis of the concentration of organic acids and pollutant dissolution products.

## 2.4 Experimental Protocol

Four experiments were performed as detailed in Table 4. Experiments 1, 2 and 3 have been carried out in the laboratory using the gas generator while experiment 4 was performed when the pilot was plugged on the industrial incinerator. For all runs, stripper column conditions are the same (100°C and 1 atm).

First, we carried out for experiment 1 a parametric study of $CO_2$ absorption without $O_2$ and pollutants in absorption gas inlet with short runs. The main variable was the $CO_2$ gas concentration on absorption inlet (5, 7, 9, 11, 13 vol. dry). Those experiments have yielded to a data set for mass balances and overall mass transfer coefficient $K_g a_W$ calculations. Second, we have studied in experiment 2 and 3 the MEA degradation and pollutants absorption on the solvent using long laboratory runs. Experiment 2 studied MEA degradation without $O_2$ and pollutants in absorption gas inlet while experiment 3 has included them in average range of waste incinerator flue gas compositions. Finally, experiment 4 has yielded data on the process performance in real industrial conditions.

This paper proposes the results about the mass balances and overall mass transfer coefficient $K_g a_W$ measurements obtained from the laboratory experiments. Moreover, MEA degradation and pollutants absorption will be discussed from laboratory and industrial experiments.

## 3 EXPERIMENTAL RESULTS

### 3.1 Mass Balances

The mass balances were calculated using $CO_2$ molar flow measurements in several points presented at Figure 1. The $CO_2$ absorption inlet is calculated with the measurements on flue gas (1) and solvent flow (2). The $CO_2$ gas inlet flow for the absorption column (1) is provided by a mass flow meter while the liquid inlet flow (2) is calculated with the solvent circulation flow and the $CO_2$ concentration at the reboiler liquid outlet. The $CO_2$ absorption outlet is figured out with the measurements at (3) and (4). The $CO_2$ gas inlet flow (3) is calculated from the total gas outlet flow and the $CO_2$ concentration measurement. At point (2), the liquid $CO_2$ inlet flow is calculated by multiplying the solvent circulation flow and the $CO_2$ concentration at the absorption liquid outlet. The $CO_2$ stripper outlet is the sum of the measurement of the mass flow meter at (5) and the flow at (2). The $CO_2$ stripper inlet is provided by the measurement at (4). Finally, the process mass balance is performed with the measurement at point (1) for the inlet and the sum of point (3) and (5) for the outlet. Influence of $CO_2$ loading on density can be neglected ($< 0.3\%$) as shown in Aouini et al. (2011).

Figures 3, 4 and 5 present the 37 mass balances provided by experiments 1, 2 and 3. Figure 3 focuses on the overall process mass balance while Figure 4 checks data for the absorption section and Figure 5 deals with the stripper section.

The results of mass balances show an uncertainty below 10% which demonstrate the accuracy of the pilot instrumentation and the analytical methods for the solvent composition.

### 3.2 Overall Mass Transfer Coefficient Measurements $K_G a_W$

The overall mass transfer coefficient ($K_G a_W$) is the most useful tool for comparing the hydrodynamic efficiency for $CO_2$ absorption of different equipments in various operating conditions. DeMontigny et al. (2001) proposed a comprehensive pilot study of $CO_2$ absorption in MEA solvents with several packing structures, $CO_2$ gas inlet concentrations, MEA solvent concentrations, and operating conditions. Dugas (2006) carried out a

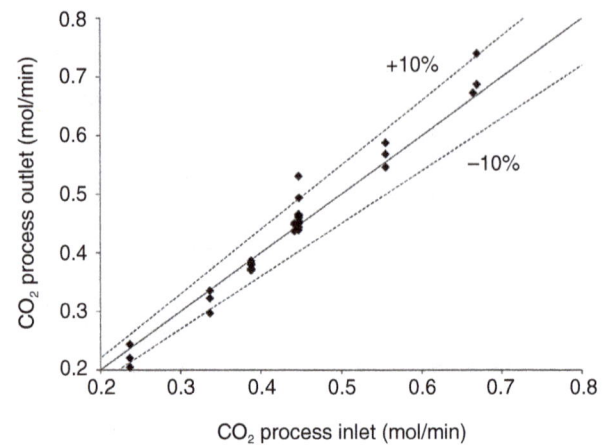

Figure 3

Overall process mass balances.

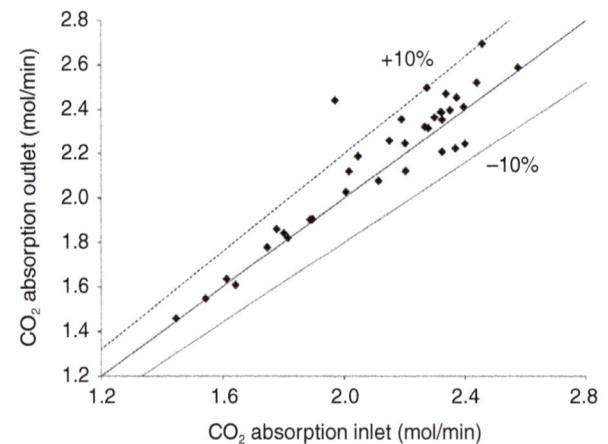

Figure 4

Absorption mass balances.

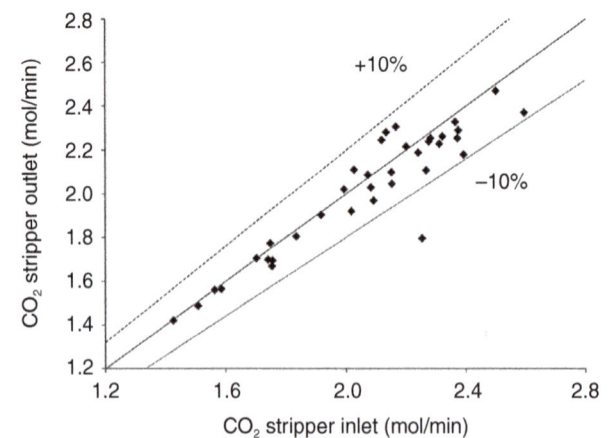

Figure 5

Stripper mass balances.

study on the $CO_2$ absorption in 30 wt% MEA solvent in a full pilot plant. The author varied the $CO_2$ gas inlet concentration, the fluid flow rates, and the desorption conditions.

The overall mass transfer coefficient ($K_Ga_W$) measurement method is based on the transfer unit theory by applying the following expression:

$$K_Ga_W = \frac{\Phi_{CO_2}}{V_{pck} \times \left(P_{CO_2} - P^*_{CO_2}\right)_{M_{ln}}}$$

Starting from experimental $CO_2$ concentration measurements in the flue gas and the solvent flow, the absorption rate $\Phi_{CO_2}$ is obtained by mass balances and the logarithm mean average driving force $\left(P_{CO_2} - P^*_{CO_2}\right)_{M_{ln}}$ is calculated using the operating and equilibrium $CO_2$ partial pressure at the top and bottom of the absorber (Aouini et al., 2012).

Nevertheless, the heat of absorption of $CO_2$ in amine solvent induces a temperature gradient along the absorption column called bulge (Kohl and Nielsen, 1997). Kvamsdal and Rochelle (2008) show that bulge has strong effect on mass transfer performance, and therefore on the overall mass transfer coefficient ($K_Ga_W$) measurements.

The runs of experiment 1, 2 and 3 (Tab. 3) were performed with a high solvent-to-gas ratio ($L/G$) (higher than 6). Heat of absorption is then dissipated in a higher liquid flow limiting temperature increase of the liquid phase. Minimizing heat accumulation in liquid phase to homogenise temperature allows to consider a uniform mass transfer along the packing. For all the experiments, the temperature gradient was less than 10°C.

Figure 6 presents the overall mass transfer coefficient ($K_Ga_W$), for experiment 1 runs, plotted as a function of $CO_2$ concentration gas inlet and the solvent-to-gas ratio ($L/G$).

The overall mass transfer coefficients ($K_Ga_W$) are similar with the DeMontigny et al. (2001), and Dugas (2006) measurements. As expected by DeMontigny et al. (2001), $K_Ga_W$ value decreases as $CO_2$ concentration increases, reflecting that hydrodynamic contribution in mass transfer is disadvantaged by the $CO_2$ partial pressure increase. Indeed, increase of $CO_2$ content in the gas phase leads to a higher $\alpha$ (loading ratio) which impact $k_G$ through the enhancement factor. Our $L/G$ range doesn't seem to have any influence on the overall mass transfer coefficients.

Nevertheless, as shown in Figure 7, the mass balances show that the absorption rate $\Phi_{CO_2}$ increases with high $CO_2$ partial pressure inlet which induce a higher thermodynamic driving force $\left(P_{CO_2} - P^*_{CO_2}\right)_{M_{ln}}$ over the

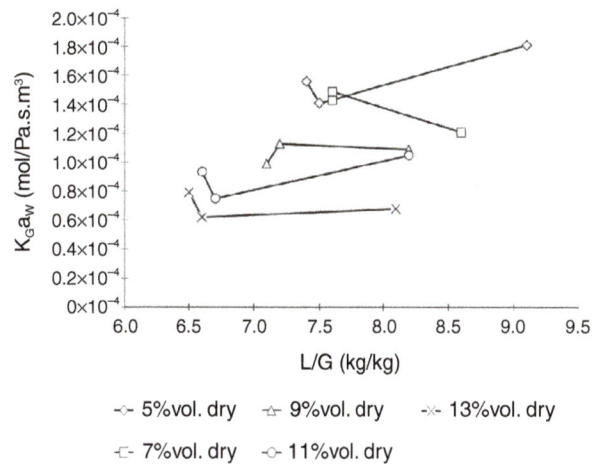

Figure 6

Overall mass transfer coefficient ($K_Ga_W$) in function of $CO_2$ gas inlet for runs of experiment 1.

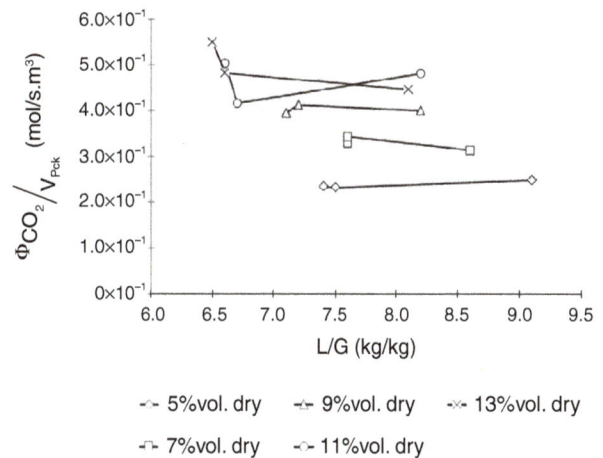

Figure 7

Absorption rate $\Phi_{CO_2}/V_{pck}$ in function of $CO_2$ gas inlet for runs of experiment 1.

absorption packing. Figure 8 presents overall higher thermodynamic driving force $\left(P_{CO_2} - P^*_{CO_2}\right)_{M_{ln}}$, for experiment 1 runs, plotted as a function of $CO_2$ concentration gas inlet and the solvent-to-gas ratio ($L/G$). In both case, the ratio $L/G$ doesn't show a great influence on the absorption rate on the driving force. The hydrodynamic behavior of the column isn't modified other the range of $L/G$ values.

Therefore, the thermodynamic driving force is predominant in $CO_2$ absorption compared with hydrodynamic, and higher the $CO_2$ partial pressure inlet is, lower the needed absorption height is.

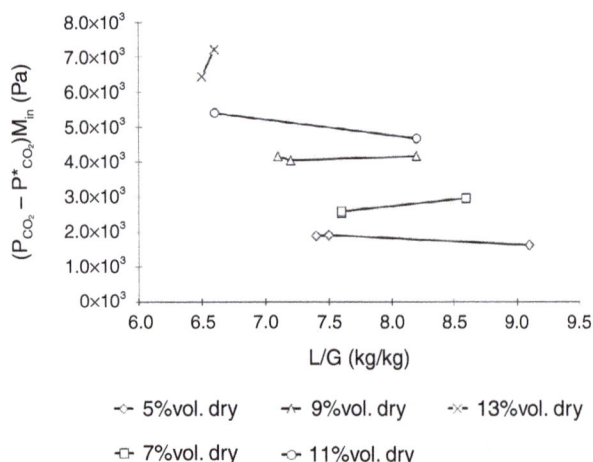

Figure 8

Absorption thermodynamic driving force $(P_{CO_2} - P^*_{CO_2})_{M_{ln}}$ in function of $CO_2$ gas inlet for runs of experiment 1.

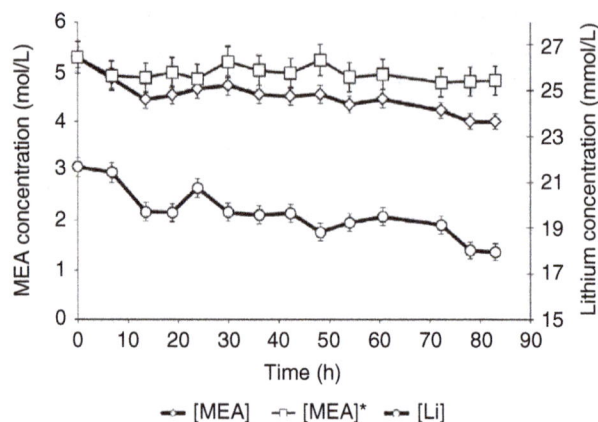

Figure 9

Evolution of MEA and Lithium concentration during experiment 2.

Finally, the overall mass transfer coefficient ($K_G a_W$) coefficient remained stable during long runs of experiments 2 and 3 which show that the accumulation of MEA degradation product was not sufficient to affect $CO_2$ mass transfer.

## 3.3 MEA Chemical Stability

The degradation of MEA in the solvent was followed by measuring its concentration variations during long runs experiments. Nevertheless, $CO_2$ capture by amine solvents is sensitive to the water evaporation or accumulation at the absorption section linked to the variation of the process operating conditions.

Therefore, the actual MEA concentration ([MEA]) is corrected with a Lithium concentration ([Li]). Lithium carbonate was added to the solvent as a non-reactive and a non-volatile compound. The lithium concentration follows the water balance of the solvent. For example an increase of Li concentration occurs when water is lost in the liquid loop. Then the corrected MEA concentration ([MEA]*) is calculated as following:

$$[MEA]^*_t = [MEA]_t \times \frac{[Li]_t}{[Li]_0}$$

Figures 9, 10 and 11 present the evolution of MEA and lithium concentrations during long runs performed in experiments 2, 3 and 4 (Tab. 3).

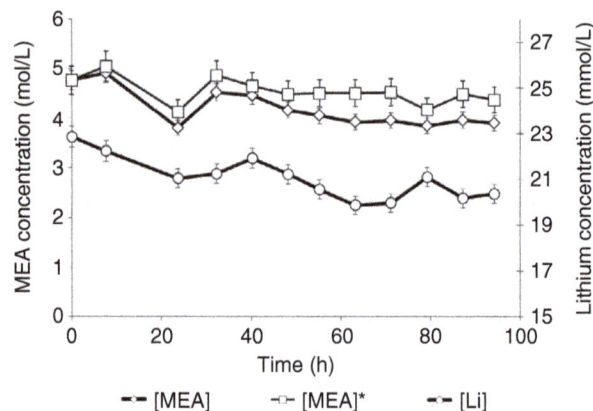

Figure 10

Evolution of MEA and Lithium concentration during a run of experiment 3.

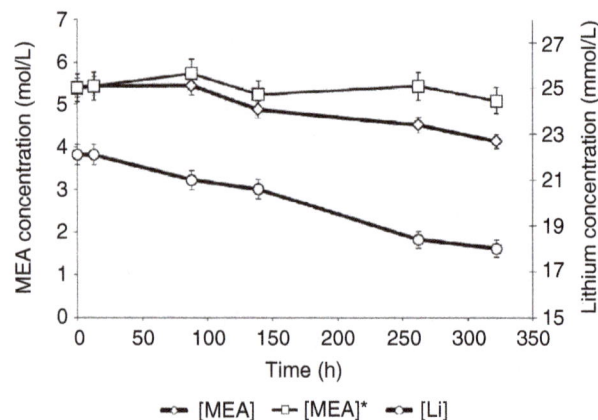

Figure 11

Evolution of MEA and Lithium concentration during a run of experiment 4.

TABLE 5

$SO_2$ and $NO_2$ concentration in the gas inlet and outlet of the absorption column in experiment 3

| Runs | $SO_2$ inlet | $SO_2$ outlet | $SO_2$ capture ratio | $NO_2$ inlet | $NO_2$ outlet | $NO_2$ capture ratio |
|------|-----------|------------|------------------|-----------|------------|------------------|
| 1 | 40 ppm | 0 ppm | 100% | 25.6 ppm | 14.1 ppm | 45% |
| 2 | 40 ppm | 0 ppm | 100% | 26.2 ppm | 13.9 ppm | 47% |

Figure 9 shows that MEA did not undergo degradation in the experiment 2 conditions. In the same manner, Figure 10 shows that $O_2$ and pollutants did not affect MEA concentration for 100 hours in the experiment 3. A minimum of concentration is clearly observed between 20 and 40 hours which could be explained by experimental error like uncertainty of analytical method. Figure 11 confirms the results for a longer run in industrial conditions.

The 30 wt% MEA solvent shows a good resistance during 300 hours against the incinerator flue gas. Nevertheless, it is important to remember that metallic materials are avoided in the pilot to minimize the oxidative degradation of MEA. Thus, this result could be projected in industrial scale in the case of a narrow control of the corrosion.

### 3.4 Absorption of $SO_2$ et $NO_2$

We carried out two runs in the experiment 3 where gas inlet composition includes sulfur dioxide ($SO_2$) and nitrogen dioxide ($NO_2$) in average range of actual waste incinerator flue gas concentrations. The inlet gas composition did not vary during those runs allowing accurate calculation for the capture ratio of the pollutants.

Table 5 gives the absorption results of $SO_2$ and $NO_2$ in experiment 3.

As Table 5 shows, the $SO_2$ was completely absorbed in less than 10 cm of packing which may correspond to an instantaneous regime for the gas-liquid transfer. $NO_2$ was partially absorbed in the solvent with an average of 46%. It corresponds to a fast regime for the gas-liquid transfer and it agrees with Hüpen and Kenig (2005) results.

### 3.5 Heat Stable Salts Accumulation

Previously, MEA reactivity review shows that it reacts into HSS with strong acid anions. Organic strong acid anions (formate, acetate, and oxalate) are produced by the oxidative degradation of MEA. All of the $SO_2$ of the absorption inlet gas leads to sulfate anions and the absorption fraction of $NO_2$ produces nitrite and nitrate anions.

Those observations were confirmed by experiment 2 where the gas inlet has excluded oxygen ($O_2$), $SO_2$, and $NO_2$ leading to the absence of accumulation of HSS.

The purpose of this section is to evaluate induced MEA losses for an industrial scale. The high reactivity of $SO_2$ makes the calculation obvious while organic and $NO_2$ HSS were evaluated starting from measurements obtained during runs of experiment 3. The two runs gas inlet composition includes $O_2$, $SO_2$, and $NO_2$ in average range of waste incinerator flue gas concentrations.

Figure 12 presents the accumulation of organic during a run of experiment 3. It shows that organic HSS concentration increases regularly which leads to calculate an average kinetic of production.

In the same way, Figure 13 shows the accumulation of nitrite and nitrate HSS during experiment 3 and leads to calculate an average kinetic of production.

The average kinetic of production of organic and $NO_2$ Heat Stable Salts was obtained by the slopes of the linear regressions of experimental measurements presented in Table 6.

The results of experiment 3 agree with industrial measurements obtained during experiment 4. Figure 14 presents an example of organic Heat Stable Salts

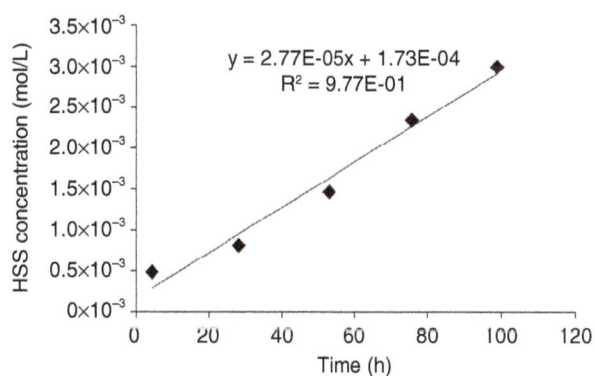

Figure 12

Organic Heat Stable Salts accumulation during experiment 3 (laboratory).

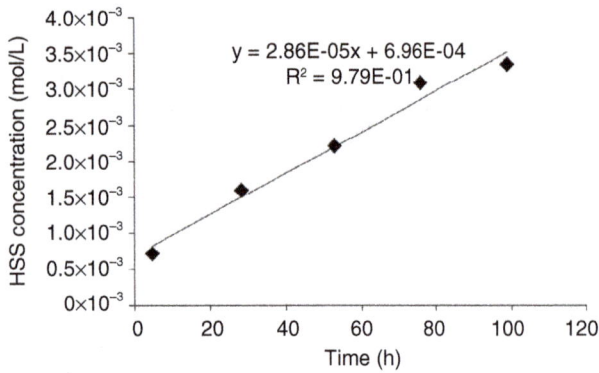

Figure 13

$NO_2$ Heat Stable Salts accumulation during a run of experiment 3 (laboratory).

TABLE 6

Average kinetic of production of organic and $NO_2$ Heat Stable Salts during the runs of experiment 3

| Runs | Organic HSS mol/(h.L) | $R^2$ | $NO_2$ HSS mol/(h.L) | $R^2$ |
|---|---|---|---|---|
| 1 | $2.08 \times 10^{-05}$ | 0.969 | $3.33 \times 10^{-05}$ | 0.933 |
| 2 | $2.77 \times 10^{-05}$ | 0.977 | $2.86 \times 10^{-05}$ | 0.979 |
| Average | $2.42 \times 10^{-05}$ | | $3.10 \times 10^{-05}$ | |

accumulation while Figure 15 illustrates the accumulation of $NO_2$ salts.

For an industrial scale, MEA losses due to Heat Stable Salts formation are estimated from previous results and with the following assumption:

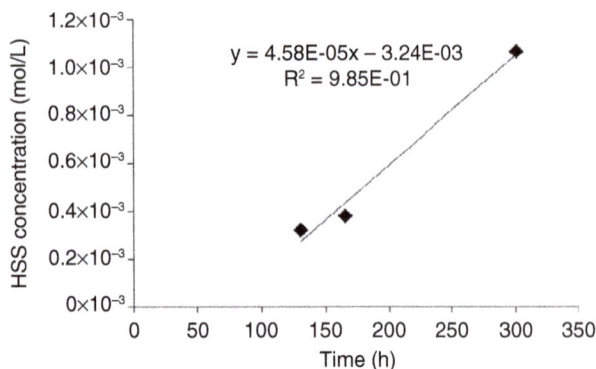

Figure 14

Organic Heat Stable Salts accumulation during experiment 4 (industrial measurements) ($L/G = 4$).

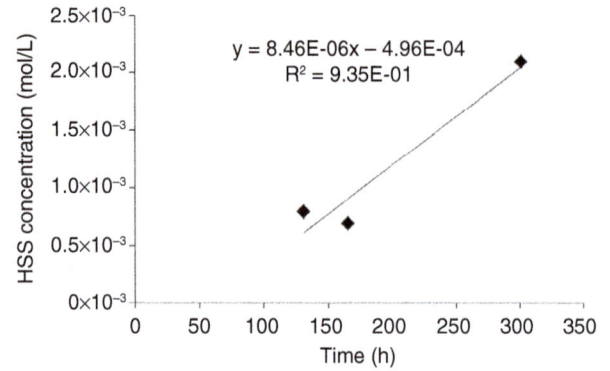

Figure 15

$NO_2$ Heat Stable Salts accumulation during experiment 4 (industrial measurements) ($L/G = 4$).

TABLE 7

MEA blocked on Heat Stable Salts during 0.7 year running using 30 wt % MEA solvent

| | Organic HSS | $NO_2$ HSS | $SO_2$ HSS | Sum |
|---|---|---|---|---|
| % MEA blocked | 2.97% | 3.80% | 5.75% | 12.52% |

- the inlet gas composition corresponds to the composition of experiment 3 (except $SO_2 = 6$ mg/m$^3$ dry),
- the capture unit works with $L/G$ ration close to 4,
- the corrosion is negligible and avoids catalysis of MEA oxidation.

Table 7 presents the proportion of MEA blocked on Heat Stable Salts for an industrial unit which has been running for 0.7 year (6 132 hours) using a 30 wt% solvent.

Table 6 shows that Heat Stable Salts accumulation is not negligible for $CO_2$ capture from industrial incinerator flue gas and could lead to loss of performance and corrosion problems. Thus for the future industrial units, it is important to implement a reclaiming unit to draw back these salts.

## CONCLUSIONS

The pilot plant presented in this paper allows studying process performances of absorption/desorption $CO_2$ capture unit using MonoEthanolAmine (MEA) solvent and feed by waste incinerator flue gas.

The pilot is fully instrumented to establish balances with an uncertainty below 10%. A laboratory parametric study allows analyzing $CO_2$ mass transfer in the absorption pilot column for various operating conditions. Experiments show that it is important to use a high gas-to-liquid ratio ($L/G$) to avoid heat accumulation in the liquid phase and to obtain a uniform mass transfer through the packing. The $CO_2$ absorption rate decreases with the reduction of $CO_2$ partial pressure gas inlet while the overall mass transfer coefficient ($K_G a_W$) increases. Therefore, the thermodynamic driving force is predominant compared with hydrodynamics in $CO_2$ absorption using MEA solvent.

Laboratory and industrial long runs were carried out to evaluate MEA chemical stability against waste incinerator flue gas. The experiments show that the 30 wt% MEA solvent has a good resistance during 300 hours against the incinerator flue gas. They also provide useful information on behavior of incinerator flue gas pollutants ($SO_2$, $NO_2$) with the 30 wt% MEA solvent. Sulfur dioxide ($SO_2$) has a high reactivity with the solvent. It is fully absorbed and leads to sulfate salts formation with MEA. Nitrogen dioxide has less reactivity with an absorption ratio close to 46% and also leads to nitrite and nitrate salts formation.

Finally, long runs permit to evaluate the Heat Stable Salts (HSS) accumulation through a MEA solvent with the waste incinerator treated flue gas. According to $SO_2$ reactivity, sulfate salt accumulation is calculated by mass balance. Organic, nitrite and nitrate salts accumulation rates were obtained by experimental data regressions. Calculations show that the proportion of MEA blocked on HSS is about 10 to 15% for an industrial unit which has been running for 0.7 year (6 132 hours) using a 30 wt% solvent. Therefore, it is important to implement a reclaiming unit to draw back these salts.

## ACKNOWLEDGMENTS

The authors thank Bruno Daronat for his technical support. The authors also thank Marie-Benedict Koko and Maria Ouboukhlik for their experimental contributions.

## REFERENCES

Aouini I., Ledoux A., Estel L., Mary S., Grimaud J., Valognes B. (2011) Study of carbon dioxide capture from industrial incinerator flue gas on a laboratory scale pilot, *Energy Procedia* **4**, 1729-1736.

Aouini I., Ledoux A., Estel L., Mary S., Evrard P., Valognes B. (2012) Experimental study of carbon dioxide capture from synthetic industrial incinerator flue gas with a pilot and laboratory measurements, *Procedia Engineering* **42**, 704-720.

Artanto Y., Jansen J., Pearson P., Do T., Cottrell A., Meuleman E., Feron P. (2012) Performance of MEA and amine-blends in the CSIRO PCC pilot plant at Loy Yang Power in Australia, *Fuel* **101**, 264-275.

Bedell S.A. (2009) Oxidative degradation mechanisms for amines in flue gas capture, *Energy Procedia* **1**, 771-778.

Bedell S.A. (2011) Amine autoxidation in flue gas $CO_2$ capture – Mechanistic lessons learned from other gas treating processes, *International Journal of Greenhouse Gas Control* **5**, 1, 1-6. doi: 10.1016/j.ijggc.2010.01.007.

Bedell S.A., Worley C.M., Darst K., Simmons K. (2011) Thermal and oxidative disproportionation in amine degradation - $O_2$ stoichiometry and mechanistic implications, *International Journal of Greenhouse Gas Control* **5**, 3, 401-404. doi: 10.1016/j.ijggc.2010.03.005.

Bello A., Idem R. (2005) Pathway for the formation of products of oxidative degradation of $CO_2$-loaded concentrated aqueous MonoEthanolAmine solutions during $CO_2$ absorption from flue gases, *Industrial and Engineering Chemistry Research* **44**, 945-969.

Carta G. (1984) Role of $HNO_2$ in the absorption of nitrogen oxides in alkaline solutions, *Industrial and Engineering Chemistry Research* **23**, 260-264.

Chapel D.G., Mariz C.L., Ernest J. (1999) Recovery of $CO_2$ from flue gases: Commercial trends, *The Canadian Society of Chemical Engineers Annual Meeting*, Saskatoon, Saskatchewan, Canada, 4-6 Oct.

Chen E. (2007) Carbon dioxide absorption into piperazine promoted potassium carbonate using structured packing, *PhD Thesis*, University of Texas at Austin.

Chi S., Rochelle G.T. (2002) Oxidative degradation of MonoEthanolAmine, *Industrial and Engineering Chemistry Research* **41**, 4178-4186.

Davis D.J. (2009) Thermal degradation of aqueous amines used for carbon dioxide capture, *PhD Thesis*, University of Texas at Austin.

Davis D.J., Rochelle G.T. (2009) Thermal degradation of MonoEthanolAmine at stripper conditions, *Energy Procedia* **1**, 327-333.

Decanini E., Nardini G., Paglianti A. (2000) Absorption of nitrogen oxides in columns equipped with low pressure drops structured packing, *Industrial and Engineering Chemistry Research* **3**, 5003-5011.

DeMontigny D., Tontiwachwuthikul P., Chakma A. (2001) Parametric Studies of Carbon Dioxide Absorption into Highly Concentrated MonoEthanolAmine Solutions, *The Canadian Journal of Chemical Engineering* **79**, 137-142.

Dugas R.E. (2006) Pilot plant study of carbon dioxide capture by aqueous MonoEthanolAmine, *M.S.E Thesis*, University of Texas at Austin.

Dupart M.S., Bacon T.R., Edwards D.J. (1993a) Understanding corrosion in Alkanolamine gas treating plants Part I, *Hydrocarbon Processing* April, 75-80.

Dupart M.S., Bacon T.R., Edwards D.J. (1993b) Understanding corrosion in Alkanolamine gas treating plants Part II, *Hydrocarbon Processing* May, 89-94.

Gabrielsen J. (2007) $CO_2$ Capture from Coal Fired Power Plants, *PhD Thesis*, Technical University of Denmark.

Goff G., Rochelle G.T. (2004) MonoEthanolAmine degradation: $O_2$ mass transfer Effects under $CO_2$ capture conditions, *Industrial and Engineering Chemistry Research* **43**, 6400-6408.

Goff G., Rochelle G.T. (2006) Oxidation inhibitors for copper and iron catalyzed degradation of MonoEthanolAmine, *Industrial and Engineering Chemistry Research* **45**, 2513-2521.

Hüpen B., Kenig E.Y. (2005) Rigorous modelling of NO$_x$ absorption in tray and packed columns, *Chemical Engineering Science* **60**, 22, 6462-6471.

Idem R., Wilson M., Tontiwachwuthikul P., Chakma A., Veawab A., Aroonwilas A., Gelowitz D. (2006) Pilot plant studies of the CO$_2$ capture performance of aqueous MEA and mixed MEA/MDEA solvents at the university of Regina CO$_2$ capture technology development plant and the Boundary Dam CO$_2$ capture demonstration plant, *Industrial and Engineering Chemistry Research* **45**, 2414-2420.

Idem R., Gelowitz D., Tontiwachwuthikul P. (2009) Evaluation of the performance of various amine based solvents in an optimized multipurpose technology development pilot plant, *Energy Procedia* **1**, 1543-1548.

Johnsen K., Helle K., Myhrvold T. (2009) Scale-up of CO$_2$ capture processes: The role of technology qualification, *Energy Procedia* **1**, 163-170.

Jou F.Y., Mather A.E., Otto F.D. (1995) The Solubility of CO$_2$ in a 30 Mass Percent MonoEthanolAmine Solution, *The Canadian Journal of Chemical Engineering* **73**, 140-147.

Kittel J., Idem R., Gelowitz D., Tontiwachwuthikul P., Parrain G., Bonneau A. (2009) Corrosion in MEA units for CO$_2$ capture: Pilot plant studies, *Energy Procedia* **1**, 791-797.

Knudsen J.N., Jensen J.N., Vilhelmsen P.J., Biede O. (2009) Experience with CO$_2$ capture from coal flue gas in pilot-scale: Testing of different amine solvents, *Energy Procedia* **1**, 783-790.

Kohl A., Nielsen R. (1997) *Gas purification*, 5th ed., Gulf Professional Publishing.

Kvamsdal H.M., Rochelle G.T. (2008) Effects of the temperature bulge in CO$_2$ absorption from flue gas by aqueous MonoEthanolAmine, *Industrial and Engineering Chemistry Research* **47**, 867-875.

Lemaire E., Bouillion P.A., Gomez A., Kittel J., Gonzalez S., Carrette P.L., Delfort B., Mongin P., Alix P., Normand L. (2011) New IFP optimized first generation process for postcombustion carbon capture: HiCapt+$^{TM}$, *Energy procedia* **4**, 1361-1368.

Lepaumier H., Picq D., Carette P.L. (2009a) New amines for CO$_2$ capture. I. Mechanisms of amine degradation in the presence of CO$_2$, *Industrial and Engineering Chemistry Research* **48**, 9061-9067.

Lepaumier H., Picq D., Carette P.L. (2009b) Degradation study of new solvents for CO$_2$ capture in post-combustion, *Energy Procedia* **1**, 893-900.

Lepaumier H., Picq D., Carette P.L. (2009c) New amines for CO$_2$ capture. II. Oxidative degradation mechanisms, *Industrial and Engineering Chemistry Research* **48**, 9068-9075.

Mangalapally H.P., Hasse H. (2010) Pilot plant experiments with MEA and new solvents for post combustion CO$_2$ capture by reactive absorption, *Distillation Absorption*, Eindhoven, The Netherlands.

Mejdell T., Knuutila H., Hoff K.A., Hallvard V.A., Svendsen F., Vassbotn T., Juliussen O., Tobiesen A., Einbu A. (2011) Novel full height pilot plant for solvent development and model validation, *Energy Procedia* **4**, 1753-1760.

Mimura T., Shimojo S., Suda T., Iijima M., Mitsuoka S. (1995) Research and development on energy saving technology for flue gas carbon dioxide recovery and steam dioxide recovery in power plant, *Energy Conversion Management* **36**, 6-9.

Notz R., Mangalapally H.P., Hasse H. (2012) Post combustion CO$_2$ capture by reactive absorption: Pilot plant description and results of systematic studies with MEA, *International Journal of Greenhouse Gas Control* **6**, 84-112.

Rochelle G.T., Bishnoi S., Chi S., Dang H., Santos J. (2001) *Research needs for CO$_2$ capture from flue gas by aqueous absorption/stripping*, Final report for DOE contract DE-AF26-99FT01029.

Rochelle G.T. (2012) Thermal degradation of amines for CO$_2$ capture, *Current Opinion in Chemical Engineering* **1**, 183-190.

Sexton A., Rochelle G.T. (2009) Catalysts and inhibitors for MEA oxidation, *Energy Procedia* **1**, 1179-1185.

Steeneveldt R., Berger B., Torp T.A. (2006) CO$_2$ capture and storage closing the knowing–doing gap, *Trans. IChemE, Part A, Chemical Engineering Research and Design* **84**, 739-763.

Supap T., Idem R., Veawab A., Aroonwilas A., Tontiwachwuthikul P., Chakma A., Kybett B.D. (2001) Kinetics of the oxidantive degradation of aqueous MonoEthanolAmine in flue gas treating unit, *Industrial and Engineering Chemistry Research* **40**, 3445-3450.

Supap T., Idem R., Tontiwachwuthikul P., Saiwan C. (2006) Analysis of MonoEthanolAmine and its oxidative degradation products during CO$_2$ absorption from flue gases: A comparative study of GC-MS, HPLC-RID, and CE-DAD analytical techniques and possible optimum combinations, *Industrial and Engineering Chemistry Research* **45**, 2437-2451.

Tatsumi M., Yagi Y., Kadono K., Kaibara K., Iijima M., Ohishi T., Tanaka H., Hirita T., Mitchell R. (2011) New energy efficient processes and improvements for flue gas CO$_2$ capture, *Energy Procedia* **4**, 1347-1352.

Tobiesen F.A., Svendsen H.F., Juliussen O. (2007) Experimental validation of a rigorous absorber model for CO$_2$ postcombustion capture, *AIChE Journal* **53**, 846-864.

Veawab A., Aroonwilas A. (2002) Identification of oxidizing agents in aqueous amine-CO$_2$ systems using mechanistic corrosion model, *Corrosion Science* **44**, 5, 967-987.

Wang M., Lawal A., Stephenson P., Sidders J., Ramshaw C. (2011) Post-combustion CO$_2$ capture with chemical absorption: A state-of-the-art review, *Chemical Engineering Research and Design* **89**, 1609-1624.

# 4

# Hollow Fiber Membrane Contactors
# for Post-Combustion CO$_2$ Capture:
# A Scale-Up Study from Laboratory to Pilot Plant

E. Chabanon[1], E. Kimball[2]*, E. Favre[1], O. Lorain[3], E. Goetheer[2], D. Ferre[4],
A. Gomez[4] and P. Broutin[4]

1 Laboratoire Réactions et Génie des Procédés, LRGP (UPR CNRS 3349), 1 rue Grandville, 54000 Nancy - France
2 TNO, Leeghwaterstraat 46, 2628 CA Delft - The Netherlands
3 Polymem, impasse du Palayré, 31100 Toulouse - France
4 IFP Energies nouvelles-Lyon, Rond-point de l'échangeur de Solaize, BP 3, 69360 Solaize - France
e-mail : elodie.chabanon@ensic.inpl-nancy.fr - erin.kimball@tno.nl - eric.favre@ensic.inpl-nancy.fr - o.lorain@polymem.fr - earl.goetheer@tno.nl
daniel.ferre@ifpen.fr - adrien.gomez@ifpen.fr - paul.broutin@ifpen.fr

* Corresponding author

**Résumé — Captage postcombustion du CO$_2$ par des contacteurs membranaires de fibres creuses : de l'échelle laboratoire à l'échelle pilote industriel** — Depuis des décennies, les contacteurs membranaires sont proposés pour intensifier les procédés de transfert de matière. Le captage post-combustion du CO$_2$ par absorption dans un solvant chimique est actuellement l'un des sujets le plus intensivement examiné. Un grand nombre d'études ont déjà été reportées dans la littérature, malheureusement, elles ne concernent pratiquement que des expériences menées à l'échelle laboratoire sur de petits modules. Étant donné les débits de fumées qui doivent être traités dans une application industrielle du captage du CO$_2$, une étude consistante à plus grande échelle est nécessaire à l'obtention d'un design rigoureux du procédé. Dans cette étude, les possibilités et les limites de l'échelle laboratoire et de l'échelle pilote industrielle ont été évaluées et seront discutées. Les expériences (absorption du CO$_2$ d'un mélange gazeux par une solution aqueuse de MEA à 30 %mass.) ont été menées à la fois sur un mini-module à l'échelle laboratoire et sur un module de taille industrielle (10 m$^2$), tous deux constitués de fibres PTFE. L'approche des résistances en série a ensuite été utilisée pour simuler les résultats. Un seul paramètre ajustable est utilisé : le coefficient de transfert de matière dans la membrane ($k_m$) qui joue logiquement un rôle clé. Les difficultés et les incertitudes des calculs liées au changement d'échelle, plus particulièrement sur la valeur de k$_m$, sont présentées et discutées.

*Abstract — Hollow Fiber Membrane Contactors for Post-Combustion CO$_2$ Capture: A Scale-Up Study from Laboratory to Pilot Plant — Membrane contactors have been proposed for decades as a way to achieve intensified mass transfer processes. Post-combustion CO$_2$ capture by absorption into a chemical solvent is one of the currently most intensively investigated topics in this area. Numerous studies have already been reported, unfortunately almost systematically on small, laboratory scale, modules. Given the level of flue gas flow rates which have to be treated for carbon capture applications, a consistent scale-up methodology is obviously needed for a rigorous engineering design. In this study, the possibilities and limitations of scale-up strategies for membrane contactors have been explored and will be discussed. Experiments (CO$_2$ absorption from a gas mixture in a 30%wt MEA aqueous solution) have been performed both on mini-modules and at pilot-scale (10 m$^2$ membrane contactor module) based on*

*PTFE hollow fibers. The results have been modeled utilizing a resistance in series approach. The only adjustable parameter is in fitting the simulations to experimental data is the membrane mass transfer coefficient ($k_m$), which logically plays a key role. The difficulties and uncertainties associated with scale-up computations from lab scale to pilot-scale modules, with a particular emphasis on the $k_m$ value, are presented and critically discussed.*

## NOMENCLATURE

| | |
|---|---|
| $c$ | Concentration (mol.m$^{-3}$) |
| $C_p$ | Heat capacity (J.K$^{-1}$.kg$^{-1}$) |
| $D$ | Diffusion coefficient (m$^2$.s$^{-1}$) |
| $H$ | Enthalpy of reaction (J.mol$^{-1}$) |
| $K_i$ | Equilibrium constant of reaction $i$ (-) |
| $k_{app}$ | Apparent constant of equilibrium (-) |
| $r$ | Radial coordinate (m) |
| $r_i$ | Inner fiber radius (m) |
| $r_o$ | Outer fiber radius (m) |
| $r_g$ | Inner shell radius (m) |
| $R$ | Reaction rate (mol.m$^{-3}$.s$^{-1}$) |
| $T$ | Temperature (K) |
| $v$ | Interstitial velocity (m.s$^{-1}$) |
| $z$ | Axial coordinate (m) |
| $\varphi$ | Packing ratio (-) |
| $\kappa$ | Thermal conductivity (W.m$^{-1}$.K$^{-1}$) |
| $\rho$ | Density (kg.m$^{-3}$) |

## ABBREVIATIONS

| | |
|---|---|
| MEA | MonoEthanolAmine |
| PP | PolyPropylene |
| PTFE | PolyTetraFluoroEthylene |
| PVDF | PolyVinyliDeneFluoride |

## INTRODUCTION

The rise of greenhouse gases in the atmosphere is commonly accepted to be caused by anthropological emissions and could be responsible for global climate change. This results in serious damage to the environment in the forms of retreating glaciers, rising sea-level and increasing storm intensity [1]. It is established that $CO_2$ is one of the most important greenhouse gases due to the immense volumes and increasing rates at which it is emitted to the atmosphere. This is mostly due to the constantly growing demand for energy, an industry which emits around 30 billion tons of $CO_2$ each year worldwide [2].

In January 2007, the European Commission recommended to the member states to cut their collective greenhouse gas emissions by 20% from the 1990 levels by 2020 [3]. Carbon Capture and Storage (CCS) is thus considered to be an essential component in the strategy to meet the ambitious emissions reduction goals and must be applied to both new power plants and as a retrofit to existing plants [4].

For existing power plants, the $CO_2$ must be removed after the fuel combustion. This results in a dilute stream of $CO_2$, about 14% vol. [5], in nitrogen for a coal power plant. The conditions of the flue gas are mild at atmospheric pressure and about 40°C, but the flow rates are immense, around 600 m$^3$/s for an 800 MW coal fired power plant. The typical systems required to separate the $CO_2$ at low concentrations involve contacting the flue gas with a solvent, which reacts chemically with the $CO_2$ and is later regenerated by either a pressure or temperature swing [6]. In order to achieve a capture rate of about 90%, huge contact surface areas are required, about 700 000 m$^2$ for the example above.

The conventional design for providing such high contact surface areas is with one or more large towers filled with a structured packing material. The flue gas is introduced at the bottom of the column and flows counter-currently to the solvent liquid introduced at the top. The size of these columns, at least 10 m in diameter and at least four times as high, makes them both expensive and difficult to install when space is limited.

As an alternative, the utilization of membrane contactors could be a promising process for $CO_2$ capture [7, 8]. With Hollow Fiber Membrane Modules (HFMM), the liquid solvent flows inside the membrane fiber tubes (lumen side) while the $CO_2$-rich gas flows around the outside of the fibers (shell side) [9-11] or *vice versa* [12-14]. The gas is transported through the (ideally) liquid-free pores of the membrane to come into contact with the liquid where it reacts with MEA. Thousands of fibers are bundled tightly together within each module, with several modules in parallel to achieve the desired capture rate of $CO_2$. In comparison to packed columns, HFMM avoid several operational problems including flooding, channeling, foaming and entrainment. Gas absorption membrane systems offer further advantages, such as independent control of the gas and liquid flows, linear scale-up [15] and a significant size reduction, with a specific surface area (m$^2$.m$^{-3}$) between two and ten times greater than with packed columns [16].

The focus of much of the work investigating the operation of HFMM has been on modeling and predicting the mass transfer characteristics of the $CO_2$ across the membrane, usually with limited experimental data for comparison. The earlier models solve the mass balances in two dimensions in order to compare the radial and axial concentration profiles at different gas flow rates, liquid flow rates and temperatures [17, 18] and also with varying solvents [19, 20]. More recent work has looked at the transient behavior of gas absorption into a liquid [21], predictions using computational fluid dynamics [22] and effects of the wetting behavior of the membrane [23-25]. In addition to $CO_2$ capture, HFMM are also being investigated for $H_2S$ absorption, by experimentally looking at the effect of different solvent concentrations [26] and by both experimentally and theoretically looking at the effects of module design and flow rates [27].

The work discussed here has extended the traditional approach to investigate the implications of scale-up for these systems. A lab-scale membrane module system was constructed in order to validate a mass transfer model similar to those discussed above. The validated model was then used for simulation of a full scale system. The results of the full scale simulation were then compared with experimental data from an industrial pilot-scale membrane system utilizing the same materials as those for the lab-scale. In general, this study intends to evaluate and discuss the possibilities and limitations of scale-up strategies for membrane contactors through these two case studies.

## 1 MATERIAL AND METHODS

### 1.1 Membrane Characterization

In the first step, a literature review combined with the partners' experience in membrane testing [28] and contactor development [29] showed that even though a very large number of studies on $CO_2$ absorption in membrane contactors have been reported [9-14, 30-33], several key issues remain largely unexplored. Among them, the stability in time of the material which is used for the membrane in the contactor appears to be of primary importance but poorly documented. Most studies report lab-scale experiments only on a short term basis (*i.e.* on an hours or days scale [34-36]). Consequently, the following priority targets are defined, in order to select the most appropriate materials for module preparation:

- a series of 10 different commercially available flat sheet membrane materials, based on 4 different polymer types — polypropylene (PP), polyvinylidenefluoride (PVDF), polytetrafluoroethylene (PTFE) and nylon — are selected for experimental tests;
- native membrane material properties, such as gas permeability and contact angle, are determined;
- the evolution of the previous properties when the samples are put in contact with a 30%wt monoethanolamine (MEA) aqueous solution are investigated on a long time basis (*i.e.* up to 2 months exposure) at temperatures of 293.15, 313.15 and 353.15 K.

In parallel, compatibility tests of the solvent solution with different glue and casing candidate materials were also performed. The results of the aging tests (evolution of contact angle and gas permeability with time) unambiguously and consistently show that:

- PTFE (5 different samples) demonstrates a remarkable stability with time in terms of gas permeability and stability of the non wetting properties, both at 313.15 and 353.15 K;
- PP (1 sample) and PVDF (3 different samples) present a fair to reasonable stability of gas permeability and contact angle with time at 313.15 K, but degrade significantly at 353.15 K;
- nylon does not show a good stability in MEA solutions, both at 313.15 and 353.15 K.

From the conclusions of this preliminary work, PTFE is selected as the most appropriate membrane material for the membrane contactor designs. A tailor made PTFE hollow fiber has been provided by Polymem and samples are characterized more fully.

Several studies are done to characterize the porosity of the fibers. The first is the Hg method (mercury intrusion technique) as it is the usual method for catalysts. The main result is that the internal structure of the sample is mainly macro-porous. The pores are comprised of sizes between 1 and 10 μm, with a mean value of 5-6 μm and a mean porosity of 0.28 mL/g. The second study is an X-rays analysis done at the European Synchrotron Radiation Facility in Grenoble (France). A typical reconstructed 3D view is shown in Figure 1. Globally, the pores form interconnected compartments, flattened in the direction of the length of the fiber (anisotropic texture). From the zoom-ins in the right part of the figure, one can see that the pores are very well connected in the axial direction and that the superficial porosity is very high, both on the inner and outer sides of the fibbers. The flattened aspect of the pores is not adequate for the usual definition of the pore sizes which supposes an isotropic distribution. In good accordance with the Hg

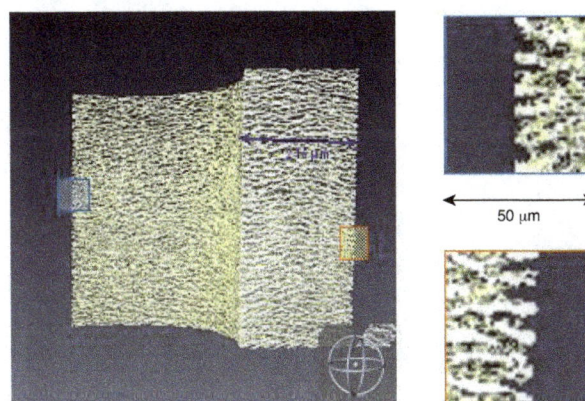

Figure 1

Synchrotron analysis of the PTFE hollow fiber tested at lab and pilot-scale in this study (image obtained at the European Synchrotron Radiation Facility).

analysis, it is again found that 5 μm is the typical dimension of the pore in the radial direction, but the straight pass length is quite scattered with an apparent mean value of about 30 μm and some passes reaching almost 100 μm.

## 1.2 Experimental Set-Up

### 1.2.1 Lab-Scale

The lab-scale membrane contactor is constructed utilizing the chosen PTFE fibers for the membrane polymer, as described above. A photo and diagram of the module setup are shown in Figure 2 and the module's geometrical properties are summarized in Table 1.

Mixtures of $CO_2$ and $N_2$ are prepared using two mass flow controllers and fed to the shell side of the membrane fibers. An aqueous 30%wt MEA solution is prepared by dilution of high purity MEA in distilled water. The liquid phase flows in the lumen side, counter-currently to the gas phase. The liquid flow rate is controlled by a high precision pump and the pressure is regulated by a valve placed at the liquid outlet.

During the experiments, the pressure and temperature are both at ambient conditions (1 bar and ~ 20°C). The inlet $CO_2$ volume fraction is 14-15%vol. with total gas flow rates of $25 \times 10^{-3} – 6.00$ L.min$^{-1}$. Fresh 30%wt MEA solution is used in each case with flow rates of $(10 \times 10^{-3}) - (50 \times 10^{-3})$ L.min$^{-1}$. An Infra-Red (IR) analyzer is used to measure the $CO_2$ volume fraction in the gas at both the inlet and outlet of the module. A bubble meter is used to measure the gas flow rate at the inlet and at the outlet of the module. Particular attention is paid to achieve steady-state conditions for each measurement, reflected by a constant $CO_2$ volume fraction at the outlet.

TABLE 1

Details of the lab-scale HFMM

| | | |
|---|---|---|
| Module | Inner diameter (m) | $1.24 \times 10^{-2}$ |
| | Effective length (m) | 0.30 |
| | Length (m) | 0.35 |
| | Number of fibers (-) | 119 |
| | Packing ratio (-) | 0.59 |
| | Specific interfacial area (m$^2$.m$^{-3}$) | 1 331 |
| Fiber | Inner diameter (m) | $4.30 \times 10^{-4}$ |
| | Outer diameter (m) | $8.70 \times 10^{-4}$ |
| | Porosity (-) | 0.336 |

### 1.2.2 Industrial Pilot-Scale

The pilot-scale module is constructed to be more than two orders of magnitude larger than the mini-module, both by increasing the length of the module and the number of fibers. Approximately equal specific interfacial areas (m$^2$.m$^{-3}$) and liquid phase Reynolds numbers of the same order of magnitude ensured dimensional similarity between the mini-module and pilot-scale. Due to the large number of fibers required for the industrial membrane module, 8 521, the bundle is supported by a plastic grid which is rolled into a spiral inside the fiber bundle. The grid is needed to support the fibers during the implementation of the resin sealing to make the bundle. An X-rays analysis showing the cross-section of the bundle and the plastic grid are given in Figures 3a and 3b, respectively. The presence of the grid does not appear to disrupt the distribution of the fibers over the length of the bundle to a large degree; however, the dark spaces in Figure 3a indicate

Figure 2

Photo and diagram of the lab-scale set-up for the $CO_2$ absorption experiments with mini HFMM.

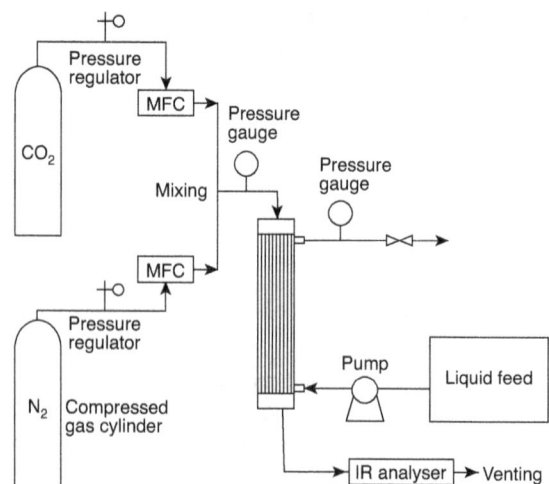

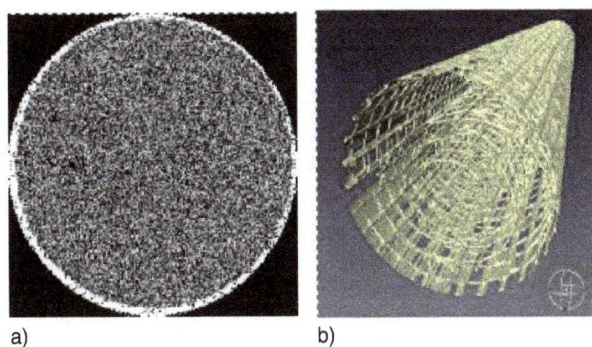

a)　　　　　　　b)

Figure 3

X-rays analysis of the pilot-scale HFMM. a) Fiber bundle cross-section. b) Plastic grid wound inside the fiber bundle.

TABLE 2

Details of pilot-scale HFMM

|  |  |  |
|---|---|---|
| Module | Inner diameter (m) | 0.105 |
|  | Length (m) | 0.88 |
|  | Effective length (m) | 1 |
|  | Number of fibers (-) | 8 521 |
|  | Packing ratio (-) | 0.648 |
|  | Specific interfacial area ($m^2 \cdot m^{-3}$) | 1 329 |
| Fiber | Inner diameter (m) | $4.30 \times 10^{-4}$ |
|  | Outer diameter (m) | $8.70 \times 10^{-4}$ |
|  | Porosity (-) | 0.336 |

some void spaces where preferential channeling of the gas flow may occur.

The final parameters of the industrial module are given in Table 2.

A diagram of the setup of the entire system is shown in Figure 4. The liquid flows axially from bottom to top and an inline transparent section (2) follows the liquid outlet, allowing for observation of possible bubbles. Keeping the counter-current mode, the gas, therefore, flows from top to bottom and an inline transparent filter (1) is installed at the gas outlet to catch any liquid that could percolate through the membrane.

The absorption of $CO_2$ is tested using a $CO_2/N_2$ mixture. The flow rate of $N_2$ and the concentration of the $CO_2$ are kept constant during each test-run using control valves. The liquid circuit is a closed loop including a 600 L storage tank open to the atmosphere. The MEA solvent reaches the module through a pump-controlled valve-flow meter system and flows back to the tank by gravity. A bend and a valve placed at the liquid outlet of the module are used to maintain a constant pressure in the liquid phase. This facility is used for the

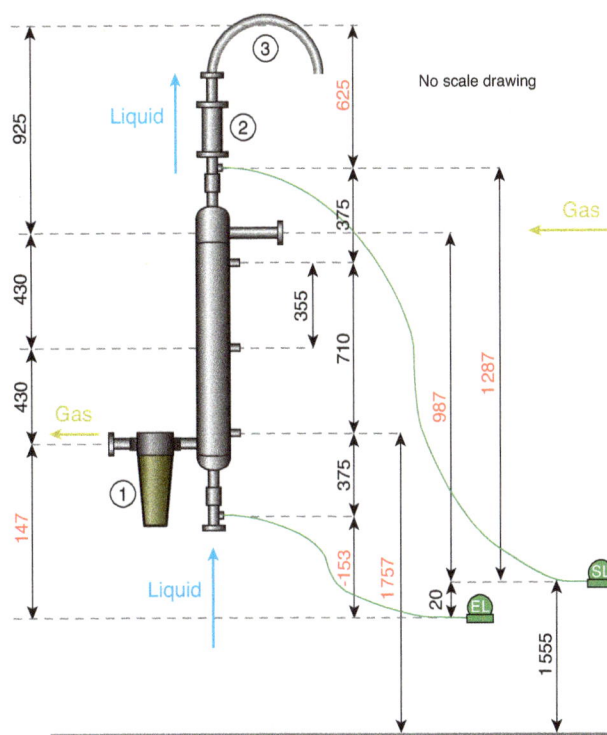

Figure 4

Diagram of the pilot-scale HFMM setup.

highest gas flow rates to prevent the gas from passing through the hollow membrane fibers.

The pressure drops of the gas phase and the liquid phase are measured by four relative-to-atmosphere pressure transducers connected to the inlet and the outlet of each phase. For both types of modules (mini and pilot), the pressure of the gas phase had to be monitored so as not to be high enough to cause bubbling into the liquid phase.

Two small lines from the gas inlet and outlet running down to the control room and made of capillary tubes are connected to IR sensors to measure the $CO_2$ concentration. Finally, the liquid inlet and outlet are equipped with bleed valves which allow for sampling of the liquid phase in order to measure the mass fraction of MEA and the $CO_2$ loading of the liquid phase over time.

As with the lab-scale experiments, each measurement is taken when steady state conditions are reached.

## 2 THEORETICAL STUDIES

In this study, a 2D mathematical model is developed for the transport of $CO_2$ from the gas phase into an aqueous MEA solution within an HFMM. This model is developed for a single hollow fiber, as shown in Figure 5. The setup of the

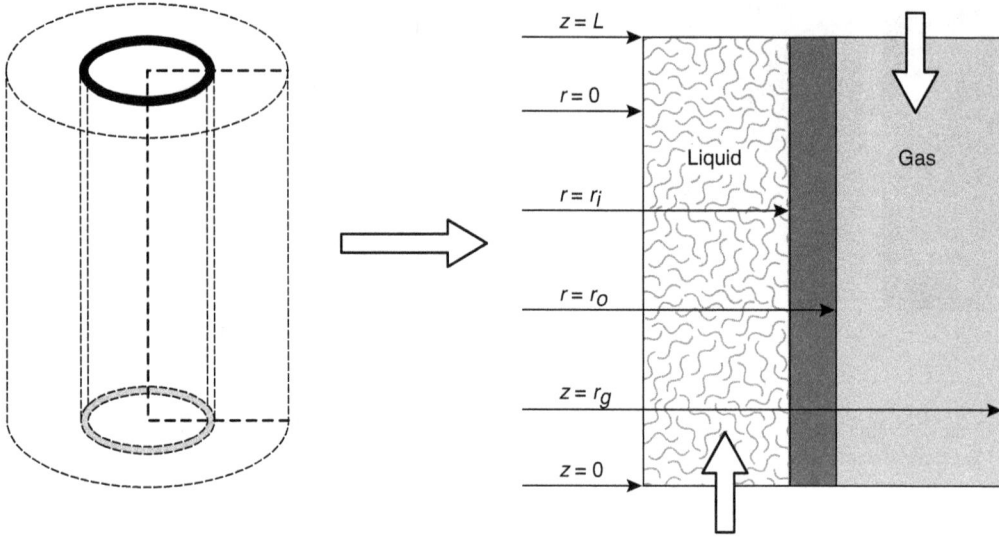

Figure 5

Schematic diagram of the membrane fiber simulated in the 2D model.

model is similar to those most commonly found in the literature [17, 19, 22-24, 26, 37] with the three layers — shell, membrane and lumen — treated separately according to a resistance-in-series approach.

The differential mass and energy balances in the axial and radial directions over a single fiber are considered:

– shell side (gas phase): diffusion and convection, no reaction ($i = CO_2, N_2$):

$$D_i^{shell}\left[\frac{\partial^2 c_i^{shell}}{\partial r^2}+\frac{1}{r}\frac{\partial c_i^{shell}}{\partial r}\right]-v_z^{shell}\frac{\partial c_i^{shell}}{\partial z}=0 \qquad (1)$$

$$\kappa^{shell}\left[\frac{\partial^2 T^{shell}}{\partial r^2}+\frac{1}{r}\frac{\partial T^{shell}}{\partial r}\right]-\rho\, C_p v_z^{shell}\frac{\partial T^{shell}}{\partial z}=0 \qquad (2)$$

– membrane: diffusion only ($i = CO_2, N_2$):

$$D_i^{mem}\left[\frac{\partial^2 c_i^{mem}}{\partial r^2}+\frac{1}{r}\frac{\partial c_i^{mem}}{\partial r}\right]=0 \qquad (3)$$

$$\kappa^{mem}\left[\frac{\partial^2 T^{mem}}{\partial r^2}+\frac{1}{r}\frac{\partial T^{mem}}{\partial r}\right]=0 \qquad (4)$$

– lumen side (liquid phase): diffusion, convection and reaction ($i = CO_2, H_2O, MEA$):

$$D_i^{lumen}\left[\frac{\partial^2 c_i^{lumen}}{\partial r^2}+\frac{1}{r}\frac{\partial c_i^{lumen}}{\partial r}\right]-v_z^{lumen}\frac{\partial c_i^{lumen}}{\partial z}+R_i=0 \qquad (5)$$

$$\kappa^{lumen}\left[\frac{\partial^2 T^{lumen}}{\partial r^2}+\frac{1}{r}\frac{\partial T^{lumen}}{\partial r}\right]-\rho\, C_p v_z^{lumen}\frac{\partial T^{lumen}}{\partial z}+R_i\sum H_i^{lumen}=0 \ (6)$$

In the equations above, $D_i$ is the diffusion coefficient of the species $i$ in each layer (m$^2$.s$^{-1}$), $c_i$ is the concentration of the species $i$ in each layer (mol.m$^{-3}$), $v$ is the interstitial velocity

defined as below (m.s$^{-1}$), $R$ is the reaction rate (mol.m$^{-3}$.s$^{-1}$), $r$ is the radial coordinate (m), $z$ is the axial coordinate (m), $T$ is the temperature (K), $\rho$ is the density (kg.m$^{-3}$), $\kappa$ is the thermal conductivity of each layer (W.m$^{-1}$.K$^{-1}$), $H_i$ is the enthalpy of reaction of the species $i$ with the MEA (J.mol$^{-1}$) and $C_p$ is the thermal capacity (J.K$^{-1}$.kg$^{-1}$).

The three layers considered are connected by the following interface relations and boundary conditions:

– axial direction (assuming counter-current flow pattern):

$$z = 0, \ T^{lumen} = T_{in}^{lumen}, \ c_i^{lumen} = 0 \ \text{for } i = CO_2,$$
$$c_i^{lumen} = c_{i,in}^{lumen} \ \text{for } i = MEA, H_2O$$
$$z = L, \ T^{shell} = T_{in}^{shell}, \ c_i^{shell} = c_{i,in}^{shell} \ \text{for } i = CO_2,$$
$$c_i^{shell} = 0 \ \text{for } i = MEA, H_2O$$

– radial direction:

$$r = 0, \ \frac{\partial T^{lumen}}{\partial r} = 0, \ \frac{\partial c_i^{lumen}}{\partial r} = 0 \quad \text{for } i = \text{all species}$$

$$r = r_i, \ T^{lumen} = T^{mem}, \ \frac{\partial c_i^{lumen}}{\partial r} = 0 \quad \text{for } i = \text{amine species},$$
$$c_i^{lumen} = m\, c_i^{mem} \ \text{for } i = CO_2, H_2O$$

$$r = r_o, \ T^{mem} = T^{shell}, \ c_i^{mem} = c_i^{shell} \ \text{for } i = CO_2, H_2O,$$
$$c_i^{mem} = 0 \ \text{for } i = MEA$$

$$r = r_g, \ \frac{\partial T^{shell}}{\partial r} = 0, \ \frac{\partial c_i^{shell}}{\partial r} = 0 \quad \text{for } i = CO_2, H_2O,$$
$$c_i^{shell} = 0 \ \text{for } i = MEA$$

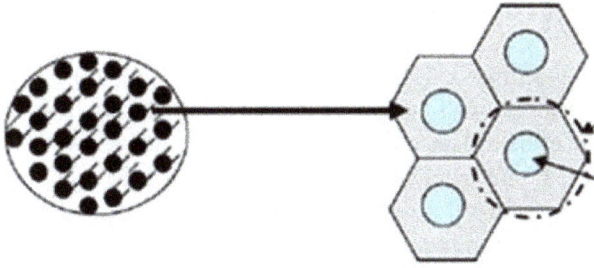

Figure 6

Membrane module cross section and approximation of the gas shell surrounding the fibers according to the Happel's free surface model [26].

The implementation is based on the method of lines and executed in Matlab. The specifications used are: gas flow on the shell side of the fibers, liquid flow on the lumen side and counter-current flows. The key assumptions are as follows:
– negligible axial diffusion (in the gas and liquid phases);
– amine compounds and water only present in the liquid phase (no evaporation);
– no wetting of the membrane pores;
– uniform fiber shapes and packing.

The flow of the liquid on the tube side of the membrane is assumed to be fully-developed laminar parabolic flow, as supported by low Reynolds numbers (on the order of 100) for the conditions tested in the experiments:

$$c_i^{lumen} = 2\langle v \rangle \left[ 1 - \left( \frac{r_i}{r} \right)^2 \right] \quad (7)$$

where $\langle v \rangle$ is the average velocity (m.s$^{-1}$) and $r_i$ is the inner radius of the fiber (m).

It should be noted that non-idealities in the fiber shapes (bends, inconsistent diameters, etc.) introduce mixing in the lumen side of the fibers, but this is not considered in the model.

The flow of gas on the shell side of the membrane is modeled using Happel's free surface model [38] (Fig. 6):

$$v_z^{shell} = 2\langle v \rangle \left[ 1 - \left( \frac{r_o}{r_g} \right)^2 \right] \left[ \frac{\left( \frac{r}{r_g} \right)^2 - \left( \frac{r_o}{r_g} \right)^2 + 2\ln\left( \frac{r_o}{r} \right)}{3 + \left( \frac{r_o}{r_g} \right)^4 - 4\left( \frac{r_o}{r_g} \right)^2 + 4\ln\left( \frac{r_o}{r_g} \right)} \right] \quad (8)$$

$$r_g = r_o \sqrt{\frac{1}{1 - \varphi}} \quad (9)$$

where $r_o$ is the outer radius of the fiber, $r_g$ is the gas shell radius (m) and $\varphi$ is the packing ratio of the module. This equation assumes uniform laminar flow around each of the fibers. It is known that the construction of HFMM

may result in a highly non-uniform distribution of the fibers within the module. Therefore, the Happel's free surface model is mostly used to calculate the effective diameter of the gas shell around each fiber and the non-uniformity of the gas flow is accounted for in the adjustment of the mass transfer across the membrane.

For the membrane, the only mechanism of mass transfer considered is diffusion.

A simplified mechanism for the kinetics is used to formulate the overall reaction rate expression, as given below:

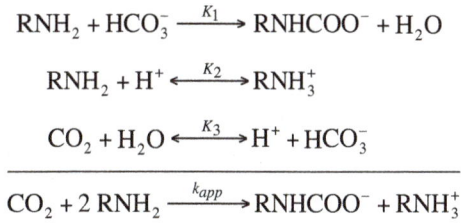

$$RNH_2 + HCO_3^- \xrightarrow{K_1} RNHCOO^- + H_2O$$

$$RNH_2 + H^+ \xleftarrow{K_2} RNH_3^+$$

$$CO_2 + H_2O \xleftarrow{K_3} H^+ + HCO_3^-$$

$$\overline{CO_2 + 2\,RNH_2 \xrightarrow{k_{app}} RNHCOO^- + RNH_3^+}$$

The full reaction mechanism for $CO_2$-MEA-$H_2O$ systems is more complex than that given here, involving several acid-base reactions [39]. These acid-base reactions are not considered here as they allow for the calculation of the concentrations of $H^+$ ions and water. In this case, experimental data are available for the equilibrium pH *versus* temperature and $CO_2$ partial pressure, allowing for direct calculation of the concentration of $H^+$ ions based on the $CO_2$ concentration, assuming a fast reaction time of the acid-base reactions relative to the reactions with $CO_2$ and amine species. Also, the concentration of $H_2O$ is assumed to be constant as there is a large excess of water (70%wt) in the solution.

The expressions for the equilibrium constants for the specific reactions above are obtained from literature, including forward and backward rate constants, and are dependent on temperature. Conservation of mass for the amine species (RNH group) is also used in deriving the final rate expression:

$$R_{CO_2} = k_3 \left[ CO_2 \right] - \frac{k_{-3} K_1 [H^+][RNHCOO^-]}{[RNH_2]} \quad (10)$$

The concentrations of $CO_2$, $RNHCOO^-$ (carbamate) and $RNH_2$ (MEA) are calculated at each point of the model grid discretization by solving the corresponding model differential equations.

## 3 RESULTS AND DISCUSSION

### 3.1 Results from Lab-Scale Setup

The membrane mass transfer coefficient, $k_m$ (m.s$^{-1}$), is defined by:

$$k_m = \frac{D^{mem} \varepsilon}{\tau \delta}$$

where $D^{mem}$ is the membrane diffusion coefficient (m$^2$.s$^{-1}$), $\varepsilon$ is the membrane porosity (-), $\tau$ is the membrane tortuosity (-) and $\delta$ is the membrane thickness (m).

By adjusting the value of $k_m$ for $CO_2$, the experimental results from the lab-scale module fit very well with the model predictions. The value of $k_m$ is fit to the data by hand using one liquid flow rate of $50 \times 10^{-3}$ L.min$^{-1}$ and an intermediate gas flow rate of 3.00 L.min$^{-1}$. The value is adjusted in the model until the calculated $CO_2$ percentage at the outlet is about equal to the experimental value. The best fit is found using a value of:

$$k_m = 2.58 \times 10^{-4} \text{ m.s}^{-1}$$

The results of the simulations from the model compared to the experimental data are shown in Figure 7, given in terms of both the volume fraction of $CO_2$ at the outlet of the gas phase *(Fig. 7a)* and as the capture rate of $CO_2$ *(Fig. 7b)*, both *versus* the gas flow rate at the inlet of the shell side. All of the points shown are with an inlet $CO_2$ volume fraction of approximately 15%, fresh 30% wt MEA and an inlet solvent flow rate of either $10 \times 10^{-3}$ or $50 \times 10^{-3}$ L.min$^{-1}$. The results for the higher liquid flow rate, $Q_l = 50 \times 10^{-3}$ L.min$^{-1}$, fit extremely well over the range of gas flow rates, from $50 \times 10^{-3}$ to 6.00 L.min$^{-1}$, where the data also follows a smooth trend. At the lower liquid flow rate, $Q_l = 10 \times 10^{-3}$ L.min$^{-1}$, there are larger differences between the experimental and model results but more fluctuations in the data are also observed.

With the close fit between the model and experimental results, the model is considered to be validated and able to describe the 2D mass transfer of $CO_2$ across the membrane and its absorption into the solvent.

## 3.2 Results from Pilot-Scale Setup

Several experiments with varying liquid and gas flow rates were conducted with the industrial pilot-scale HFMM. The results considered here are for a range of liquid flow rates of 0.50-3.33 L.min$^{-1}$ and gas flow rates of 5-30 L.min$^{-1}$. The MEA is recirculated and so is already partially loaded with $CO_2$, which is also measured at the start of each experiment. The inlet volume fraction of $CO_2$ in the $N_2$ is kept approximately constant at about 15%. The experimental results are then compared with simulation results from the 2D mass transfer model.

Following the validation of the 2D model with the data from the lab-scale module, the parameters of the model describing the module design and MEA conditions are adjusted to simulate the pilot-scale module. The fibers in both modules are made from the same material, allowing for the membrane mass transfer coefficient determined from the fit with the lab-scale module experimental data to also be used in the simulations of the larger module. However, the results of the pilot-scale module simulations using the previously determined $k_m$ value show a significant over-prediction of the $CO_2$ capture for all of the gas and liquid flow rates, indicated by very low values for the $CO_2$ volume fraction at the outlet of the gas phase, as shown in Figure 8.

In order to have the model fit the experimental data at an intermediate gas flow rate of 10 L.min$^{-1}$, the membrane mass transfer coefficient is adjusted to:

$$k_m = 5.31 \times 10^{-5} \text{ m.s}^{-1}$$

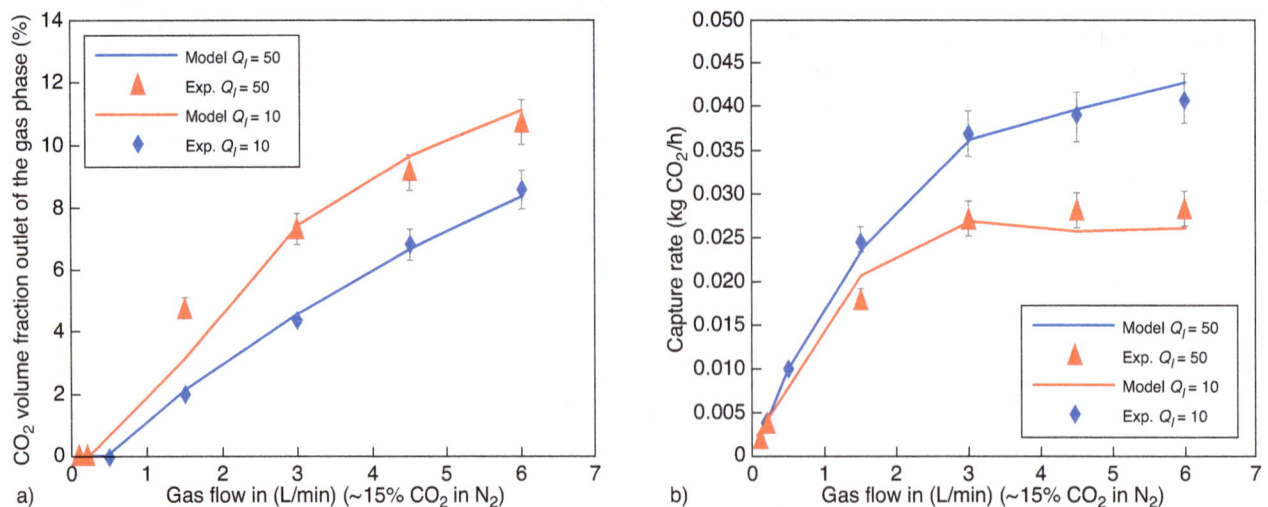

Figure 7

Comparison of simulation results and experimental data for the mini HFMM for gas inlet flow rates of $50 \times 10^{-3}$ to 6.00 L.min$^{-1}$ and two liquid flow rates, $Q_l = 10 \times 10^{-3}$ and $50 \times 10^{-3}$ L.min$^{-1}$. a) Results for $CO_2$ volume fraction at the outlet of the shell side. b) Results for calculation of the capture rate (kg $CO_2$ captured per hour).

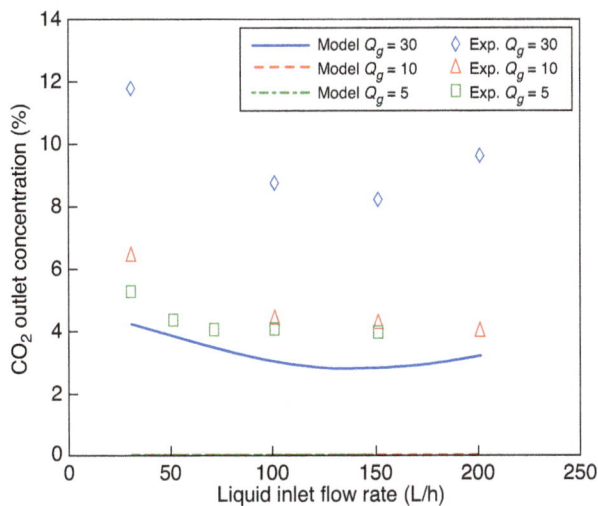

Figure 8

Comparison of simulation results and experimental data for the pilot-scale module utilizing the $k_m$ value from the fit with the mini-module data. Values are given as $CO_2$ volume fraction at the outlet of the gas phase for liquid inlet flow rates of 0.50-3.33 L.min$^{-1}$ and gas inlet flow rates of 5, 10 or 30 L.min$^{-1}$. The inlet volume fraction of $CO_2$ is about 15% in every case and the accuracy of the $CO_2$ detector is ± 0.01%.

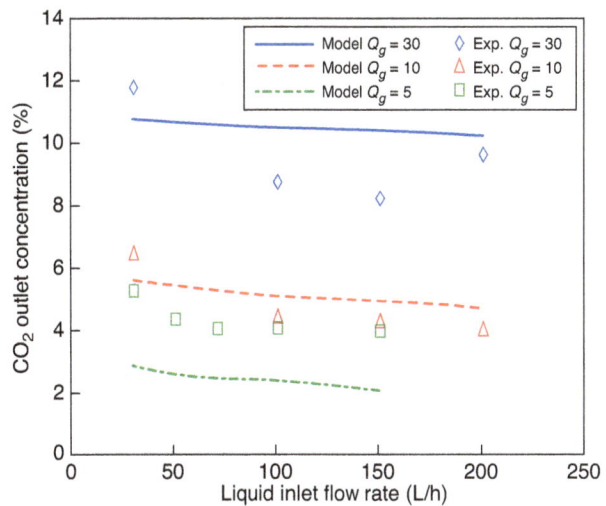

Figure 9

Comparison of simulation results and experimental data for the pilot-scale module utilizing a $k_m$ value of $5.31 \times 10^{-5}$ m.s$^{-1}$. Values are given as $CO_2$ volume fraction at the outlet of the shell side for liquid inlet flow rates of 0.50-3.33 L.min$^{-1}$ and gas inlet flow rates of 5, 10 or 30 L.min$^{-1}$. The inlet fraction of $CO_2$ is about 15%vol. in every case and the accuracy of the $CO_2$ detector is ± 0.01% (same experimental data as in *Fig. 8*).

and gives good results over the range of liquid flow rates, as shown in Figure 9 (red markers). However, for the lower gas flow rate of 5 L.min$^{-1}$ (green markers), the simulations show a lower prediction of the $CO_2$ volume fraction at the outlet and at a higher gas flow rate of 30 L.min$^{-1}$ (blue markers), the prediction of the percent $CO_2$ is higher than what is observed experimentally. The effects of the different parameters on the results are investigated separately but the simulation results cannot be made to fit well with all of the data sets with one set of parameter values. This indicates that the non-idealities in the larger pilot-scale system are much more significant than with the lab-scale module, as logically expected. The 2D mass transfer model with the stated assumptions cannot account for the different phenomena occurring in the pilot-scale module.

### 3.3 Comparison between Lab-Scale and Pilot-Scale

The significant differences in the results between the lab- and pilot-scale modules show the difficulties in predicting the performance of larger scale systems with simple models. It is important to stress that the modeling approach tested in this study corresponds to the basic 2D equations classically proposed in membrane contactor studies. The comparison performed here shows that significant improvements need to be incorporated into the 2D mass transfer model before the operation of a full scale module will be acceptably simulated

over a wide range of conditions. When simulating the pilot-scale module, the over-prediction of the $CO_2$ capture may be due to errors in the mass transfer parameters, kinetics or assumptions of uniformity. Given that the $k_m$ value is set by fitting the data from the small lab-scale module, it is thought that the maldistribution of gas flow around the fibers of the pilot-scale module, especially with the mesh grid that is incorporated into the fiber bundle, is the most significant source of error. This would effectively make the resistance of the gas flow through the membrane appear higher. The mass transfer resistance of the gas flow close to the fibers is still relatively low, but not all of the gas flows in the shell of a fiber, as defined by Happel's free surface model. The resistance to mass transfer through the membrane for some fraction of the $CO_2$ in the gas phase would then be essentially infinite.

Another source of error that would be much more significant with the pilot-scale module than with the lab-scale module is the effect of the temperature rise due to the heat of the absorption reaction. The model assumed no evaporation of water from the lumen of the fibers to the gas phase. However, evaporation may be a significant factor, given that the gas is dry at the inlet and the simulations predicted about a 10°C temperature rise over the length of the fiber, for the pilot-scale module. This is supported by the fact that a large amount (approximately 10 L but not precisely measured) of condensed water is present in the outlet of the experimental pilot-scale system between tests. In contrast, the predicted temperature rise for the much smaller lab-scale module is less than 1°C and no water condensation in the experimental setup is reported. The effect that the water evaporation can have on the $CO_2$ capture performance is to decrease the capture ratio. As water evaporates into the gas phase, the partial pressure of $CO_2$ in the gas phase decreases; this in turn decreases the driving force for $CO_2$ transport through the membrane and into the absorption liquid, thus resulting in a lower $CO_2$ capture ratio (higher $CO_2$ volume fraction at the outlet) than expected with no evaporation. It has to be noted that for power plant capture systems, the flue gas treated by the HFMM is usually hydrated to some degree. Consequently, the evaporation of water from the liquid can be expected to be smaller than in the experimental system.

### 3.4 Ability to Predict Operation of Large Scale Modules

With the construction of large scale HFMM, the performance becomes harder to predict. The plastic grid, as shown in Figure 3b, is one example of the non-uniformities that can exist within the module which cause maldistribution of flows. Others include bends in the fibers, variations in the local fiber packing and varying thickness of the membranes. The ability to take into account these effects in a rigorous model is a challenge. Attempts have been made to capture the effects of non-uniformities, such as by considering a random distribution of fibers [37], however experiments at several different scales of HFMM are needed. The two orders of magnitude difference in size between the two HFMM considered here caused large differences in the ability to predict the operation. In order to understand how the non-uniformities evolve with each step increase in size and each variation in module construction, experimental data from many more systems are needed. This will then likely allow for correlations based on scale to be derived for the several parameters needed to accurately simulate HFMM operation.

## CONCLUSIONS

Two different scales of HFMM are constructed and operated to provide insight into the ability of predicting the behavior of HFMM at full scale. The results of the two-dimensional model based on mass and energy balances showed a good prediction of the performance of HFMM. For the lab-scale HFMM, with less non-ideality in the system, the model and experimental results fit very well for varying gas and liquid flow rates. When simulating the larger pilot-scale module, the prediction of the performance over a range of gas flow rates is not as good. The maldistribution of the flows around and inside the fibers is not accounted for in the 2D model and proved to be much more important in the pilot-scale module.

The assumption of uniform flow is one of the key assumptions used in the development of the model. Another is the hypothesis of a negligible evaporation of water. Although only modest temperature rises of up to 15°C are predicted by the simulations, this may have an effect in several ways. For one, the evaporation will cause a cooling effect in the liquid and decrease the temperature rise. This has implications for the diffusion of the compounds in the liquid, the density of both the gas and liquid and the reaction rates. Another effect of water in the gas phase is to decrease the partial pressure of $CO_2$. This will decrease the driving force for $CO_2$ to be absorbed into the liquid and thus decrease the capture ratio. It has not been possible to incorporate evaporation into the mass transfer model for this project, but this should clearly be a major focus of future work.

More generally, this contribution is one of the first attempts to experimentally evaluate the scale-up possibilities and limitations of membrane contactors from laboratory- to pilot-scale. Publications almost systematically refer to laboratory scale membrane contactors, while large scale applications (such as CCS) obviously require adequate simulation tools and extrapolation strategies for much larger units. This study has shown that the scale-up issue of membrane contactors is far from straightforward. Consequently, research and development efforts are urgently needed in this direction.

## ACKNOWLEDGMENTS

This work was completed as part of the larger CESAR Project, part of the European Commission 7th Framework Program.

## REFERENCES

1 IPCC (2007) 4th Assessment Report (AR4), Climate Change 2007: Synthesis Report.

2 Energy-Related Carbon Dioxide Emissions, *International Energy Outlook 2010*, available at: http://www.eia.doe.gov/oiaf/ieo/emissions.html, accessed on 9 March 2012.

3 European Commission, Analysis of options to move beyond 20% greenhouse gas emission reductions and assessing the risk of carbon leakage, available on http://ec.europa.eu/clima/policies/package/index_en.htm, accessed on 9 March 2012.

4 Wang M., Lawal A., Stephenson P., Sidders J., Ramshaw C. (2010) Post-combustion $CO_2$ capture with chemical absorption: A state-of-the-art review, *Chem. Eng. Res. Design* **89**, 1609-1624.

5 Tobiesen F.A., Svendsen H.F., Juliussen O. (2007) Experimental Validation of a Rigorous Absorption Model for $CO_2$ postcombustion capture, *AIChE J.* **53**, 4, 846-865.

6 Abu-Zahra M.R.M., Schneiders L.H.J., Niederer J.P.M., Feron P.H.M., Versteeg G.F. (2007) $CO_2$ capture from power plants. Part I. A parametric study of the technical performance based on monoethanolamine, *Int. J. Greenhouse Gas Control* **1**, 37-46.

7 Gabelman A., Hwang S.T. (1999) Hollow fiber membrane contactors, *J. Membr. Sci.* **159**, 61-106.

8 Mansourizadeh A., Ismail A.F. (2009) Hollow fiber gas-liquid membrane contactors for acid gas capture: A review, *J. Hazardous Mater.* **17**, 38-53.

9 Mansourizadeh A., Ismail A.F., Matsuura T. (2010) Effect of operating conditions on the physical and chemical $CO_2$ absorption through the PVDF hollow fiber membrane contactor, *J. Membr. Sci.* **353**, 192-200.

10 Dindore V.Y., Brilman D.W.F., Versteeg G.F. (2005) Hollow fiber membrane contactor as a gas-liquid model contactor, *Chem. Eng. Sci.* **60**, 467-479.

11 Qi Z., Cussler E.L. (1985) Microporous Hollow fibers for gas absorption, *J. Membr. Sci.* **23**, 321-332.

12 Bottino A., Capannelli G., Comite A., Di Felice R., Firpo R. (2008) $CO_2$ removal from a gas stream by membrane contactor, *Sep. Purifi. Technol.* **59**, 85-90.

13 Lu J.G., Zheng Y.F., Cheng M.D., Wang L.J. (2007) Effects of activators on mass-transfer enhancement in a hollow fiber contactor using activated alkanolamine solutions, *J. Membr. Sci.* **289**, 138-149.

14 Mavroudi M., Kaldis S.P., Sakellaropoulos G.P. (2003) Reduction of $CO_2$ emissions by a membrane contacting process, *Fuel* **82**, 2153-2159.

15 Li J.L., Chen B.H. (2005) $CO_2$ absorption using chemicals solvents in hollow fiber membrane contactors, *Sep. Purifi. Technol.* **41**, 109-122.

16 Karoor S., Sirkar K.K. (1993) Gas absorption studies in microporous hollow fiber membranes modules, *Ind. Eng. Chem. Res.* **32**, 674-684.

17 Al-Marzouqi M., El-Naas M.H., Marzouk S.A.M., Al-Zarooni M.A., Abdullatif N., Faiz R. (2008) Modeling of $CO_2$ absorption in membrane contactors, *Sep. Purifi. Technol.* **59**, 286-293.

18 Lee Y., Noble R.D., Yeom B.Y., Park Y.I., Lee K.H. (2001) Analysis of $CO_2$ removal by hollow fiber membrane contactors, *J. Membr. Sci.* **194**, 57-67.

19 Wang R., Li D.F., Liang D.T. (2004) Modelling of $CO_2$ capture by three typical amine solutions in hollow fiber membrane contactors, *Chem. Eng. Process.* **43**, 849-856.

20 Boucif N., Favre E., Roizard D. (2008) $CO_2$ capture in HFMM contactor with typical amine solutions: A numerical analysis, *Chem. Eng. Sci.* **63**, 5375-5385.

21 Porcheron F., Drozdz S. (2009) Hollow fiber membrane contactor transient experiments for the characterization of gas/liquid thermodynamics and mass transfer properties, *Chem. Eng. Sci.* **64**, 265-275.

22 Shirazian S., Moghadassi A., Moradi S. (2009) Numerical simulation of mass transfer in gas-liquid hollow fiber membrane contactors for laminar flow conditions, *Simul. Modell. Pract. Theory* **17**, 708-718.

23 Zhang H.Y., Wang R., Liang D.T., Tay J.H. (2008) Theoretical and experimental studies of membrane wetting in the membrane gas-liquid contacting process for $CO_2$ absorption, *J. Membr. Sci.* **308**, 162-170.

24 Malek A., Teo W.K. (1997) Modeling of microporous hollow fiber membrane modules operated under partially wetted conditions, *Ind. Eng. Chem. Res.* **36**, 784-793.

25 Lu J.G., Zheng Y.F., Cheng M.D. (2008) Wetting mechanism in mass transfer process of hydrophobic membrane gas absorption, *J. Membr. Sci.* **308**, 180-190.

26 Faiz R., Al-Marzouqi M. (2009) Mathematical modeling for the simultaneous absorption of $CO_2$ and $H_2S$ using MEA in hollow fiber membrane contactors, *J. Membr. Sci.* **342**, 269-278.

27 Boucif N., Favre E., Roizard D., Belloul M. (2008) Hollow fiber membrane contactor for hydrogen sulfide odor control, *AIChE J.* **54**, 122-131.

28 Nguyen P.T., Roizard D., Thomas D., Favre E. (2010) Gas permeability: A simple, novel and efficient method for testing membrane material/solvent compatibility for membrane contactors applications, *Desalination Water Treatment* **14**, 7-14.

29 Dindore V.Y., Brilman D.W.F., Feron P.H.M., Versteeg G.F. (2004) $CO_2$ absorption at elevated pressures using a hollow fiber membrane contactor, *J. Membr. Sci.* **235**, 99-109.

30 Chen S.C., Lin S.H., Wang Y.H., Hsiao H.C. (2011) Chemical absorption of carbon dioxide with asymmetrically heated polytetrafluoroethylene membranes, *J. Environ. Manage.* **92**, 1083-1090.

31 Takahashi N., Furuta Y., Fukunaga H., Takatsuka T., Mano H., Fujioka Y. (2011) Effects of membrane properties on $CO_2$ recovery performance in a gas absorption membrane contactor, *Energy Procedia* **4**, 693-698.

32 Hedayat M., Soltanieh M., Mousavi S.A. (2011) Simultaneous separation of $H_2S$ and $CO_2$ from natural gas by hollow fiber membrane contactor using mixture of alkanolamines, *J. Membr. Sci.* **377**, 191-197.

33 Boributh S., Assabumrungrat S., Laosiripojana N., Jiraratananon R. (2011) A modelling study on the effects of membrane characteristics and operating parameters on physical absorption of $CO_2$ by hollow fiber membrane contactor, *J. Membr. Sci.* **380**, 21-33.

34 deMontigny D., Tontiwachwuthikul P., Chakma A. (2005) Using polypropylene and polytetrafluoroethylene membranes in a membrane contactor for $CO_2$ absorption, *J. Membr. Sci.* **277**, 99-107.

35 Nishikawa N., Ishibashi M., Ohta H., Akutsu N., Matsumoto H., Kamata T., Kitamura H. (1995) $CO_2$ removal by hollow fiber gas liquid contactor, *Energy Convers. Manage.* **36**, 415-418.

36 Khaisri S., deMontigny D., Tontiwachwuthikul P., Jiraratananon R. (2010) Comparing membrane resistance and absorption performance of three different membranes in a gas absorption membrane contactor, *Sep. Purifi. Technol.* **65**, 290-297.

37 Keshavarz P., Ayatollahi S., Fathikalajahi J. (2008) Mathematical modeling of gas-liquid membrane contactors using random distribution of fibers, *J. Membr. Sci.* **325**, 98-108.

38 Happel J. (1959) Viscous flow relative to arrays of cylinders, *AIChE J.* **5**, 174-177.

39 Blauwhoff P.M.M., Versteeg G.F., van Swaaij W.P.M. (1982) A study on the reaction between $CO_2$ and alkanolamines in aqueous solutions, *Chem. Eng. Sci.* **39**, 2, 207-225.

# 5

# Aqueous Ammonia (NH$_3$) Based Post Combustion CO$_2$ Capture: A Review

Nan Yang[1,2], Hai Yu[2]*, Lichun Li[2,3], Dongyao Xu[1], Wenfeng Han[3] and Paul Feron[2]

[1] School of Chemical & Environmental Engineering, China University of Mining & Technology (Beijing), Beijing 100083 - P.R. China
[2] CSIRO Energy Technology, 10 Murray Dwyer Circuit, Mayfield West, NSW 2304 - Australia
[3] Institute of Catalysis, Zhejiang University of Technology, Hangzhou, Zhejiang 310014 - P.R. China
e-mail: hai.yu@csiro.au

* Corresponding author

**Résumé — Capture de CO$_2$ en postcombustion par l'ammoniaque en solution aqueuse (NH$_3$) : synthèse** — L'ammoniaque en solution aqueuse (NH$_3$) est un solvant novateur, prometteur et un challengeur pour la capture du CO$_2$ émis par les centrales électriques au charbon et au gaz et autres sources industrielles. Au cours des dernières années, d'intenses activités de recherche ont été menées pour tenter de comprendre les propriétés thermochimiques et physiques du système CO$_2$-NH$_3$-H$_2$O, ainsi que la chimie et la cinétique associées, afin d'obtenir plus d'informations sur le taux d'absorption, la perte en ammoniaque, les énergies de régénération et les problèmes pratiques, et aussi pour évaluer la faisabilité technique et commerciale des technologies de capture en postcombustion par l'ammoniaque en solution aqueuse. Cet article offre une synthèse de ces activités de recherche, apporte des comparaisons avec le monoéthanolamine (MEA, le solvant de référence) et identifie des opportunités de recherche pour de futures améliorations.

*Abstract — Aqueous Ammonia (NH$_3$) Based Post Combustion CO$_2$ Capture: A Review — Aqueous ammonia (NH$_3$) is an emerging, promising and challenging solvent for capture of CO$_2$ emitted from coal and gas fired power stations and other industrial sources. Over the last few years, intensive research activities have been carried out to understand the thermo-chemical and physical properties of the CO$_2$-NH$_3$-H$_2$O system and the chemistry/kinetics involved, to obtain information on absorption rate, ammonia loss, regeneration energies, and practical issues, and to assess the technical and economical feasibility of the aqueous ammonia based post combustion capture technologies. This paper reviews these research activities, makes comparisons with MonoEthanolAmine (MEA, the benchmark solvent) and identifies the research opportunities for further improvements.*

## INTRODUCTION

With the rapid industry development globally, excessive amounts of the greenhouse gases in particular $CO_2$ have been released to the atmosphere, which is believed to be the major contributor to global warming. It is generally agreed that Carbon Capture and Storage (CCS) is the only technology available to make deep cuts in greenhouse gas emissions while still using fossil fuels and much of today's energy infrastructure. Developing an economic and efficient method to capture carbon dioxide is an important objective for successful deployment of effective CCS process chains.

Post Combustion Capture (PCC) is a process that uses an aqueous absorption liquid incorporating compounds such as ammonia or amine to capture $CO_2$ from power station flue gases and many other industrial sources. It is the leading capture technology as a result of the potential benefits, such as,

- it can be retrofitted to existing power plants or integrated with new infrastructure to achieve a range of $CO_2$ reductions, from partial retrofit to full capture capacity;
- it has a lower technology risk compared with other competing technologies;
- renewable technologies can be integrated with PCC, for example, low cost solar thermal collectors can provide the heat required to separate $CO_2$ from solvents;
- PCC can be used to capture $CO_2$ from a range of sources – smelters, kilns and steel works, as well as coal- and gas-fired power stations.

Currently, the commercially available PCC technology is mainly based on alkanol/alkyl amine solutions. A study by Dave et al. (2009) shows that retrofitting a MEA based PCC plant to the existing/new mechanical draft water cooled black coal fired plants will reduce the power plant efficiency by 10 absolute percentage points and involves significant capital investment costs. Moreover, there is some concern about the formation of carcinogenic nitrosamines from the use of amines in PCC and their possible spread into the environment (MacDowell et al., 2010).

Aqueous $NH_3$, as a promising solvent for $CO_2$ capture, has been receiving increasing attention recently. Compared to traditional amines, $NH_3$ is a low cost solvent that is not easily decomposed by other gas present in flue gases. Moreover, it has a high $CO_2$ removal capacity, low absorption heat and low regeneration energy and is less corrosive to instrument. It also has the potential of capturing multiple flue gas components ($NO_x$, $SO_x$ and $CO_2$) and producing value added chemicals, such as ammonium sulphate and ammonium nitrate, which are commonly used as fertilizers. However, ammonia is volatile and hence it slips to flue gas

and $CO_2$ product, thus requiring additional energy and capital costs to reduce ammonia evaporation and/or recover ammonia. This could offset the advantages offered by ammonia.

Over the last few years, there has been a significant amount of research work on the aqueous ammonia based $CO_2$ capture technologies including experimental investigations at various scales and process modeling to evaluate energy consumption and economics. This paper will briefly review these research activities and make comparisons with other solvents in particular MonoEthanolAmine (MEA), the benchmark solvent, and identify the research opportunities for further improvements.

## 1 AQUEOUS AMMONIA BASED CO₂ CAPTURE PROCESSES

A number of pilot and demonstration plants has been constructed and operated to evaluate the technical and economic feasibility of $NH_3$ based PCC processes to remove $CO_2$ from flue gas being emitted from various industrial sources including both coal and natural gas fired power stations. *Alstom*, *Powerspan*, *CSIRO* and *KIST* (Korean Research Institute of Industrial Science & Technology) are the major players in this area and have tested the technologies at pilot-plant scale (McLarnon and Duncan, 2009; Rhee et al., 2011; Telikapalli et al., 2011; Yu et al., 2011a).

*Powerspan* developed an aqueous $NH_3$ based $CO_2$ capture process, called $ECO_2^®$, in which the absorption takes place at ambient temperatures and no slurry is involved in the absorber (McLarnon and Duncan, 2009). The ammonia concentration used was not given and believed to be between 5 and 15%. Desorption occurs at low pressures. An illustration of how the $ECO_2$ system is integrated with the ECO multi-pollutant control system is shown in Figure 1. The ECO system uses a barrier discharge reactor, wet scrubber and wet electrostatic precipitator to remove high levels of $NO_x$, $SO_2$, mercury and fine particulate matter. The wet scrubber uses aqueous ammonia for $SO_2$ scrubbing, producing ammonium sulphate fertilizer as a co-product. *Powerspan* initiated a pilot test program with FirstEnergy at the R.E. Burger Plant to demonstrate $CO_2$ capture through integration with the ECO® multi-pollutant control process. A 1-MW pilot demonstration was designed to produce approximately 20 metric tonnes of sequestration ready $CO_2$ per day. However pilot plant results have not been reported in public domain and there is no further development of this technology reported in the literature.

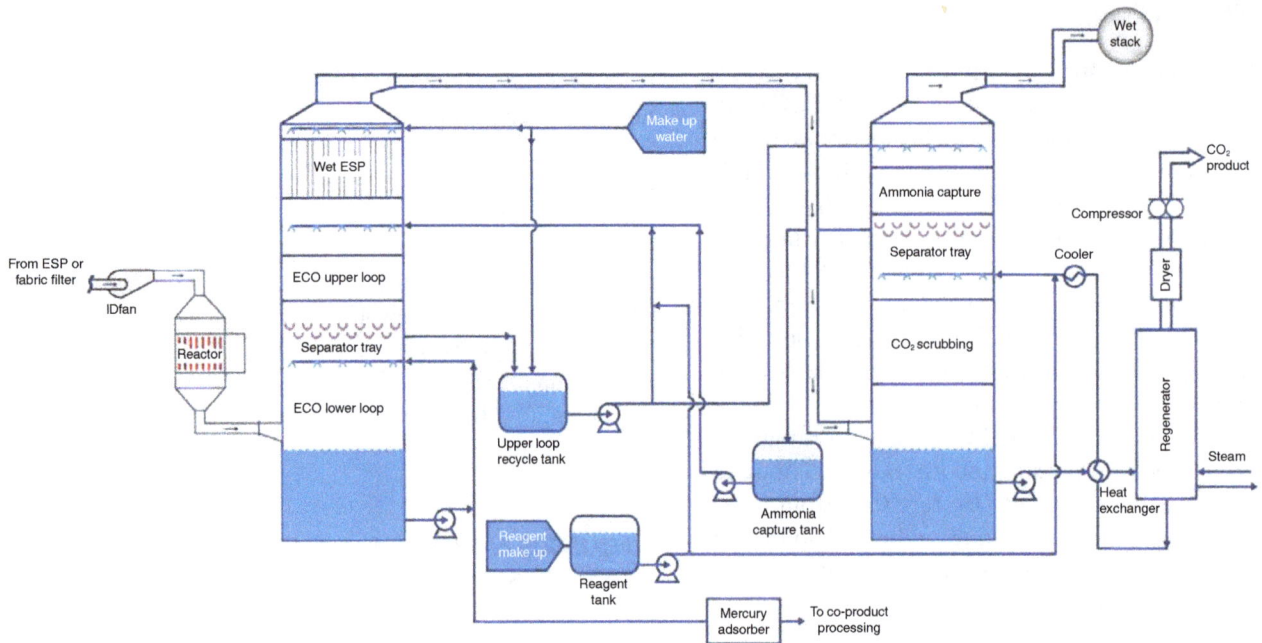

Figure 1

Incorporation of the ECO$_2$ scrubbing process with *Powerspan's* commercially available multi-pollutant control ECO process. Figure is copied from McLarnon and Duncan (2009).

Alstom has been playing a leading role in developing and advancing a Chilled Ammonia Process (CAP). In this process as described in the open literature (Darde *et al.*, 2009; Telikapalli *et al.*, 2011), CO$_2$ is absorbed in a highly ammoniated solution at low temperatures (0-10°C), producing a slurry containing ammonium bicarbonate. In the stripper, ammonium bicarbonate is converted to ammonium carbonate at temperatures above 100°C and pressures of 20-40 bar. Figure 2 shows the simplified chilled ammonia process flow diagram.

*Alstom* conducted a number of field tests of CAP and completed the demonstration project at American Electric Power's Mountaineer Plant at end of May 2011 after a 21 month period (Jönsson and Telikapalli, 2012). The technology is ready for the next major validation on a gas turbine exhaust as well as industrial (refinery) off-gas with the installation of demonstration plant at Technology Centre Mongstad (TCM). TCM is the world's largest facility for evaluating carbon capture technologies.

The processes investigated by CSIRO and under development at KIST are similar to *Powerspan's* ECO$_2$: they both operate at close to ambient temperatures and use ammonia concentrations below 10%.

The major difference is that in the CSIRO research, the desorption takes place under medium pressure and the process combines SO$_2$ removal and ammonia slip reduction (Yu *et al.*, 2011a, 2012a) while RIST is developing technology for CO$_2$ capture from Blast Furnace Gas (BFG) generated from iron and steel making process. The process utilises the unrecovered waste heat from a steelmaking process for solvent regeneration and ammonia recovery at low temperatures (Rhee *et al.*, 2011).

## 2 PROCESS CHEMISTRY

Thermo-chemical properties for the CO$_2$-NH$_3$-H$_2$O system have been reasonably understood thanks to intensive experimental and modeling studies over the years, in particular the last few years. These studies include experimental determination of vapour-liquid-solid equilibrium, speciation, heat of absorption, reaction kinetics, and the thermodynamic model development. Research work in this area has been reviewed by Maurer (2011) and Zhao *et al.* (2012). This section only briefly reviews vapour–liquid-solid equilibrium and reaction mechanism involved in the liquid phase.

Figure 2

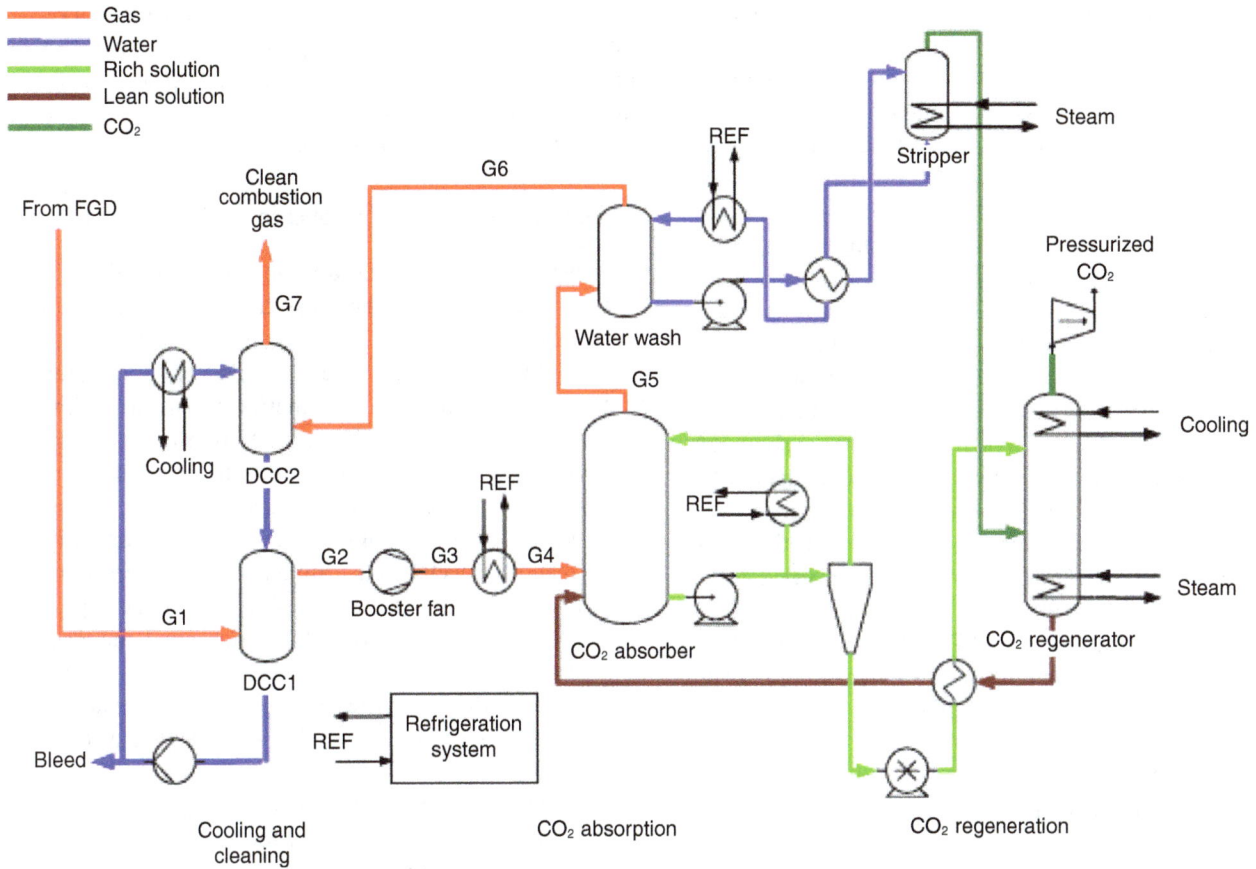

Chilled ammonia process flow diagram. Figure is copied from Jönssona and Telikapalli (2012).

## 2.1 Vapour-Liquid-Solid Equilibrium (VLSE)

The $CO_2$-$NH_3$-$H_2O$ equilibria relevant for the application of aqueous ammonia based $CO_2$ capture processes are described in Figure 3 (Thomsen and Rasmussen, 1999).

As shown in Figure 3, despite the fact that ammonia can be regarded as the simplest amine, its interaction with $CO_2$ is quite complex involving the evaporation of ammonia to the gas phase, the formation of carbonate, bicarbonate and carbamate in the liquid phase and solid precipitation. Ammonium bicarbonate is believed to be the major specy in the solid, but not all authors agreed on the possible formation and combination of the salts made of $(NH_4)_2CO_3$, $NH_2COONH_4$, $2NH_4HCO_3$. The complex nature of the $CO_2$-$NH_3$-$H_2O$ system makes ammonia an intriguing solvent and it imposes some challenges for its application on $CO_2$ capture, as discussed in the following sections.

Figure 3

Vapour liquid solid equilibria in the $CO_2$-$NH_3$-$H_2O$ system.

A number of thermodynamic models has been developed for the NH₃-CO₂-H₂O system in particular models developed by Thomsen and Rasmussen (1999) and Que and Chen (2011) which satisfactorily predict the vapour-liquid-solid equilibria over a wide range. Thomsen and Rasmussen developed a thermodynamic model which includes the aforementioned equilibrium processes (*Fig. 3*) under the 0-110°C range of temperature and 1-100 bar of the pressure, the ammonia concentration up to 80 molal. This model is not only for the vapour-liquid system but also for vapour-liquid-solid equilibrium including formation of ammonium bicarbonate: $NH_4HCO_3$; ammonium carbonate: $(NH_4)_2CO_3 \cdot H_2O$; ammonium carbamate: $NH_2COONH_4$ and ammonium sesqui-carbonate: $(NH_4)_2CO_3 \cdot 2NH_4HCO_3$. The thermal properties of the aqueous electrolyte solution are calculated on the basis of the extended UNIQUAC model and the Soave-Redlich-Kwong equation of state. Que and Chen (2011) developed the electrolyte NRTL (Non-Random Two Liquid) activity coefficient model which can satisfactorily represent the thermodynamic properties of the NH₃-CO₂-H₂O system at temperatures up to 473 K, pressures up to 7 MPa, NH₃ concentrations up to 30 wt%, and CO₂ loadings up to unity. However, the model considers ammonium bicarbonate as the only species in the solid.

The availability of these models allows the reliable calculation of thermochemical properties of the CO₂-NH₃-H₂O system under various conditions, assessment of the energy performance of the capture process and even the definition of the best range of operation conditions. For example, Figure 4 shows the predicted solubility curve for ammonium bicarbonate as a function of ammonia concentration and CO₂ loading at various temperatures

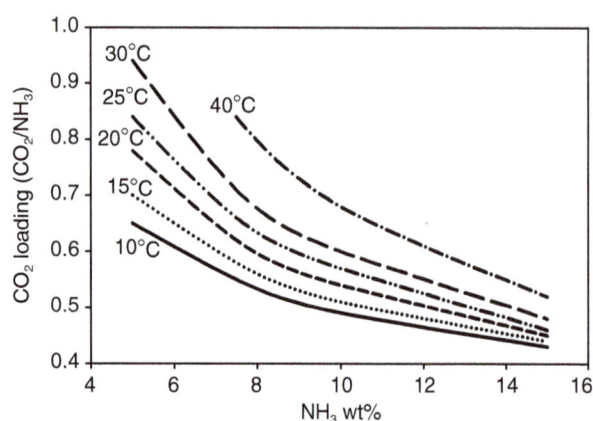

(Yu *et al.*, 2012a). In the region above each curve, ammonium bicarbonate precipitates from the liquid phase. To avoid solid precipitation in the absorber, the ammonia concentration, CO₂ loading and absorption temperatures should be considered carefully.

## 2.2 Reaction Mechanism/Kinetics in the Liquid Phase

CO₂ absorption in aqueous ammonia is a relatively slow process and limited by the mass transfer in the liquid phase (Darde *et al.*, 2011; Liu *et al.*, 2011; Qi *et al.*, 2013). So it is critical to have a detailed understanding of the reaction mechanism/kinetics involved in the absorption chemistry in order to develop a process model for process evaluation, optimisation and scale-up. A number of researchers investigated the reactions involved in the absorption process and put forward different reaction mechanisms. Similar to the case for the reaction of CO₂ with primary amines, the most important reaction in the presence of free ammonia is the reaction of NH₃ with CO₂. The reaction mechanisms proposed include the zwitterion mechanism (Derks and Versteeg, 2009) and the ter-molecular mechanism (Qin *et al.*, 2010).

A more detailed model was developed by Wang *et al.* (2011) who studied the kinetics of the reversible reaction of CO₂ with ammonia (instead of irreversible reaction proposed by other researchers) in aqueous solution using stopped-flow techniques and established a complete reaction mechanism, as shown in Figure 5 (Wang *et al.*, 2011).

The interactions of CO₂ (aq) with water and hydroxide ions are indicated in dashed lines. Equations (1-6) summarize this part of the reaction scheme as well as the protonation equilibrium of ammonia:

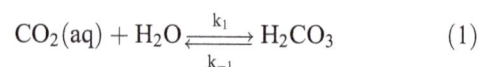

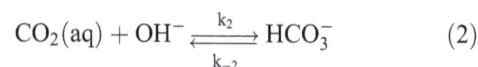

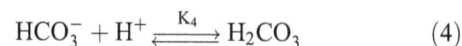

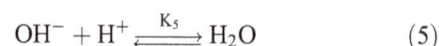

$$CO_2(aq) + H_2O \underset{k_{-1}}{\overset{k_1}{\rightleftharpoons}} H_2CO_3 \qquad (1)$$

$$CO_2(aq) + OH^- \underset{k_{-2}}{\overset{k_2}{\rightleftharpoons}} HCO_3^- \qquad (2)$$

$$CO_3^{2-} + H^+ \overset{K_3}{\rightleftharpoons} HCO_3^- \qquad (3)$$

$$HCO_3^- + H^+ \overset{K_4}{\rightleftharpoons} H_2CO_3 \qquad (4)$$

$$OH^- + H^+ \overset{K_5}{\rightleftharpoons} H_2O \qquad (5)$$

$$NH_3 + H^+ \overset{K_6}{\rightleftharpoons} NH_4^+ \qquad (6)$$

The formation reactions of the carbamate/carbamic acid included in the mechanism are shown in Equation (7),

Figure 4

Predicted ammonium bicarbonate solubility operation line.

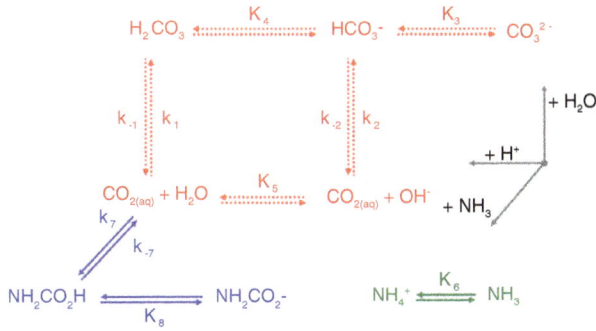

Figure 5

Reaction scheme, including all reactions between ammonia, the $CO_2$/carbonate group, and protons (Wang et al., 2011).

while Equation (8) describes the protonation of carbamate to the carbamic acid:

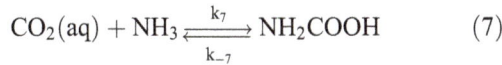

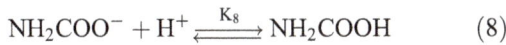

$$CO_2(aq) + NH_3 \underset{k_{-7}}{\overset{k_7}{\rightleftharpoons}} NH_2COOH \qquad (7)$$

$$NH_2COO^- + H^+ \overset{K_8}{\rightleftharpoons} NH_2COOH \qquad (8)$$

The temperature dependence of all rate and equilibrium constants including the protonation constant of $NH_3$ between 15 and 45°C were reported by Wang et al.

In a $CO_2$ capture process aqueous ammonia reacts with $CO_2$ (aq) to form carbamic acid which deprotonates instantly to give a carbamate. Due to the pH changes with continued absorption of $CO_2$, the equilibrium is reversed and the carbamate eventually decomposes to generate protonated ammonia and bicarbonate. Table 1 shows the carbamate equilibrium constants and pKa of amines including ammonia. The carbamate equilibrium constants for MEA is much higher than that for ammonia, suggesting that ammonia-derived carbamate is formed with a lower yield than the equivalent Mono-EthanolAmine carbamate, which contributes to the higher $CO_2$ loading capacity for aqueous ammonia compared to MonoEthanolAmine.

## 3 PERFORMANCES OF AQUEOUS AMMONIA BASED POST COMBUSTION CAPTURE PROCESSES

Pilot plant trials and demonstration projects have confirmed the technical feasibility of the aqueous ammonia based $CO_2$ capture processes and some of the anticipated benefits (Jönsson and Telikapalli, 2012; Rhee et al., 2011; Yu et al., 2011b). The benefits include high $CO_2$ removal efficiency and high purity of $CO_2$ product; high stability; high $CO_2$ loading capacity and high effectiveness for capturing $SO_2$ (Yu et al., 2011b). This section focuses on absorption rate, regeneration energy, ammonia loss and overall performance.

### 3.1 CO$_2$ Absorption Rate

The $CO_2$ absorption rate is an important parameter which can determine the size of column and hence determine capital cost. It is related to the mass transfer coefficient, the effective surface area for mass transfer and the driving force as described in Equation (9):

$$N_{CO_2} = K_G A_{int}(P_{CO_2} - P_{CO_2}^*) \qquad (9)$$

where $N_{CO_2}$ is the $CO_2$ absorption rate (mol/s), $K_G$ is the overall gas phase mass transfer coefficient (mol/(s.m$^2$.kPa)), $P_{CO_2}$ is the partial pressure of $CO_2$ in the flue gas (kPa), $P_{CO_2}^*$ is the $CO_2$ equilibrium partial pressure in the solvent (kPa) and $A_{int}$ is the effective interfacial surface area of packing (m$^2$).

Mass transfer coefficients are used to assess the $CO_2$ absorption rates in the solvents. The bench scale studies using wetted wall column and pilot plant studies all suggest that $CO_2$ mass transfer coefficient in aqueous ammonia are lower than those in MEA (Yu et al., 2011a, 2012b). Figure 6 shows the comparison of overall

TABLE 1

Carbamate equilibrium constants and pKa of amines at 25°C (Fernandes et al., 2012; Puxty et al., 2009)

| | | NH$_3$ | MEA | AMP | DEA | PIPD | 4-PIPDM | 4-PIPDE |
|---|---|---|---|---|---|---|---|---|
| $HCO_3^- + RNH_2 \overset{K_1}{\rightleftharpoons} RNHCOO^- + H_2O$ | Log $K_1$ | 0.33 | 1.76 | Very unstable carbamate formed | 0.92 | 1.38 | 1.39 | 1.38 |
| $RNH_3^+ \overset{K_2}{\rightleftharpoons} RNH_2 + H^+$ | pKa | 9.24 | 9.44 | 8.84 | 8.88 | 11.2 | 10.9 | 10.6 |

AMP: 2-amino-2-methyl-1-propanol; DEA: Diethanolamine; PIPD: Piperidine; 4-PIPDM: 4-piperidinemethanol; 4-PIPDE: 4-piperidineethanol.

$K_1 = \frac{[RNHCOOH]}{[RNHCOO^-][H^+]}$, M$^{-1}$; $K_2 = \frac{[RNH_2][H^+]}{[RNH_3^+]}$, M; pKa $= -\log_{10}K_2$

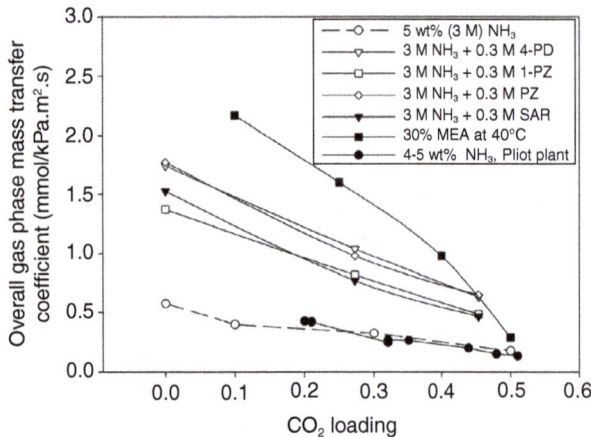

**Figure 6**

Mass transfer coefficients for $CO_2$ at various $CO_2$ loadings obtained from Wetted Wall Column experiments (WWC) and pilot plant trials. WWC: for 30 wt% MEA, absorption temperature = 40°C; for other solvents, absorption temperature = 15°C. Pilot plant: absorption temperature = 10-15°C. 4-PD: 4 amino piperidine. PZ: piperazine. 1-PZ: 1-methyl piperazine. SAR: Sodium Sarcosinate.

gas phase mass transfer coefficient as a function of $CO_2$ loading in aqueous ammonia and the standard 30 wt% MEA (Yu *et al.*, 2012b). Good agreement between results from pilot plant and wetted wall column experiments for the ammonia solvent was observed. This suggests that the mass transfer coefficients obtained from the wetted wall column can be used to estimate those in large-scale packed columns, at least under the conditions reported here.

The mass transfer coefficients for $CO_2$ in 3 M aqueous ammonia alone are 3 to 5 times lower than those in MEA solutions at the $CO_2$ loading range of 0.2-0.4 mole $CO_2$ per mole amine. The low values for the mass transfer coefficients are due to the low rate constant for the reaction of $NH_3$ and $CO_2$ in comparison with that for the reaction between MEA and $CO_2$. This reaction is the key step for $CO_2$ absorption in aqueous ammonia (Wang *et al.*, 2011). The reaction of $CO_2 + OH^-$, can contribute to $CO_2$ absorption but its contribution is small due to the low concentration of $OH^-$. According to surface renewal theory of mass transfer with chemical reactions, for a pseudo first order reaction and assuming resistance in the gas phase is negligible, the liquid side mass transfer coefficient can be estimated by Equation (10) (Yu *et al.*, 2012b):

$$K_G = \frac{\sqrt{k\, C_{NH_3} D_{CO_2}}}{H_{CO_2}} \qquad (10)$$

$k$ is the rate constant for forward reaction of $CO_2$ with $NH_3$, $C_{NH_3}$ is the concentration of free ammonia in the liquid bulk, and $D_{CO_2}$ and $H_{CO_2}$ are diffusivity coefficients and Henry's law constants in aqueous ammonia, respectively.

$k$ values are typically around 200 $M^{-1} \cdot s^{-1}$ at the temperature of 15°C (Wang *et al.*, 2011). The values are significantly lower than the rate constants for the reaction of MEA with $CO_2$ at 40°C which are more than 10 000 $M^{-1} \cdot s^{-1}$ (Versteeg *et al.*, 1996). Both temperature and ammonia concentration can be increased to enhance $K_G$, but the ammonia loss would increase as well.

There is no publicly available report on the $CO_2$ absorption rate in the chilled ammonia process. Recently, *Alstom* power published a paper in which described the rate based model for the chilled ammonia process (Lia *et al.*, 2012). The model used the zwitterion mechanism to explain the reaction between $CO_2$ and $NH_3$ and the kinetics parameters in equations were derived by fitting them against experimental data reported by Derks (Derks and Versteeg, 2009). So it is speculated that the mass transfer coefficients for $CO_2$ in the chilled ammonia process are low, if no promoter/additive is used.

## 3.2 Regeneration Energy

Table 2 lists the solvent regeneration energies reported in the literature. The regeneration energies refer to the heat requirement for regeneration of solvents (reboiler duty) in the stripper. The values are quite different and vary from less than 1 000 KJ/kg $CO_2$ to more than 4 200 KJ/kg $CO_2$. The significant discrepancy is due to under-estimation of heat of desorption in the previous studies and the large variation in operational conditions in particular the solvent concentrations.

During the solvent regeneration, heat is required for:
- $CO_2$ desorption reaction;
- steam regeneration;
- sensible heat.

Due to the different process conditions, the contribution of the above three parts of energy is various, resulting in the various heat requirements in the $CO_2$ capture process.

During the stripper process, $CO_2$ can be desorbed *via* many pathways including the following overall reactions (Jilvero *et al.*, 2012).

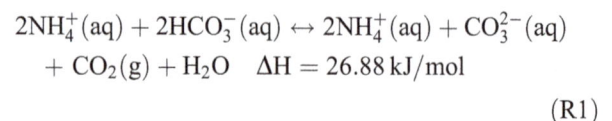

$$2NH_4^+(aq) + 2HCO_3^-(aq) \leftrightarrow 2NH_4^+(aq) + CO_3^{2-}(aq)$$
$$+ CO_2(g) + H_2O \quad \Delta H = 26.88\,kJ/mol$$

$$(R1)$$

TABLE 2

Heat requirement of desorption processes that use aqueous ammonia

| Source | Year | Method | Heat requirement (KJ/kg $CO_2$) | Ammonia concentration (wt%) | $CO_2$ loading | |
|---|---|---|---|---|---|---|
| | | | | | Lean | Rich |
| EPRI (2006) | 2006 | Estimate | 930 | - | - | - |
| McLarnon and Duncan (2009) | 2009 | Estimate | 1 000 | - | - | - |
| Darde et al. (2009) | 2009 | Modeling | 2 050 | 28 | 0.33 | 0.67 |
| Dave et al. (2009) | 2009 | Modeling | 2 900 | 7 | - | - |
| Mathias et al. (2010) | 2009 | Modeling | 2 377 | 26 | 0.35 | 0.92 |
| Yu et al. (2011a) | 2010 | Pilot plant | 4 000-4 200 | 4-5 | 0.17-0.19 | 0.35-0.40 |
| Valenti et al. (2012) | 2012 | Modeling | 2 460 | 15 | 0.33 | - |
| Darde et al. (2012) | 2012 | Modeling | ~2 500 | 7.8 | 0.33 | 0.66-0.68 |
| Jilvero et al. (2012) | 2012 | Modeling | 2 200 | 10 | 0.3 | - |
| Alstom power (Jönsson and Telikapalli, 2012) | 2012 | Demonstration project | 2 200 | - | - | - |

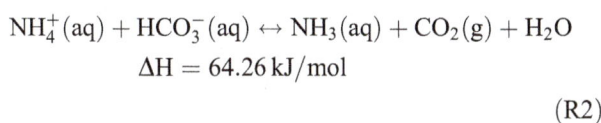

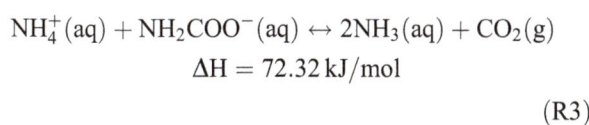

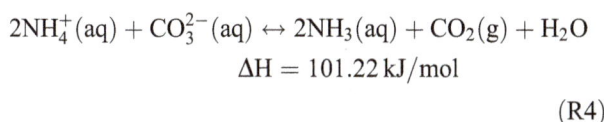

$$NH_4^+(aq) + HCO_3^-(aq) \leftrightarrow NH_3(aq) + CO_2(g) + H_2O$$
$$\Delta H = 64.26\,kJ/mol$$

(R2)

$$NH_4^+(aq) + NH_2COO^-(aq) \leftrightarrow 2NH_3(aq) + CO_2(g)$$
$$\Delta H = 72.32\,kJ/mol$$

(R3)

$$2NH_4^+(aq) + CO_3^{2-}(aq) \leftrightarrow 2NH_3(aq) + CO_2(g) + H_2O$$
$$\Delta H = 101.22\,kJ/mol$$

(R4)

The previous research reported a low regeneration energy requirement based on the assumption that the regeneration only involves the decompose reaction of ammonia bicarbonate to $CO_2$ and ammonia carbonate (R1) and the heat of reaction for this reaction is low ($\Delta H = 26.88$ kJ/mol). The equilibrium based modeling suggested that one cannot single out only one particular pathway for the solvent regeneration. The experimental and modeling investigations show that the heat of absorption is a function of $CO_2$ loading and a weak function of temperature and ammonia concentration. Figure 7 shows the experimental and predicted heat of absorption of $CO_2$ in 5% ammonia solution at the

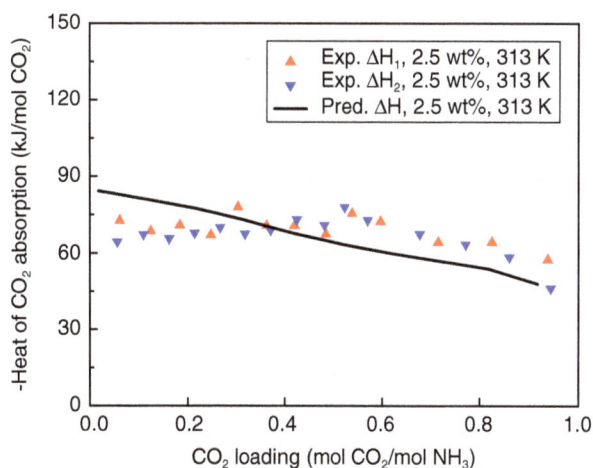

Figure 7

Predicted and experimental heat of $CO_2$ absorption in 5 wt% $NH_3$ at $T = 40°C$.

temperature of 40°C (Qi et al., 2013). The predicted results were obtained using the thermodynamic model for $CO_2$-$NH_3$-$H_2O$ developed by *Aspentech* (Que and Chen, 2011). Over the large $CO_2$ loading range (0-0.6), the heat of absorption is more than 60 kJ/mol, suggesting that the overall absorption/decomposition pathway

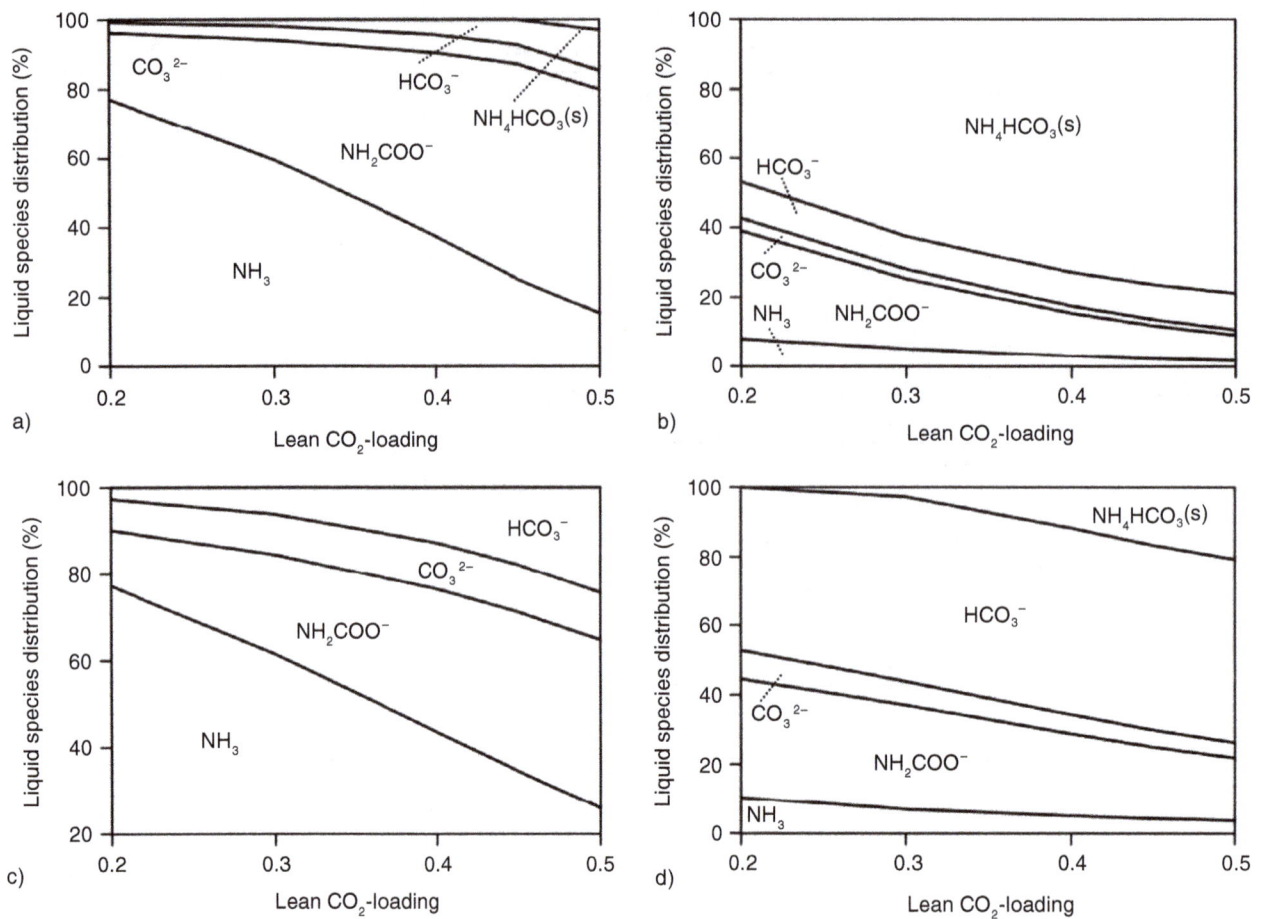

Figure 8

The $CO_2$-lean solution (a and c) and the $CO_2$-rich solution (b and d) liquid composition at 15% ammonia (a and b) and 5% ammonia (c and d) are shown as a function of the lean $CO_2$-loading. The figures exclude water and trace element, such as $N_2$, $OH^-$, $H^+$, and $CO_2$. The process is conducted at a temperature of 10°C and at atmospheric pressure. The figure is copied from Jilvero et al. (2012).

is mainly *via* the formation of ammonium bicarbonate (R2) and carbamate (R3) which is consistent with the conclusion drawn by Jilvero et al. (2012). Through the thermodynamic analysis of the speciation distribution in both lean and rich solvent at $NH_3$ concentrations 5 and 15% and lean loading of 0.2 to 0.5, it has been found that a large proportion of the $CO_2$ is absorbed through reactions with free ammonia. As shown in Figure 8, for lean and rich solvents, there are negligible differences with respect to the fractions of carbamate and the carbonate, while the level of free ammonia is decreasing and the amount of bicarbonate is increasing. This suggests that the main reaction pathway for the absorption of $CO_2$ was identified as the formation of bicarbonate through the reaction of ammonia with $CO_2$. In other words, the decomposition of bicarbonate to ammonia

and $CO_2$ is the main reaction for the $CO_2$ desorption. The heat of reaction required to desorb $CO_2$ of ammonia was similar to that required for MonoEthanolAmine (MEA). The simulation results show that the heat requirement of ammonia regeneration could be less than 2 500 kJ/kg $CO_2$ captured. (Jilvero et al., 2012). The main reason for the lower heat requirement for ammonia compared to MEA is that the thermodynamic properties of the $NH_3$-$CO_2$-$H_2O$ system allow for pressurized regeneration, which results in significantly lower water vaporization.

All investigations based on equilibrium-based thermodynamic process modeling are in agreement regarding the potential to achieve a regeneration heat requirement that ranges between 2 000 and 3 000 kJ/kg $CO_2$ for an aqueous ammonia process. The recent publication from *Alstom* shows a heat requirement of 2 200 KJ/kg $CO_2$ captured for the chilled ammonia process (Jönsson and Telikapalli, 2012). This result is in line with the modeling prediction. These recent studies show that the aqueous ammonia based $CO_2$ capture process requires much less heat for the solvent regeneration than MEA based capture processes (Jönsson and Telikapalli, 2012; Jilvero et al., 2012; Valenti et al., 2012; Darde et al., 2012).

The pilot plant trials conducted by *CSIRO* and *Delta Electricity* at Delta Electricity's Munmorah power station showed a much higher heat requirement (Yu et al., 2011b). One reason for the high heat requirement is that a diluted ammonia solution was used and the amount of $CO_2$ captured/released is low. As a result, more than 50% of energy is used to heat up the solvent (sensible heat). The results also show the heat of desorption is around 70-80 kJ/mol, suggesting under the pilot plant conditions where $CO_2$ loading in stripper varied between 0.4 and 0.1, the major pathway is the formation of carbamate *via* the reaction of $CO_2$ with free ammonia.

### 3.3 Ammonia Loss

Ammonia is very volatile and characterised by high equilibrium pressure even in the $CO_2$ loaded aqueous ammonia solution. Figure 9 shows predicted partial and total pressure at the ammonia concentration of 28% and temperature of 8°C; Figure 10 predicted equilibrium $CO_2$ and $NH_3$ partial pressure as a function of $CO_2$ loading at the ammonia concentration of 5 wt% and

the temperature of 10, 20 and 30°C; Figure 11 predicted equilibrium $NH_3$ and $CO_2$ partial pressure as a function of $CO_2$ loading at the ammonia concentration of 2.5 and 5% and the temperature of 10°C. The conditions for Figure 9 are relevant for the chilled ammonia process in which precipitation of products is taken into account while the conditions for Figures 10 and 11 are for processes at ambient temperatures, used by *Powerspan*, *CSIRO* and *KIST* at lower ammonia

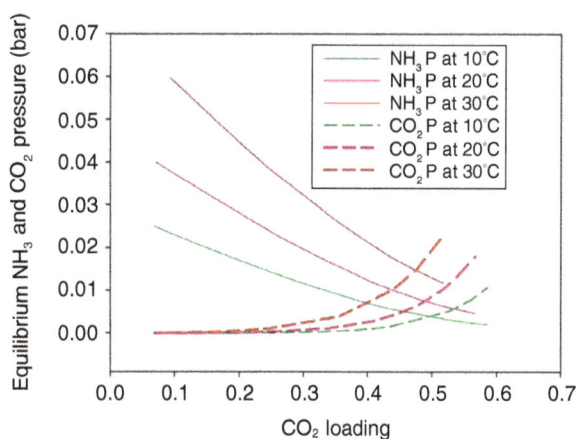

Figure 10

Predicted equilibrium $CO_2$ and $NH_3$ partial pressure as a function of $CO_2$ loading at the ammonia concentration of 5 wt% and the temperature of 10°C. Aspen Plus® V7.3 with a built-in thermodynamic model for $CO_2$- $NH_3$ – $H_2O$ was used to obtain equilibrium pressure in Figure 10 and Figure 11 (AspenTechnology, 2012; Que and Chen, 2011).

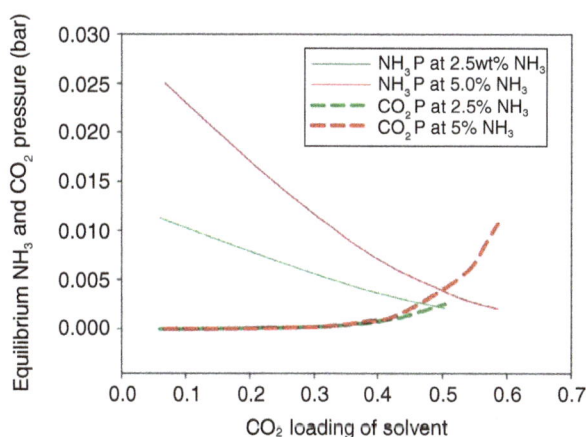

Figure 9

Predicted partial and total pressure at the ammonia concentration of 28 wt% and the temperature of 8°C (Darde et al., 2009).

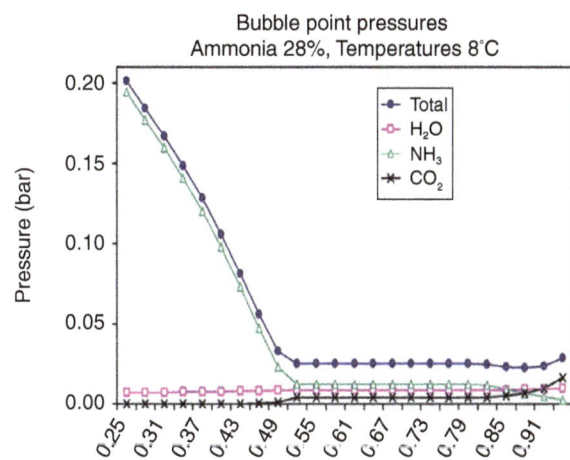

Figure 11

Predicted equilibrium $NH_3$ and $CO_2$ partial pressure as a function of $CO_2$ loading at the ammonia concentration of 2.5 and 5% and the temperature of 10°C.

concentration in which precipitation did not take place. It is clear in all cases, that $NH_3$ vapour pressures are significant at $CO_2$ loading below 0.5.

As a result, ammonia will inevitably slip to flue gas in the absorber and to the $CO_2$ product stream from the stripper in the capture process and measures will have to be taken to recover the ammonia. The operational and capital costs for an ammonia recovery process will be determined by the amount of ammonia slip and the impact of the operational parameters. Despite the fact that it is widely acknowledged that ammonia loss is a big challenge to the aqueous ammonia based $CO_2$ capture process, there are limited studies on the ammonia loss rate in the literature.

For the chilled ammonia processes, the results available in the literature are all from the modeling work. Mathias et al. (2010) studied the effect of ammonia concentration in the gas phase at the outlet of absorber. They found out that the $NH_3$ slip from the absorber is only weakly dependent on the $NH_3$ concentration in the range of 15-26% and depends mainly on temperature. At the absorber temperature of 10°C and atmospheric pressure, the $NH_3$ concentration is approximately constant at 2 230 ppmv at ambient pressure, the ammonia concentration of 26% and $CO_2$ loading of ca. 0.4 in the lean solvent and reduced to 242 ppm if absorber temperature can drop to −1.1°C. The equilibrium $NH_3$ partial pressure for a solvent at the ammonia concentration of 28% and $CO_2$ loading of 0.4 and 8°C is more than 0.05 bar equivalent to 50 000 ppm at atmospheric pressure (*Fig. 9*), according to Darde et al. (2009). The large discrepancy between the two references is most likely indicative of the uncertainty around the thermodynamics of liquid-solid systems.

Versteeg and Rubin (2011) did a thorough analysis of effect of ammonia concentration (0-30%), $CO_2$ loading (0.25 to 0.67), absorber temperature (5 to 20°C) on ammonia slip in a chilled ammonia process in which rich solvent is not recycled to absorber. An increase in ammonia concentration and absorber temperature as well as a decrease in $CO_2$ loading can lead to an increase in ammonia slip. Darde et al. (2012) also modeled the chilled ammonia process at the low ammonia concentrations up to 12% and investigated the effect of ammonia concentration, $CO_2$ loading, absorber temperature and process configuration. In a configuration close to the configuration proposed by *Alstom* (*Fig. 2*), it has been found that with increase in ammonia concentration in the solvent from 4 to 11%, the ammonia concentration in the gas stream leaving the absorber varies between 4 500 and 19 000 ppm. In Darde's work, a murphy tray efficiency was introduced to account for deviation from the equilibrium.

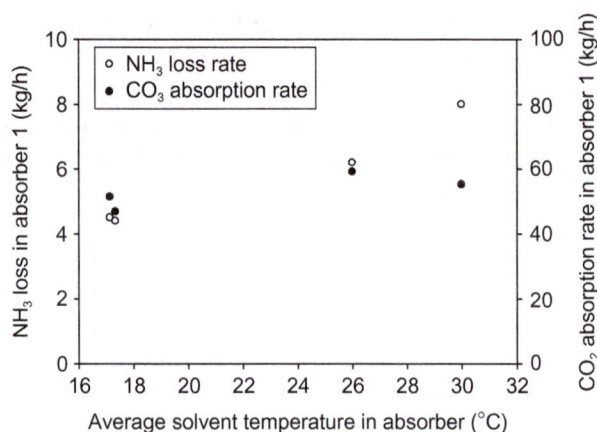

Figure 12

Effect of temperature on the $CO_2$ absorption rate and $NH_3$ loss rate in absorber 1 under similar conditions. $NH_3$ wt% = 3.8-4.2; $CO_2$ loading in lean solvent = 0.22-0.25. Solvent flow-rate = 134 L/min; Gas flow-rate = 640-820 kg/h; $CO_2$ in the flue gas = 8-10 vol%. Inner diameter of absorber 1 = 0.6 m; packing height = 3.8 m (Yu et al., 2011a).

Budzianowski (2011), Niu et al. (2012) and Yu et al. (2012c) conducted experiments to investigate ammonia loss at low ammonia concentrations and relatively high absorber temperatures. All results from these studies observed the trends similar to the modeling studies described above, but the ammonia loss is much higher. For example, Niu et al. (2012) studies the absorption at the ammonia concentration of 1-11 wt%, the $CO_2$ loading of 0.12 and the temperature of 24°C and the ammonia concentrations at outlet were 1-14 vol%. Similar to research work by Niu et al., Yu et al. used a pilot scale research facility to investigate the effect of a number of parameters. The pilot plant trials show that the conditions which typically enhance $CO_2$ absorption rate such as an increase in ammonia concentration and absorption temperature and a decrease in $CO_2$ loading tend to increase ammonia loss significantly. Among many parameters which affect ammonia loss, temperature is one of the most sensitive parameters. As shown in Figure 12, with the increase in solvent temperature, $CO_2$ absorption rate in absorber 1 remains similarly but ammonia loss rate increases significantly. From the ammonia loss point of view, the absorption temperature should be as low as possible. However, it is expected that a considerable amount of additional energy is required to produce chilled water. Dave et al. (2009) simulated the aqueous ammonia based process by keeping the flue gas and lean solvent temperatures to the absorber at 10°C and solvent concentration at 2.5 and 5 wt%.

The electricity consumption is estimated to be more than 1 400 KJ/kg $CO_2$.

In addition, the research by Budzianowski (2011) and Qi et al. (2013) suggested that the $NH_3$ desorption process is a fast process and controlled by the vapour liquid equilibrium which could guide future research on process modification to reduce ammonia slip in absorber.

Ammonia is also present in the $CO_2$ product and requires removal. Yu et al. (2012a) investigated the effect of operation conditions on ammonia concentration in the $CO_2$ product. To reduce ammonia concentration, the high pressure operation and low temperature in the over-head condenser will help reduce ammonia concentration in the $CO_2$-product. However, this can cause solid precipitation in the stripper where solvent regeneration leads to generation of vapour which contains significant amount of ammonia and $CO_2$ gas apart from water vapour. When the temperature of vapour drops in the condenser/reflux line or even in stripper when the regeneration stops, part of vapour will condense, leading to formation of the solution/droplet in which ammonia concentration is very high (more than 10 wt%). Since partial pressure of $CO_2$ is high in stripper, this facilitates formation of a highly carbonated ammonia solution which reaches the ammonium bicarbonate solubility limit. The precipitation of solid will block the stripper condenser and reflux line, causing a shut-down of the plant. Blockage also occurs in instrument tubing, leading to false readings and affecting the operation. For ammonia processes which operate at higher ammonia concentrations, the blockage could also occur in the absorber packing.

### 3.4 Overall Performance

The aqueous ammonia based $CO_2$ capture process is quite complex as it involves flue gas cooling, $CO_2$ absorption and stripping, ammonia capture and recovery. The technology is still under development. So far there are no published results available from optimised pilot plant/demonstration plants, which prevent the rigorous economical assessment of the technologies. In the literature, the assessment of the overall performance was carried out using the equilibrium based process model with the estimated $CO_2$ absorption rates in the absorber.

Previous studies indicated significant improvements in performance over traditional amine technologies but the analysis was preliminary and based on limited understanding of the system (Bai and Yeh, 1997; Ciferno et al., 2005; Gal et al., 2011). As described above, the understanding of the system has improved significantly over the last few years. Based on improved understanding of the system

thermodynamics model, Versteeg and Rubin (2011) modeled a chilled ammonia-based post-combustion $CO_2$ capture system processing flue gas from a supercritical coal-fired power plant. The estimated performance and cost were compared to an amine-based capture system. For the ammonia system the absorber $CO_2$ capture efficiency, $NH_3$ slip, and solids precipitation were evaluated for changes in lean solution $NH_3$ concentration, $NH_3/CO_2$ ratio, and absorber temperature. For 90% $CO_2$ capture the levelized cost of electricity generation (annual revenue requirement) for the plant with ammonia-based capture was estimated at \$US 105/MWh, which is comparable to the levelized cost of electricity generation for the plant with an amine-based capture system. The ammonia-based $CO_2$ system benefits from lower steam loads and reduced compressor power requirements, but the chilling loads and associated costs offset these benefits. Versteeg and Rubin (2011) suggest that the viability of ammonia-based $CO_2$ capture may be location dependent. A plant located in a characteristically cold climate, or with direct access to a large cooling sink such as a deep water lake, would have lower parasitic energy demands for the process chillers.

Linnenberg et al. (2012) evaluated the impact of an ammonia-based post-combustion $CO_2$ capture process on a steam power plant. To study the influence of the cold end of the process, two power plants with different cooling water temperatures are analysed. Additionally, two different process configurations of the capture plant, with either one single absorber or two absorbers connected in series where the first absorber captures the majority of the $CO_2$ and the second limits the $NH_3$ slip, are evaluated. The study shows that the configuration of the process with absorption at low temperature (approximately 10°C) with or without precipitation of ammonium bicarbonate compounds leads to a lower net efficiency penalty than an MEA-based process, assuming that low temperature cooling water is available. An estimate of the size of the absorber shows that the absorber columns of an ammonia-based process are significantly higher than the ones required for an MEA-based process. The conclusions are consistent with those by Versteeg and Rubin (2011).

### 4 OUTLOOK

As highlighted by Zhuang et al. (2012), three inherent hurdles with the aqueous ammonia based capture technology have not been fully overcome yet: low $CO_2$ absorption rate resulting in large absorption equipment; high ammonia vapour pressure leading to ammonia slip into the cleaned flue gas, and ammonium bicarbonate solid formation in the absorption tower (in stripper condenser as well) which may plug the tower packing, stripper condenser and reflux line. Further research efforts should be focused on these areas in order to improve the technical and economic feasibility of the technologies. Recent research has shown that introduction of additives can enhance $CO_2$ absorption rate or suppress ammonia vapour (Ahn et al., 2012; Gal et al., 2011; Salentinig et al., 2012; Yu et al., 2012b). For example, a PZ-promoted ammonia solution, which has been patented by Alstom (Gal et al., 2011), is a highly effective promoter of $CO_2$ absorption in ammonia, as shown in Figure 6. The study by Yu et al. (2012b) has shown that alternative additives, which are environmentally friendly, have the potential to significantly increase the $CO_2$ mass transfer coefficients to a level comparable with MEA at high $CO_2$ loadings relevant to industrial applications. However, it is not clear about possible side effect of introduction of additives. New processes such as high pressure absorption, rich solvent split, re-engineering of the industrial ammonium bicarbonate fertilizer production process and combined capture of $CO_2$, $SO_2$ and other impurities could offer potential for improvement (Yu et al., 2011a; Zhuang et al., 2012). All these research efforts are the early stage of the development.

## CONCLUSIONS

The aqueous ammonia based post combustion $CO_2$ capture processes have gained intensive attention over the last few years thanks to many advantages over amine based solvents. A number of pilot plants and demonstration plants have been constructed and used to assess the technical and economical feasibility of the processes and identify opportunities to further advance the technologies. This paper has reviewed the research work available in the public domain including understanding of the thermo-chemical and physical properties of the $CO_2$-$NH_3$-$H_2O$ system, and absorption chemistry/kinetics involved during the absorption, information on absorption rate, ammonia loss, regeneration energies, and practical issues, and assessment of the technical and economical feasibility of the aqueous ammonia based post combustion capture technologies.

Thermo-chemical properties for the $CO_2$-$NH_3$-$H_2O$ system have been well understood and an extended UNIQUAC model, described by Thomsen and Rasmussen (1999) and an electrolyte NRTL activity coefficient model by Aspentech (Que and Chen, 2011) can satisfactorily represent the thermodynamic properties of the $CO_2$-$NH_3$-$H_2O$ system over a wide range. A comprehensive reaction mechanisms involved in the reaction of $CO_2$ with ammonia in the aqueous medium has been developed by Wang et al. (2011).

Bench scale and pilot plant scale investigations show that aqueous ammonia absorbs $CO_2$ at a slower rate than the standard MEA solution under their respective operational conditions. Process modeling and experimental results both indicate that ammonia losses can be substantial. All investigations based on equilibrium-based thermodynamic process modeling are in agreement regarding a heat requirement that ranges between 2 000 and 3 000 kJ/kg $CO_2$. The recent publication from Alstom shows a regeneration heat requirement of 2 200 kJ/kg $CO_2$ captured for the chilled ammonia process. The assessment of the overall performance using the equilibrium based process models shows that the chilled ammonia process leads to a lower net efficiency penalty than an MEA-based process, assuming that low temperature cooling water is available. An estimate of the size of the absorber shows that the absorber columns of an ammonia-based process will be significantly higher than the ones required for an MEA-based process.

## ACKNOWLEDGMENTS

The authors acknowledge the support of CSIRO's Advanced Coal Technology and Nan Yang is grateful to China Scholarship Council for supporting her PhD studies in CSIRO.

## REFERENCES

Ahn C.K., Han K., Lee M.S., Kim J.Y., Chun H.D., Kim Y., Park J.M. (2012) Experimental studies of additives for suppression of ammonia vaporization in the ammonia based $CO_2$ capture process, 11st International Conference on Greenhouse Gas Control Technologies (GHGT-11), Kyoto, Japan, 18-22 Nov.

AspenTechnology (2012) Aspen Plus® v7.3, Cambridge MA, USA.

Bai H., Yeh A.C. (1997) Removal of $CO_2$ greenhouse gas by ammonia scrubbing, Ind. Eng. Chem. Res. 36, 2490-2493.

Budzianowski W.M. (2011) Mitigating $NH_3$ vaporization from an aqueous ammonia process for $CO_2$ capture, Int. J. Chem. React. Eng. 9, A58.

Ciferno J.P., Di Pietro P., Tarka T. (2005) An economic scoping study for $CO_2$ capture using aqueous ammonia, DOE/NETL Final Report.

Darde V., Thomsen K., van Well W.J.M., Stenby E.H. (2009) Chilled ammonia process for $CO_2$ capture, Energy Procedia 1, 1035-1042.

Darde V., van Well W.J.M., Fosboel P.L., Stenby E.H., Thomsen K. (2011) Experimental measurement and modeling of the rate of absorption of carbon dioxide by aqueous ammonia, Int. J. Greenhouse Gas Control 5, 1149-1162.

Darde V., Maribo-Mogensen B., van Well W.J.M., Stenby E. H., Thomsen K. (2012) Process simulation of $CO_2$ capture with aqueous ammonia using the extended UNIQUAC model, *Int. J. Greenhouse Gas Control* **10**, 74-87.

Dave N., Do T., Puxty G., Rowland R., Feron P.H.M., Attalla M.I. (2009) $CO_2$ capture by aqueous amines and aqueous ammonia – A Comparison, *Energy Procedia* **1**, 949-954.

Derks P.W.J., Versteeg G.F. (2009) Kinetics of absorption of carbon dioxide in aqueous ammonia solutions, *Energy Procedia* **1**, 1139-1146.

EPRI (2006) Chilled ammonia post combustion $CO_2$ capture system – Laboratory and economic evaluation results, Palo Alto, CA.

Fernandes D., Conway W., Burns R., Lawrance G., Maeder M., Puxty G. (2012) Investigations of primary and secondary amine carbamate stability by H-1 NMR spectroscopy for post combustion capture of carbon dioxide, *J. Chem. Thermodyn.* **54**, 183-191.

Gal E., Bade O.M., Jayaweera I., Gopala K. (2011) Promoter enhanced chilled ammonia based system and method for removal of $CO_2$ from flue gas, USP 7,862,788, *ALSTOM Technology Ltd.*

Jilvero H., Normann F., Andersson K., Johnsson F. (2012) Heat requirement for regeneration of aqueous ammonia in post-combustion carbon dioxide capture, *Int. J. Greenhouse Gas Control* **11**, 181-187.

Jönsson S., Telikapalli V. (2012) Chilled Ammonia Process installed at the Technology Center Mongstad, *11st International Conference on Greenhouse Gas Control Technologies (GHGT-11)*, Kyoto, Japan, 18-22 Nov.

Lia M., Chen X., Hiwale R., Vitse F. (2012) Rate based modeling of chilled ammonia process (CAP) Absorber in Aspen Plus, *11st International Conference on Greenhouse Gas Control Technologies (GHGT-11)*, Kyoto, Japan, 18-22 Nov.

Linnenberg S., Darde V., Oexmann J., Kather A., van Well W.J.M., Thomsen K. (2012) Evaluating the impact of an ammonia-based post-combustion $CO_2$ capture process on a steam power plant with different cooling water temperatures, *Int. J. Greenhouse Gas Control* **10**, 1-14.

Liu J., Wang S., Qi G., Zhao B., Chen C. (2011) Kinetics and mass transfer of carbon dioxide absorption into aqueous ammonia, *Energy Procedia* **4**, 525-532.

MacDowell N., Florin N., Buchard A., Hallett J., Galindo A., Jackson G., Adjiman C.S., Williams C.K., Shah N., Fennell P. (2010) An overview of $CO_2$ capture technologies, *Energ. Environ. Sci.* **3**, 1645-1669.

Mathias P.M., Reddy S., O'Connell J.P. (2010) Quantitative evaluation of the chilled-ammonia process for $CO_2$ capture using thermodynamic analysis and process simulation, *Int. J. Greenhouse Gas Control* **4**, 174-179.

Maurer G. (2011) Phase equilibria in chemical reactive fluid mixtures, *J. Chem. Thermodyn.* **43**, 147-160.

McLarnon C.R., Duncan J.L. (2009) Testing of ammonia based $CO_2$ capture with multi-pollutant control technology, *Energy Procedia* **1**, 1027-1034.

Niu Z.Q., Guo Y.C., Zeng Q., Lin W.Y. (2012) Experimental studies and rate-based process simulations of $CO_2$ absorption with aqueous ammonia solutions, *Ind. Eng. Chem. Res.* **51**, 5309-5319.

Puxty G., Rowland R., Allport A., Yang Q., Bown M., Burns R., Maeder M., Attalla M. (2009) Carbon dioxide postcombustion capture: A novel screening study of the carbon dioxide absorption performance of 76 amines, *Environ. Sci. Technol.* **43**, 6427-6433.

Qi G., Wang S., Yu H., Wardhaugh L., Feron P., Chen C. (2013) Development of a rate based model for $CO_2$ absorption using aqueous ammonia $NH_3$ in a packed column, Submitted to *International Journal of Greenhouse Gas Control* **17**, 450-461.

Qin F., Wang S.J., Hartono A., Svendsen H.F., Chen C.H. (2010) Kinetics of $CO_2$ absorption in aqueous ammonia solution, *Int. J. Greenhouse Gas Control* **4**, 729-738.

Que H.L., Chen C.C. (2011) Thermodynamic modeling of the $NH_3$-$CO_2$-$H_2O$ system with electrolyte NRTL model, *Ind. Eng. Chem. Res.* **50**, 11406-11421.

Rhee C.H., Kim J.Y., Han K., Ahn C.K., Chun H.D. (2011) Process analysis for ammonia-based $CO_2$ capture in ironmaking industry, *Energy Procedia* **4**, 1486-1493.

Salentinig S., Jackson P., Attalla M. (2012) Strategic Vapor Suppressing Additives for Ammonia Based $CO_2$ Capture Solvent, *11st International Conference on Greenhouse Gas Control Technologies (GHGT-11)*, Kyoto, Japan, 18-22 Nov.

Telikapalli V., Kozak F., Francois J., Sherrick B., Black J., Muraskin D., Cage M., Hammond M., Spitznogle G. (2011) CCS with the Alstom chilled ammonia process development program – Field pilot results, *Energy Procedia* **4**, 273-281.

Thomsen K., Rasmussen P. (1999) Modeling of vapor–liquid–solid equilibrium in gas–aqueous electrolyte systems, *Chem. Eng. Sci.* **54**, 1787-1802.

Valenti G., Bonalumi D., Macchi E. (2012) A parametric investigation of the Chilled Ammonia Process from energy and economic perspectives, *Fuel* **101**, 74-83.

Versteeg G.F., Van Dijck L.A.J., Van Swaaij W.P.M. (1996) On the kinetics between $CO_2$ and alkanolamines both in aqueous and non-aqueous solutions. An overview, *Chem. Eng. Commun.* **144**, 113-158.

Versteeg P., Rubin E.S. (2011) A technical and economic assessment of ammonia-based post-combustion $CO_2$ capture at coal-fired power plants, *Int. J. Greenhouse Gas Control* **5**, 1596-1605.

Wang X., Conway W., Fernandes D., Lawrance G., Burns R., Puxty G., Maeder M. (2011) Kinetics of the reversible reaction of $CO_2$(aq) with ammonia in aqueous solution, *J. Phys. Chem. A* **115**, 6405-6412.

Yu H., Morgan S., Allport A., Cottrell A., Do T., McGregor J., Wardhaugh L., Feron P. (2011a) Results from trialling aqueous $NH_3$ based post-combustion capture in a pilot plant at Munmorah power station: Absorption, *Chem. Eng. Res. Des.* **89**, 1204-1215.

Yu H., Morgan S., Allport A., Cottrell A., Do T., McGregor J., Feron P. (2011b) Results from trialling aqueous ammonia based post combustion capture in a pilot plant at Munmorah, *Energy Procedia* **4**, 1294-1302.

Yu H., Qi G., Wang S., Morgan S., Allport A., Cottrell A., Do T., McGregor J., Wardhaugh L., Feron P. (2012a) Results from trialling aqueous ammonia-based post-combustion capture in a pilot plant at Munmorah Power Station: Gas purity and solid precipitation in the stripper, *Int. J. Greenhouse Gas Control.* **10**, 15-25.

Yu H., Xiang Q., Fang M., Yang Q., Feron P. (2012b) Promoted $CO_2$ absorption in aqueous ammonia, *Greenhouse Gas Sci. Technol.* **2**, 200-208.

Yu H., Qi G., Xiang Q., Wang S., Fang M., Yang Q., Wardhaugh L., Feron P. (2012c) Aqueous Ammonia Based Post Combustion Capture: Results from pilot plant operation, challenges and further opportunities, *11st International Conference on Greenhouse Gas Control Technologies (GHGT-11)*, Kyoto, Japan, 18-22 Nov.

Zhao B.T., Su Y.X., Tao W.W., Li L.L., Peng Y.C. (2012) Post-combustion $CO_2$ capture by aqueous ammonia: A state-of-the-art review, *Int. J. Greenhouse Gas Control* **9**, 355-371.

Zhuang Q., Clements B., Li Y. (2012) From ammonium bicarbonate fertilizer production process to power plant $CO_2$ capture, *Int. J. Greenhouse Gas Control* **10**, 56-63.

# 6

# Development of HiCapt+™ Process for CO₂ Capture from Lab to Industrial Pilot Plant

Eric Lemaire*, Pierre Antoine Bouillon and Kader Lettat

*IFP Energies nouvelles, Rond-point de l'échangeur, BP 3, 69360 Solaize, France*
*e-mail: eric.lemaire@ifpen.fr - pierre-antoine.bouillon@ifpen.fr - kader.lettat@ifpen.fr*

* Corresponding author

**Résumé — Développement du procédé HiCapt+™ pour le captage du CO₂ : du laboratoire au pilote industriel** — Il est aujourd'hui connu que les procédés de captage du CO₂ en postcombustion dits de "première génération" à base de MEA nécessitent une importante énergie de régénération en plus de connaître des problèmes de dégradation du solvant dus à la présence d'oxygène dans les fumées de combustion. Cependant, ce procédé est le seul disponible aujourd'hui à l'échelle industrielle, c'est pourquoi *IFP Energies nouvelles* a développé le procédé HiCpat+, basé sur un solvant conventionnel à la MEA mais mis en œuvre à haute concentration avec des additifs antioxydants. Ces additifs permettent au procédé HiCapt+ de mettre en œuvre un solvant concentré sans connaître de problèmes de corrosion et de dégradation, ce qui engendre des gains notables en terme de consommation énergétique. De plus, la baisse du taux de dégradation du solvant permet de réduire les problèmes opératoires liés à la régénération du solvant ainsi que de limiter les émissions de produits de dégradation légers comme le NH₃ dans le gaz traité.

*Abstract — Development of HiCapt+™ Process for CO₂ Capture from Lab to Industrial Pilot Plant — It is now well known that "first generation" MEA based post-combustion carbon capture processes require high energy consumption and have problems with solvent degradation due to oxygen. Nevertheless, it is the only available process for first industrial units. That is why IFP Energies nouvelles, has developed HiCapt+ process, based on a conventional MEA solvent but using high performance oxidative inhibitors and higher amine concentration. These oxidative inhibitors enable HiCapt+'s process to use high solvent concentrations without corrosion or degradation problems and lead to reduction of the regeneration energy demand due to solvent flow rate reduction. Moreover, the huge reduction of solvent degradation rate avoids the difficult management of reclaiming unit, as well as avoiding high concentration of light degradation products in the treated flue gas, like NH₃.*

## INTRODUCTION: THE REFERENCE PROCESS MEA AT 30 WT%

It is now well known that "first generation" MEA (MonoEthanolAmine) based post-combustion carbon capture processes require high energy consumption and have problems with solvent degradation due to the presence of oxygen in the inlet flue gas. Nevertheless, it is the only available and proven technology for first industrial units based on an inexpensive and widely available chemical solvent. That is why *IFP Energies nouvelles* and *PROSERNAT*, have developed the HiCapt+$^{TM}$ process, based on a conventional MEA solvent but using high performance oxidative inhibitors and higher amine concentrations.

Aqueous solution of MEA is the most widely investigated solvent for a carbon dioxide post-combustion capture process. MEA is cheap, largely available, non toxic and highly effective because of its high capacity for $CO_2$ capture and its fast reaction kinetic. $CO_2$ scrubbing by an aqueous amine solution of MEA 30 wt% is a widely proven technology to capture $CO_2$. The European project Castor [1] has demonstrated the good operability, security, flexibility, stability and reliability of this process during long run tests (more than 500 h operation without stop). These tests have been done in the pilot plant located at the Dong power plant (Esbjerg, Denmark). This pilot gave realistic data because the flue gas treated came directly from the power plant and the pilot capacity was very large (approximately 1 t/h of $CO_2$ captured).

However, the high energy consumption is a major drawback of this type of processes. The energy used for the stripping of rich amine in the reboiler was measured around 3.7 GJ/t$CO_2$ avoided. It corresponds to a penalty for the power plant around 10.5 points on the net efficiency (for a power plant with a high yield of 40% net).

Moreover another critical point with this reference technology concerns the amine degradation by oxidation with the $O_2$ contained in the flue gas. This degradation has been evaluated in the Castor pilot around 1.4 kg MEA degraded/t $CO_2$ captured. Different troubles result from this degradation:

– the first one is a high consumption of MEA, approximately 2 batches of solvent per year. The economical impact on the operating costs is really important – more than 7 M€/year for a $CO_2$ capture unit installed to remove 90% of the $CO_2$ emissions of a 600 MWe power plant. An other drawback is the management of this huge solvent volume;

– the second one is coming from the degradation products formed during the degradation reactions, which stay in the liquid phase. These products, named HSS (Heat Stable Salt), are mainly organic acids (formic acid, oxalic acid, acetic acid). These acids are very corrosive regarding carbon steel and create a decrease in the solvent reactivity. Thus a reclaiming unit is needed to purify the solvent and remove all the degradation products. This adds some complexity to the operation of the plant and of course increases the CAPEX and the OPEX;

– the third one is the emission of light degradation products in the treated flue gas and in the $CO_2$ produced because MEA oxidation by $O_2$ generates products such as $NH_3$. For example in the Castor pilot plant approximately 25 mg/Nm$^3$ of $NH_3$ were measured in the treated flue gas.

So MEA 30 wt% is a proven process and could be operated for $CO_2$ capture but has some important drawbacks.

## 1 THE HICAPT+$^{TM}$ PROCESS DEVELOPMENT

### 1.1 From MEA 30 wt% to MEA 40 wt%

Based on all the knowledge acquired during pilot tests of the 30% MEA reference process and in order to improve it and to develop a realistic industrial technology, *IFPEN* and *PROSERNAT* have developed HiCapt+$^{TM}$ process. It is based on a conventional MEA solvent but using high performance oxidative inhibitors to limit as much as possible oxidative degradation of the solvent and by-products formation and using MEA at higher amine concentration (40 wt%) to lower the energy required at the reboiler. This paragraph focuses on the developments and results achieved with the HiCapt+$^{TM}$ process, in particular the adjustments made to the models (thermodynamics, hydrodynamics and kinetics) to predict the process performances with a high concentrated solvent.

#### 1.1.1 Process Simulation – State of the Art

The performances and design criteria of post-combustion processes using MEA 30 wt% can be evaluated through a process simulation tool like *AspenTech*'s AspenPlus. Based on its knowledge of MEA 30 wt% process, *IFPEN* implemented under this simulation tool all the in-house correlations and models available for MEA 30 wt%.

Starting from these correlations, it is possible to simulate a process running with high MEA concentrations like the HiCapt+ process. Unfortunately, pilot plant data have shown the inaccuracy of the models with MEA 40 wt% tests as shown in Figure 1.

To predict correctly the performances and to perform good designs of the HiCapt+ process, it has been necessary to modify some aspects of the correlations used in the AspenPlus models.

### 1.1.2 Thermodynamics

The thermodynamic model used in AspenPlus simulation is the Non Random Two Liquids (NRTL)-electrolyte model. Lab scale experiments have been performed in *IFPEN* labs with MEA 40 wt% and the results have been compared to the AspenPlus model (absorption isotherms and enthalpy) as well as to literature data [2-4]. Figure 2 illustrates the comparison between AspenPlus model and literature data at different temperatures.

From this work, it has been concluded that the NRTL-electrolyte model used in the AspenPlus simulation is suitable to predict the equilibrium with MEA 40 wt%.

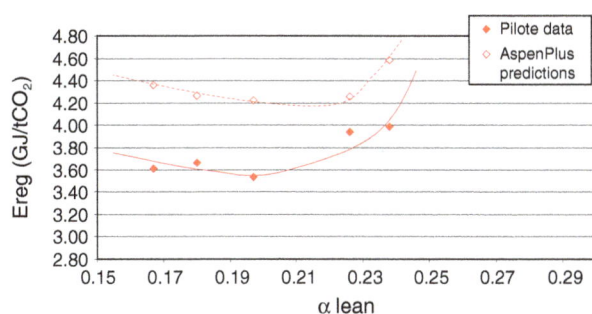

Figure 1

The AspenPlus predictions without models modifications.

### 1.1.3 Hydrodynamic and Mass Transfer

An other critical point for the optimization of the process concerns the design of the absorber and stripper columns which may represent 30 to 50% of the CAPEX of the unit and the packing of the absorber, up to 50% of the absorber itself. As a matter of fact, in post combustion technologies, because of dilution by nitrogen, the gas flow rate is really huge (for example approximately 1 700 000 $Nm^3$/h for a 600 MWe coal power plant). To reduce the size of future post-combustion capture plants and the cost of the columns, high capacity packings are highly needed. Moreover, it is really important to limit the pressure drop generated by the absorber column as it must be compensated by a blower which electric consumption can represent between 5 to 10% of the global utility cost. For the stripper column, as the $CO_2$ captured must be compressed for the transport before storage, a low pressure drop in the column is also needed.

In order to design properly the columns, absorber and stripper, it is required to have a complete characterization of each packing in term of:

- liquid hold-up,
- liquid distribution,
- effective area (this is the area available for the mass transfer, different from the geometric area of the packing),
- pressure drop.

More over the mass transfer coefficients must be known:

- $k_g$ (mass transfer coefficient in gas phase),
- $k_l$ (mass transfer coefficient in liquid phase).

To reach this goal, *IFPEN* has done a complete characterization of different commercial structured and ran-

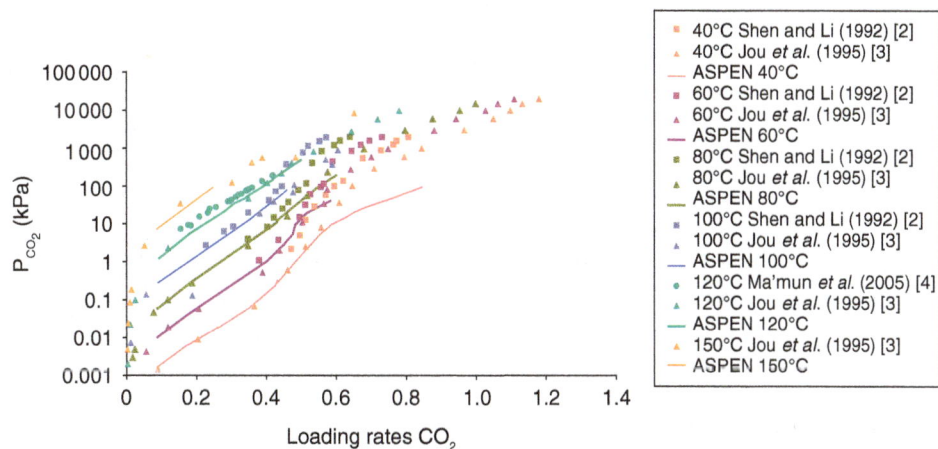

Figure 2

The AspenPlus thermodynamic correlations.

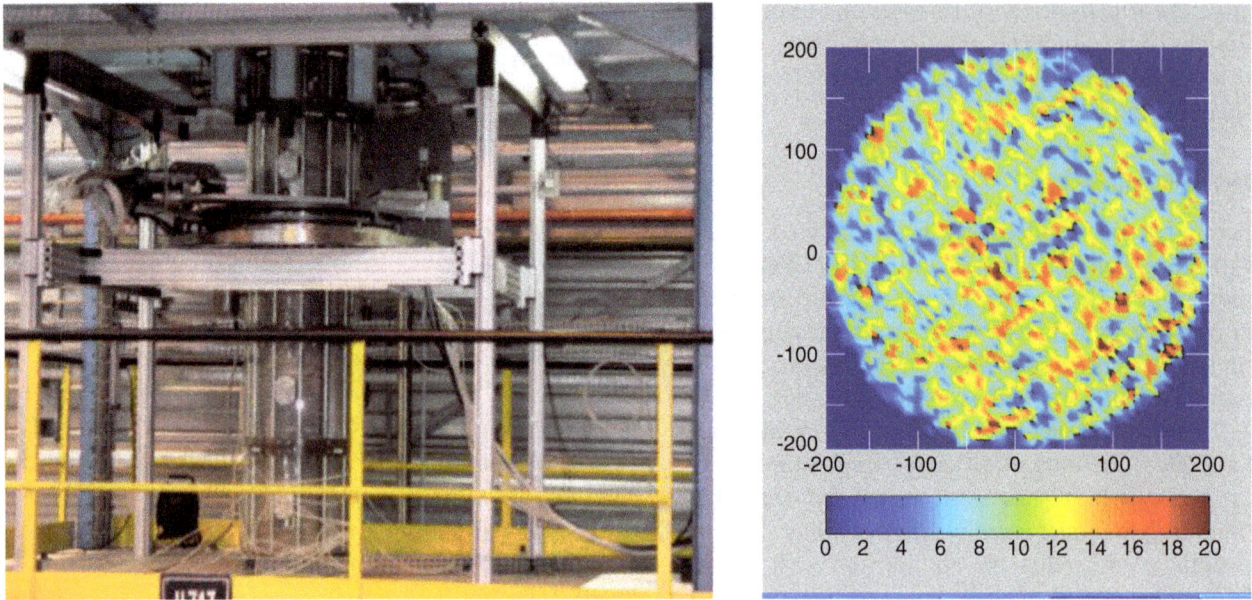

Figure 3

400 mm diameter column and liquid distribution map obtained with gamma tomograph.

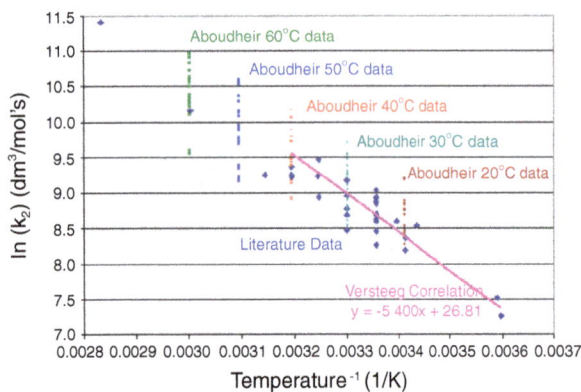

Figure 4

Aboudheir data [6] and Versteeg *et al.* [7] correlations.

dom packings, like IMTP50 developed by *Koch Glitsch* or Mellapack 250X developed by *Sulzer* [5]. This work was done using different pilot units available at *IFPEN*, for example in Figure 3 is presented a 400 mm diameter column equipped with a gamma tomography and the liquid map obtained.

From this work, correlations were obtained for effective area of packing (ae), mass transfer coefficient in gas phase ($k_g$) and mass transfer coefficient in liquid phase ($k_l$). These correlations were obtained as function of the solvent parameters such as viscosity and surface tension to take into account the increase in solvent concentration in the HiCapt+ process.

All the obtained correlations were implemented in the *AspenTech*'s AspenPlus simulation tool.

### 1.1.4 Kinetics

To increase the accuracy of the AspenPlus model, it has been decided to use Aboudheir data [6], which are validated from 20 to 40 wt% MEA, loading between 0.1 and 0.5, temperature between 20 to 60°C. It has been found out that the Aboudheir data were not in agreement with the Versteeg *et al.* [7] correlation, which was obtained in diluted unloaded solutions, as shown in Figure 4.

It has been chosen to refine the raw Aboudheir data to make them compatible with the AspenPlus thermodynamics (NRTL-electrolyte) and formalism. Moreover, correction factors had been used to take into account:
– the activity coefficient in highly concentrated solutions,
– the viscosity of the solution in highly loaded solutions.

The new kinetic correlation has been implemented in the AspenPlus RateSep model. The extrapolation of this in-house model to unloaded solutions is consistent with the Versteeg correlation for different MEA concentrations, which demonstrates the validity of the methodology used in this work.

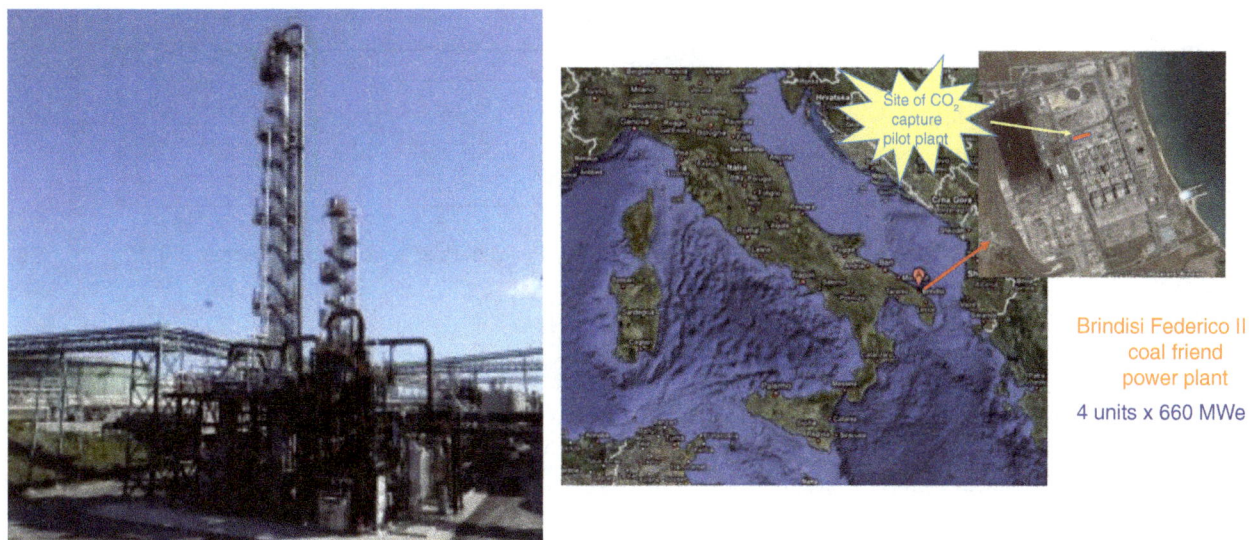

Figure 5

Picture and location of the 2.25 t/h post-combustion $CO_2$ capture pilot plant in *ENEL* power plant of Brindisi.

### 1.1.5 Models Validation at Industrial Pilot Scale

#### 1.1.5.1 Pilot Plant Description

In 2009, an agreement between *ENEL* and *IFPEN* was signed in order to collaborate in the field of $CO_2$ capture processes. This collaboration was mainly focused around the future operation of the industrial pilot plant built by *ENEL* (*Fig. 5*). This pilot which started up during summer 2010, is located in Brindisi in the south of Italy, on the site of a 4 × 660 MWe coal power plant. It has a capacity of 2.25 t/h of $CO_2$ captured for 12 000 Nm$^3$/h of flue gas.

Different tests have been done with MEA at 20, 30 wt% and also with MEA 40 wt% which is the basis of the HiCapt+$^{TM}$ process (but no additive was tested). These tests performed on this semi industrial unit allowed *IFPEN* to optimize and validate the process performances at high MEA concentration (40 wt%) in real and continuous operation. Moreover, long run tests gave the reference figures for MEA 40 wt%.

The pilot plant engineering phase started in 2008, construction was done from March 2009 to June 2010 and finally start up phase and validation of the unit were done from June to September 2010 using a solvent at 20 wt% MEA.

Detailed description of the pilot plant is given here below. First, a fan compresses the flue gas just after the DeSOx unit of the power plant. Flue gas flow rate could vary from 3 000 up to 15 000 Nm$^3$/h and the fan could compensate the pressure losses resulting from the pilot units.

Next, the flue gas is sent to a dedicated DeSOx unit that enables to operate at very low value for $SO_2$ content in flue gas. The unit uses a spray tower with a limestone slurry. Efficiency of SOx removal is higher than 95%. Right after the DeSOx unit, two WESP (Wet Electro-Static Precipitator) remove the entrained particles. They can be used in series, only one at a time or bypassed totally. After this pre-treatment, flue gas arrives in the $CO_2$ removal loop composed of the absorber and the stripper columns.

The different materials composing the $CO_2$ capture section are the following:
– absorber column:
  – diameter = 1 500 mm/height = 45 m,
  – absorption zone (structured packing): 3 beds × 7.35 m, washing section packing: 1 bed × 4 m, others: 2 demisters in the top,
  – 3 possible feeds for liquid (operation with 1 or 2 or 3 beds of packing);
– stripping column:
  – diameter = 1 300 mm/height = 31 m,
  – packing below feed (random packing): 3 beds × 3.7 m, washing section packing (up to the feed): 1 bed × 3 m, 2 possible feeds for liquid (operation with 2 or 3 beds of packing);
– rich solvent/lean solvent heat exchanger:
  – plate fine type;
– filtration section:

- 2 mechanical filters;
- 1 carbon filter;
- others:
  - solvent storage $2 \times 100$ m³ tanks;
  - cooling water section: external loop sea water;
  - industrial cooling water loop: sea water/cooling water heat exchanger.

The pilot is also fully instrumented with more than 400 acquisition points. All the regulation and control loop are exactly the same than for industrial unit.

The pilot is equipped with 5 corrosion monitoring points, each containing 6 corrosion probes. Analysis are done on line for $CO_2$ content in the gas (on line regulation of plant efficiency) and daily analysis are done for all the values regarding liquid phase (loadings, solvent concentration, etc.).

### 1.1.5.2 AspenTech *AspenPlus* Model Validation

In order to collect and use all the data obtained during the development (thermodynamic, kinetic, mass transfer, hydrodynamic, etc.) a dedicated predictive model was created. This model was created inside the *AspenTech* AspenPlus environment, using the special Aspen RateSep model. As discussed previously:

- physical properties such as density and viscosity were adjusted in Aspen's properties to match data correlations produced by Weiland et al. [8] and also produced at *IFPEN* lab. Heat of formation and heat capacity data were adjusted;
- highly concentrated and highly loaded MEA reaction rate data were shown to match unloaded, dilute literature data when activity coefficient corrections were properly considered. The temperature dependence of the Versteeg rate constant correlation was shown to be valid up to 60°C with an acceptable extrapolation to 80°C. The effect of ionic strength on the kinetics was quantified and implemented into the model;
- specific hydrodynamic and mass transfer properties were also implemented into the model for the packings characterized at *IFPEN*. Correlations developed by *IFPEN* were used to calculate the liquid holdup, interfacial area and the liquid film mass transfer coefficients.

It is important to state that there are no fitting parameters in the model which force it to match experimental data. The thermodynamic, kinetic, hydrodynamic and other aspects of the model were defined independently.

The model was tested with data coming from the Castor pilot plant experiments and appeared to represent correctly all the cases tested. In Figure 6, dots are temperature measurements and triangles are $CO_2$ concentrations in gas phase, all measured in the absorber of Castor

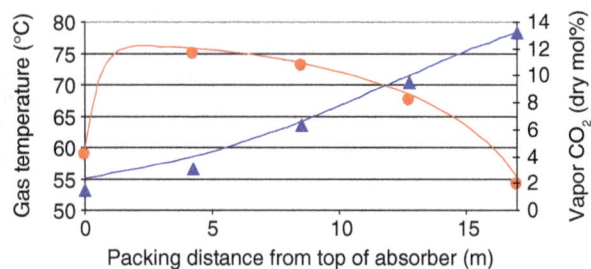

Figure 6

Comparison of HiCapt+™ model results with data of Castor pilot plant.

pilot plant during a test using MEA at 30 wt%. The lines show the model predictions. It is clearly shown in this figure that the model representation is really good.

During the MEA 40 wt% pilot plant campaign, the AspenPlus model had been extended with additional lab experiments to cover the new operating conditions (mean temperature, densities, viscosities, etc.). Pilot plant data were directly compared to the model prediction with success. All the parametric studies made on the pilot plant with MEA 40 wt% as well as the long run test were pretty well predicted by the model. Figure 7 illustrates the model prediction for all the parametric tests (more than 50 operating conditions with MEA 40 wt%), before and after model modifications made for MEA 40 wt%.

All the results of the new AspenPlus model are in the range of ±5% with respect to industrial pilot plant data. The developed simulation tool as well as results from *ENEL* pilot plants can be used for design of industrial scale $CO_2$ capture plant.

## 1.2 Corrosion

The removal of carbon dioxide from industrial gas streams by amine treating units is a well known process, extensively used for many decades in natural gas treatment. One of the most severe operational difficulties encountered is the corrosion of the process equipments. Based on extensive R&D and operational industrial feed back, *PROSERNAT* and *IFPEN* have already developed a large expertise for the design and operation of natural gas Amine units [9, 10]. Of course, corrosion risks are extremely dependent on the nature of the amine solution, MEA, DEA, MDEA, formulated MDEA (respectively from the most corrosive to the lowest

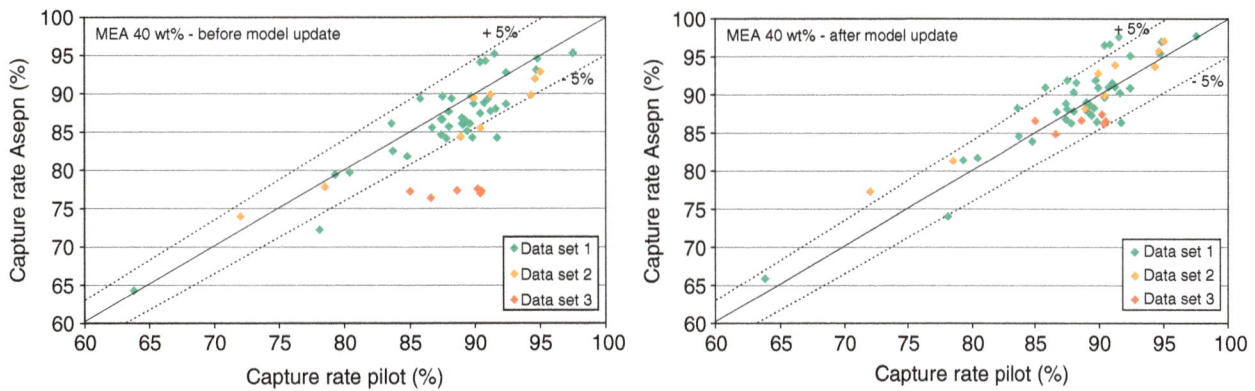

Figure 7

Comparison of HiCapt+$^{TM}$ model with data of MEA 40 wt% pilot plant.

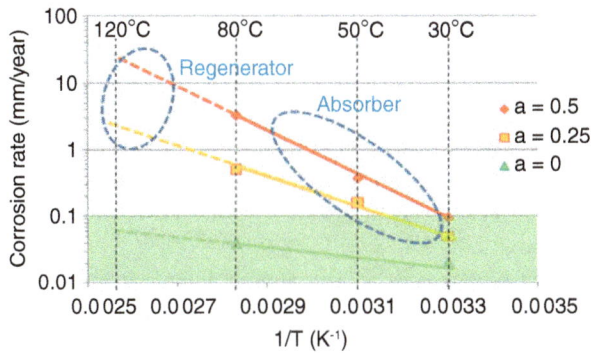

Figure 8

Corrosion speed for carbon steel in aqueous 30 wt% MEA solvent *versus* process parameters (temperature and loading (*a*)).

corrosive). Other important parameters that influence the corrosion are the temperature, the concentration of the amine solution, the solvent loading conditions, the liquid velocity and the composition of the gas to be treated. It is also well known that amine solvent degradation often increases corrosivity. Except for the use of oxidative inhibitors which decrease drastically solvent degradation (see below), in the case of HiCapt+$^{TM}$ process and post-combustion $CO_2$ capture of the flue gas, all these parameters are unfavourable: MEA is a primary amine which is very corrosive; the high amine concentration as well as the high content of oxygen in the inlet flue gas promote degradation and increase corrosion; the $CO_2$ loading which is never lowered down to zero increases also the corrosion. That is why appropriate corrosion experiments and particular design and operational attention are therefore required to take into account such parameters, to predict the risks of corrosion and to select the optimized material for unit.

To solve these questions *IFPEN* has made various tests of corrosion in its lab for all the conditions that could be found in the process and with different types of carbon steel and stainless steel. Moreover, *IFPEN* was responsible of the corrosion monitoring during the Castor project. The corrosion evaluation for the Castor project consisted in implementing corrosion monitoring tools to the Castor pilot plant. Weight loss corrosion coupons were chosen as the most convenient method, and 6 insertion points were selected. For each insertion point, one carbon steel and one stainless steel (AISI 316) corrosion coupons were installed. After the first 500 hours run of the pilot plant with 5M MEA, the coupons were removed for visual observation and corrosion rate evaluation.

With all these experiments (internally at *IFPEN* or during Castor project) we have now a clear view of corrosion in the HiCapt+$^{TM}$ process and more generally in MEA based processes. For corrosion of carbon steel in MEA at 30 wt%, results are shown in Figure 8. This figure shows the estimated speed of corrosion for carbon steel *versus* temperature for different loadings of the solvent (*i.e.* mol of $CO_2$/mol of MEA in the liquid phase). The bottom green part of the figure represents an area in which corrosion speed is lower than 0.1 mm/year, which could be considered as an acceptable corrosion rate for an industrial unit. The blue dashed circles represent the conditions of temperature and loading existing in the absorber and stripper of a MEA 30 wt% standard process. From this figure, it clearly appears that the use of carbon steel is not possible for a MEA 30 wt% process, and by extension for all MEA based processes.

Identical tests done with stainless steel showed corrosion speed lower than 10 µm/year in all the process conditions.

But the combination of MEA and anti-oxydative agent changes the corrosivity of raw MEA solvent. Considering there may be a risk of corrosion in hot parts of the *ENEL* industrial pilot plant during long test runs, it was thus decided not to use HiCapt+$^{TM}$ solvent containing inhibitor in the *ENEL* pilot plant. Nevertheless, others materials like duplex steels are identified and have been tested with really good results, less than 5 or 10 µm/an of corrosion speed with the full HiCapt+ solvent composition.

## 1.3 Degradation of MEA by Oxidation with O₂

It is well known that when MEA is exposed to oxygen, oxidative degradation occurs in a significant amount. When the water wash section is used on top of the contactor and the temperature is well controlled at the reboiler and the stripper, this oxidative degradation of the MEA molecule represents almost all the solvent losses of the unit, which would be about 2 batches of solvent per year for a typical 30 wt% MEA. Oxidation of MEA is not only a source of solvent consumption but also leads to formation of volatile compounds and ammonia which need to be removed from the treated gas. Moreover many carboxylic acids are formed and trapped as salts in the solvent. These acids can also promote corrosion phenomena.

For these reasons, the control of degradation is a major challenge in MEA based technologies. As the increase in the solvent concentration will end with an increase of the degradation issues, solving this problem will allow to design processes using solvents with an increased MEA concentration, enhancing the performance in $CO_2$ capture. Oxidative degradation, which is a critical point for the development of an industrial process, can be minimized by the use of antioxidant additives in the HiCapt+$^{TM}$ process.

A lab scale evaluation test of MEA degradation associated with a dedicated analysis of degradation products and then an evaluation of different antioxidant additives were done by *IFPEN* [11, 12]. More than 150 products have been tested and compared in regard to their oxidation inhibition capacity. Conventional antioxidant additives were found to be poorly active or inactive. But new classes of additives have been found to be effective and considerably reduce the degradation issues in MEA processes.

Some results of this investigation could be shown in Figures 9 and 10. Figure 9 represents the concentration of the main HSS resulting from degradation in the sample collected after 12 days of degradation in lab test performed at *IFPEN*. It clearly appears that with 0.25 wt% of inhibitor (U2 or V1 or V2 or Y1) the level of HSS stays at a negligible concentration, near the detection limit of the analytical method. In Figure 10, we could see the ammonia analysis of the gas exiting the lab reactor used for degradation tests. The results are in line with those obtained for HSS detection and it is shown that some inhibitors are really efficient.

As a consequence, with the very efficient inhibitors found, HiCapt+$^{TM}$ process can operate at high MEA concentration (40 wt%) without any trouble linked to oxidative degradation. Compared to the reference 30 wt% MEA process, the use of efficient oxidative inhibitors results in three direct advantages:

– the consumption of MEA would be reduced by more than a factor 10;

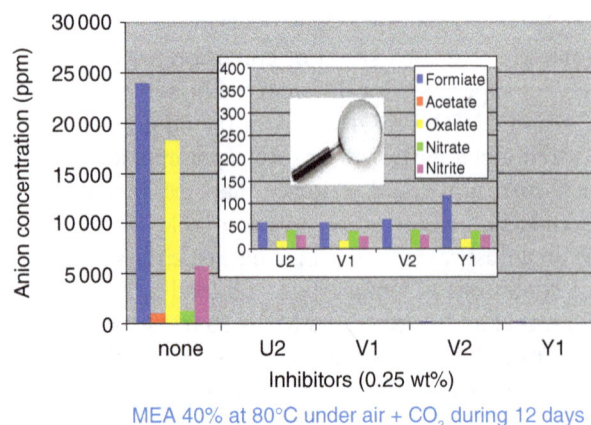

MEA 40°C at 80°C under air + CO₂ during 12 days

Figure 9

Concentration of HSS in solvent using different inhibitors.

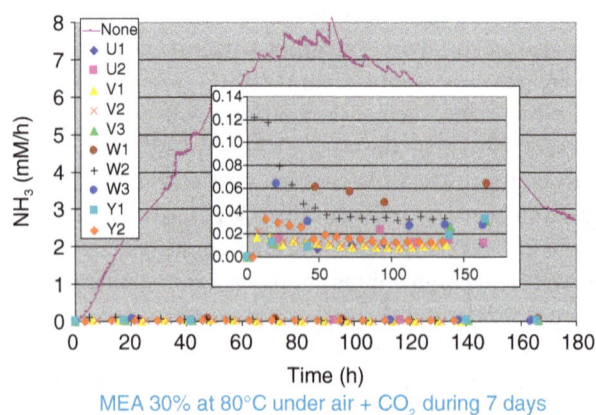

MEA 30% at 80°C under air + CO₂ during 7 days

Figure 10

Concentration of NH₃ in the outlet gas for different inhibitors.

- the reclaiming unit will be smaller than in conventional units,
- the ammonia concentration in the treated flue gas would meet the environmental specifications without additional treatment.

The use of a high concentration MEA solvent (40 wt%) results in three main advantages:

- reduced circulation flow rate of solvent, decreasing total hold-up of solvent,
- reduced heat requirement for solvent regeneration,
- reduced size of some equipments such as pumps and regenerator column as well as size of bulk material and piping.

Figure 11

Lean loading and stripper pressure optimization for MEA 40 wt% at 90% capture rate.

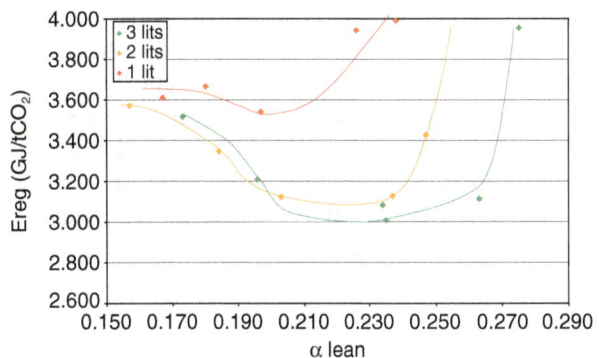

## 2 THE HICAPT+™ PROCESS OPTIMIZATION

### 2.1 Results of Pilot Plant Tests – Optimization of MEA at 40 wt% Process

A first campaign on the *ENEL*'s pilot plant was done from June to September 2010 with MEA at 20 wt%. This campaign enables to start up the unit, validate all the methods and procedures and to perform the guarantee tests. Following this start-up phase, the tests campaigns with MEA 30 wt% and 40 wt% started. This paper will only focus on MEA 40 wt% campaign. But we could say that results obtained with MEA 30 wt%, which is considered as a reference, are totally coherent with literature. This is an other proof of the pilot plant representativity.

The campaign with MEA at 40 wt%, performed without anti oxidation additives, took place from February 2011 to June 2011. This campaign represented a challenge because MEA 30 wt% is considered as the reference process and a process using MEA at 40 wt% has not been operated at such a big scale by any competitors in the field of $CO_2$ post-combustion capture technologies.

A first part of the test campaign corresponded to a parametric optimization for the key process parameters such as:

- stripper pressure – 1.6/1.8 and 2.0 bar (a),
- lean loading variation,
- capture rate 80%, 90% and 95%,
- flue gas flow rate variation – from 3 000 to 12 000 $Nm^3$/h,
- packing height variation in absorber and stripper columns.

For all these parametric studies, the process has been optimized with respect to energy consumption and techno-economic analysis including Capex evaluation, specially for packing height experiments. Some of the results obtained are illustrated in Figures 11, 12 and 13.

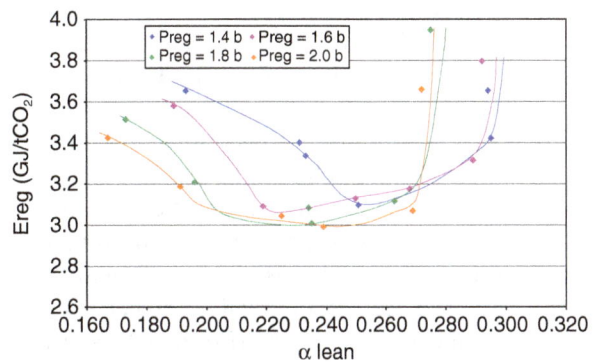

Figure 12

Packing height optimization in the absorber column for MEA 40 wt% at 90% capture rate.

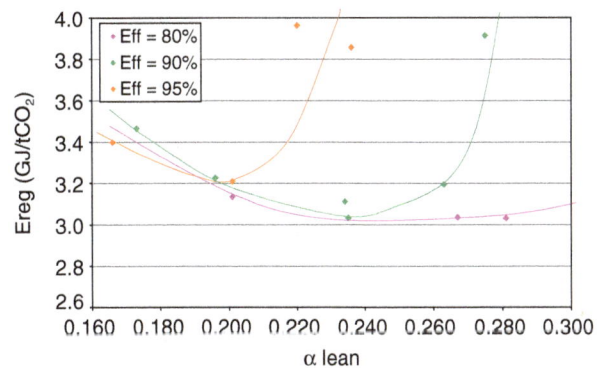

Figure 13

Lean loading optimization for different process efficiencies for MEA 40 wt%.

Figure 11 shows a series of tests performed at 90% capture rate to optimize the lean loading of the solvent for different stripper conditions (pressure). These tests allowed to determine an optimum lean loading around 0.22 and 0.24 depending of the stripper pressure. Moreover it can be seen on the graph that the higher the pressure is in stripper, the larger will be the stable operation range for the process. The drawback of operation at high stripper pressure will be an increase in the degradation rate of the solvent as temperature will be higher (over 125°C the thermal degradation for MEA 40 wt% starts to be not negligible).

Figure 12 shows another series of experiments performed still at 90% capture rate as the reference efficiency for the process. These tests aimed at optimizing the packing height in the absorber column, meaning optimizing the investment costs of the technology. For 3 different packing heights, the lean loading was optimized to minimize the energy consumption of the unit.

As expected, it is observed that for a high packing height, the lean loading is still optimum around 0.22-0.24. When the height of packing were decreased to an intermediate value, no variation of the lean loading optimum was observed and no significant increase on the energy requirement at the optimum was observed. When the height of packing was decreased to the smallest value, we observed first an increase of the energy requirement, meaning that such a gain in investment costs starts to increase operating costs. Then we observed a slight move of the lean loading around 0.20, meaning that it is possible to compensate slightly the decrease of effective area by increasing the reactivity of the solvent by a better regeneration.

In Figure 13, the lean loading was again optimized but for different capture rates. We can observe that for efficiency, the optimum lean loading increased to 0.26 (80% capture rate) whereas for high capture rates (95%) the optimum lean loading decreased around 0.2. One can also notice on this graph that the optimized energy consumption is higher for high capture rates.

## 2.2 Results of Pilot Plant Tests – Process Performances Validation

After the parametric campaign, the optimized operating conditions were fixed and validated during a long run test. The test conditions and results are the following (see also *Fig. 14* and *15* for main parameters trend):

Parameters:
- flue gas flowrate: 12 033 $Nm^3$/h,
- solvent flowrate: 32.3 $m^3$/h,
- stripper pressure: 1.8 bar(a),
- test duration: 380 h.

Results:
- $CO_2$ production: 2 327 kg/h,
- efficiency (capture rate): 89.7%,
- energy consumption: $\sim$3.02 $GJ/tCO_2$,
- solvent loadings: $\alpha$ lean = 0.23, $\alpha$ rich = 0.48.

These results proved the interest of this MEA 40 wt% process, with a reduced energy penalty at the reboiler down to around 3 $GJ/tCO_2$. It can be considered today as one of the best proven and simple solution in terms of energy consumption per tons of $CO_2$ avoided (proven during 380 h with a total $CO_2$ captured around 900 tons). Moreover, this MEA 40 wt% process is reliable, well known (same operation than standard MEA 30 wt% process), easy to operate. However, the major drawback of this process is

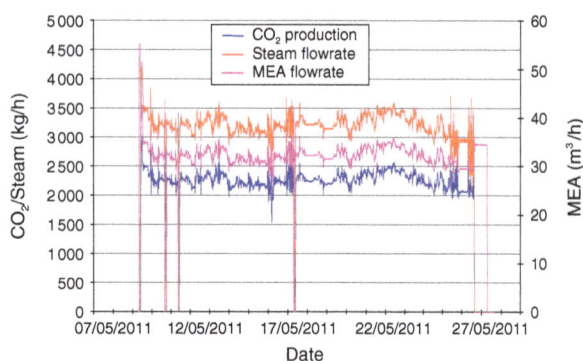

Figure 14

Parameters trend for MEA 40 wt% – long run test.

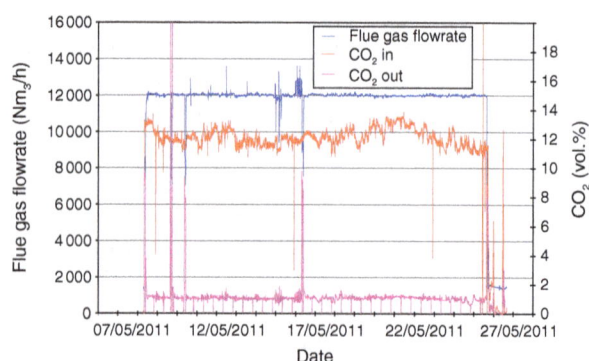

Figure 15

Parameters trend for MEA 40 wt% – long run test.

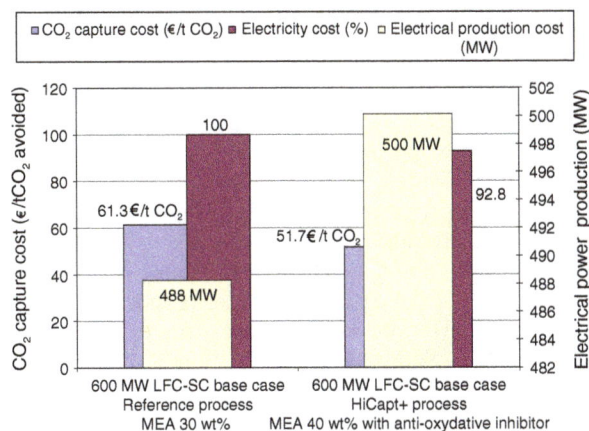

Figure 16

Technico-economic comparison between standard MEA 30 wt% process and HiCapt+$^{TM}$ process.

the high solvent degradation, significantly more than for MEA 30 wt%. That is why *IFPEN* and *PROSERNAT* have developed degradation inhibitors with the HiCapt+$^{TM}$ process.

## 2.3 Process and Techno-Economical Evaluation of the HiCapt+$^{TM}$ Process

Computed simulations with the HiCapt+$^{TM}$ model have been done with a flue gas coming from a 630 MWe coal power plant and with a full integration with the electricity production unit. The simulations showed an energetic penalty around 9 points (compare to 10.5 for the reference process) and an energetic consumption between 3.1 to 3.3 GJ/tCO$_2$ avoided which places HiCapt+$^{TM}$ among the most energy effective process technologies for CO$_2$ removal from flue gases. A technico-economic evaluation of HiCapt+$^{TM}$ compared to classical MEA 30 wt% process shows an improvement around 15% in the cost of CO$_2$ captured. A part of the results could be seen in Figure 16. It has to be noted that prices indicated are functions of many parameters like coal price, reference year for investment estimation (2008), limit of the process (included CO$_2$ compression), so the absolute value is difficult to compare with others studies. It should be noted as well that HiCapt+$^{TM}$ technology is based on cheap and widely available chemicals as solvent. And even with duplex steel (a family of stainless steel) the CAPEX is still competitive because mechanical properties are better than for austenitic stainless steel and enable the reduction of total weight of material.

## 3 A NEW AND EFFICIENT SOLUTION FOR CO$_2$ CAPTURE: HICAPT+$^{TM}$ PROCESS

As explained previously, the Brindisi pilot plant tests confirmed the interest of the process in terms of energy penalty and operability. However solvent degradation must be controlled in order to operate this process at full scale and decrease degradation and by products emissions. After an extensive research and tests of a large number and types of chemicals, *IFPEN* has developed and selected the most efficient inhibitors. Among all tests, corrosion tests have shown that most inhibitors are quite corrosive in hot conditions for some steels such as carbon steels or stainless steels but they have also demonstrated what specific material to be used in some specific hot parts of the process with the HiCapt+$^{TM}$ solvent. Nevertheless, specific materials are identified and have been tested with really good results, that is to say less than 5 or 10 µm/an of corrosion speed.

The degradation inhibitors used by the HiCapt+$^{TM}$ process being not fully compatible with the material of hot parts of the Brindisi industrial pilot plant which is built in austenitic 316L stainless steel, only the "process" part corresponding to the increase of concentration to 40 wt% MEA has been tested in Brindisi. Complete test of HiCapt+$^{TM}$ process in the Brindisi pilot plant would require some modifications in some specific hot parts of the unit, which is not yet planed.

Regarding the experiments done in the *IFPEN* lab, we are totally confident concerning the degradation limitation, and for the performance of the future HiCapt+ process. Moreover we could add that the HiCapt+$^{TM}$ solvent will be tested shortly in a mini pilot at *IFPEN* including a complete absorber/stripper loop and with correct material.

## CONCLUSION

Robustness, stability, reliability and process performance of HiCapt+$^{TM}$ process (using MEA at 40 wt%) have been proven during the pilot tests with *ENEL*. Among other performance, it was demonstrated that an energy at the reboiler as low as 3 GJ/tCO$_2$ avoided can be steadily achieved with 90% CO$_2$ capture. A simulation tool including in-house models has been successively cross checked and validated by the huge number and large range of experimentations obtained on the industrial pilot unit.

Regarding degradation, anti-oxidative additive efficiency is clearly proven by lab tests done at *IFP Energies nouvelles*. Reliability of corrosion lab tests is confirmed by good agreement between lab results and pilots results (Castor or Brindisi). Corrosion *IFP Energies nouvelles* lab tests have identified materials in order to operate HiCapt+$^{TM}$ process safely.

HiCapt+$^{TM}$ process could offer an efficient solution for $CO_2$ capture:
- low energy consumption, around 3 GJ/t$CO_2$,
- low solvent degradation with efficient inhibitor (*IFP Energies nouvelles* and *PROSERNAT* tests only),
- safe material design,
- very large flexibility,
- reliable and easy to operate.

Based on all this R&D as well as *PROSERNAT*'s industrial experience on similar technologies for natural gas sweetening, HiCapt+$^{TM}$ technology is now ready to be proposed for demonstration unit.

## ACKNOWLEDGMENTS

The authors would like to thank all the team involved in this development; J. Kittel (PhD), S. Gonzalez (Ing.), P.L. Carrette (PhD), P.A. Bouillon (Ing.), A. Gomez (PhD), B. Delfort (PhD), P. Alix (Ing.), R. Dugas (PhD), L. Raynal (PhD), K. Lettat (PhD) and all the others. The authors would like to thank also *ENEL* for the really good collaboration that is done with *IFP Energies nouvelles* and will demonstrated the HiCapt +$^{TM}$ process performance at industrial pilot scale.

## REFERENCES

1 www.co2-castor.com.

2 Shen K.P., Li M.H. (1992) Solubility of Carbon-Dioxide in Aqueous Mixtures of Monoethanolamine with Methyldiethanolamine, *J. Chem. Eng. Data* **37**, 1, 96-100.

3 Jou F.Y., Mather A.E., Otto F.D. (1995) The Solubility of $CO_2$ in A 30-Mass-Percent Monoethanolamine Solution, *Can. J. Chem. Eng.* **73**, 1, 140-147.

4 Ma'mun S., Nilsen R., Svendsen H.F., Juliussen O. (2005) Solubility of carbon dioxide in 30 mass% monoethanolamine and 50 mass% methyldiethanolamine solutions, *J. Chem. Eng. Data* **50**, 2, 630-634.

5 Alix P., Raynal L. (2008) Pressure drop and mass transfer of a high capacity random packing. Application to $CO_2$ post-combustion capture, *GHGT9 congress*, Washington.

6 Aboudheir A.A. (2002) Kinetics, Modeling, and Simulation of Carbone Dioxide Absorption into Highly Concentrated and Loaded Monoethanolamine Solutions, *PhD Dissertation*, University of Regina.

7 Versteeg G.F., van Dijck L.A., van Swaaij W.M. (1996) On the Kinetics between $CO_2$ and Alkanolamines both in Aqueous and Non-aqueous Solutions. An Overview, *Chem. Eng. Commun.* **144**, 113-158.

8 Weiland R.H., Dingman J.C., Cronin D.B., Browning G.J. (1998) Density and Viscosity of Some Partially Carbonated Aqueous Alkanolamine Solutions and Their Blends, *J. Chem. Eng. Data* **43**, 378-382.

9 Bonis M.R., Ballaguet J.P., Rigaill C. (2004) A Critical Look ar Amines: A Practical Review of Corrosion Experience over four Decades, *GPA Convention*, New Orleans, US, 14-17 March, Corrosion GPA/SOGAT *PROSERNAT*.

10 Kittel J., Bonis M.R., Perdu G. (2008) Corrosion Control on Amine Plants: New Compact Unit Design for High Acid Gas Loading, *SOGAT*, Abu Dhabi, 28 April-1 May.

11 Delfort B., Carrette P.L., Bonnard L. (2009) Additives for inhibiting MEA oxidation in a post-combustion capture process, *8th Conf. on Carbon Capture & Sequestration*, Pittsburgh, 4-7 May.

12 Carrette P.L., Delfort B., Bonnard L. (2009) Oxidation inhibitors for aqueous MEA solutions used in a post-combustion $CO_2$ capture process, *IEA Greenhouse Gas R&D's 12th Int. Post Comb Network Meeting*, U. Regina, 29 Sept-1 Oct.

# 7

# Blowout in Gas Storage Caverns

H. Djizanne[1]*, P. Bérest[1], B. Brouard[2] and A. Frangi[3]

[1] LMS, École Polytechnique ParisTech, École des Mines de Paris, École des Ponts et Chaussées,
UMR 7649 CNRS, route de Saclay, 91120 Palaiseau - France
[2] Brouard Consulting, 101 rue du temple, 75003 Paris - France
[3] Politecnico di Milano, Piazza Leonardo da Vinci, 32, 20133 Milano - Italy
e-mail: djakeun@lms.polytechnique.fr - berest@lms.polytechnique.fr - contact@brouard-consulting.com - attilio.frangi@polimi.it

* Corresponding author

*Abstract* — *A small number of blowouts from gas storage caverns has been described in the literature. Gas flow lasted several days before the caverns were emptied. In this paper, we suggest simplified methods that allow for computing blowout duration, and evolution of gas temperature and pressure in the cavern and in the well. This method is used to compute air flow from an abandoned mine, an accident described by Van Sambeek in 2009, and a natural gas blowout in an underground storage facility in Kansas. The case of a hydrogen storage cavern also is considered, as it is known that hydrogen depressurization can lead, in certain cases, to an increase in hydrogen temperature.*

*Résumé* — **Éruption en cavités de stockage de gaz** — Un petit nombre d'éruptions en cavités salines de stockage de gaz a été décrit dans la littérature. L'écoulement de gaz dure plusieurs jours avant que les cavités ne se vident complètement. Dans cet article, nous proposons une méthode de calcul de la durée de l'éruption et de l'évolution des paramètres majeurs du système tels que la température, la pression et la vitesse du gaz dans la cavité ou dans le puits. Cette méthode est utilisée pour calculer le débit d'air expulsé par un puits d'accès à une mine de sel abandonnée, un accident décrit par Van Sambeek en 2009, et une éruption suivie d'une combustion du gaz naturel sur un site de stockage souterrain de gaz naturel au Kansas. Le cas d'une cavité de stockage d'hydrogène est également examiné, avec le souci de vérifier si la détente de l'hydrogène peut conduire, dans certains cas, à une augmentation de sa température.

## LIST OF SYMBOLS

| | |
|---|---|
| $a$ | van der Waals gas coefficient, measures attraction between particles, $J.m^3/kg^2$ |
| $b$ | van der Waals gas coefficient, excluded volume per unit of mass of a gas, $m^3/kg$ |
| $C_p$ | Heat capacity at constant pressure, $J/kg.K$ |
| $C_v$ | Heat capacity at constant volume, $J/kg.K$ |
| $D$ | Well diameter, m |
| $e$ | Internal energy, J |
| $F$ | Friction coefficient, /m |
| $f$ | Friction factor |
| $g$ | Gravity acceleration, $m/s^2$ |
| $H$ | Borehole length, m |
| $h$ | Gas enthalpy, $J/kg$ |
| $K$ | Thermal conductivity of salt, $W/m.°C$ |
| $k$ | Thermal diffusivity of salt, $m^2/s$ |
| $M$ | Molar mass, $g/mol$ |
| $m$ | Gas mass, kg |
| $P$ | Pressure, Pa |
| $P_{atm}$ | Atmospheric pressure, Pa |
| $P_c$ | Cavern pressure, Pa |
| $P_{wh}$ | Wellhead cavern pressure, Pa |
| $Q$ | Heat flux at cavern wall, W |
| $S$ | Gas entropy, $J/K$ |
| $t$ | Time, s |
| $T$ | Temperature, K |
| $T_c$ | Cavern temperature, K |
| $T_{cr}$ | Critical temperature, K |
| $T_{wh}$ | Wellhead gas temperature, K |
| $u_c$ | Gas velocity at casing-shoe depth, m/s |
| $u_{wh}$ | Wellhead gas velocity, m/s |
| $V_0$ | Cavern volume, $m^3$ |
| $z$ | Depth, m |

## GREEK LETTERS

| | |
|---|---|
| $\gamma$ | Ratio of specific heats, $\gamma = C_p/C_v$ |
| $\varepsilon$ | Absolute roughness of the well, m |
| $\eta$ | Kinematic viscosity of gas, $m^2/s$ |
| $\dot{\mu}$ | Mass flow, $kg/s.m^2$ |
| $v$ | Specific volume of gas, $m^3/kg$ |
| $v_c$ | Specific volume of cavern gas, $m^3/kg$ |
| $v_{wh}$ | Specific volume of wellhead gas, $m^3/kg$ |
| $\rho$ | Gas density, $kg/m^3$ |
| $\Sigma$ | Cross-sectional area of well, $m^2$ |
| $\Sigma_c$ | Actual surface of cavern walls, $m^2$ |

## INTRODUCTION

A blowout is the uncontrolled release of crude oil and/or natural gas from an oil or gas well after pressure control systems have failed. It often is a dramatic accident when it affects a conventional reservoir, as the amount of gas or oil that can be released can be huge; blowout duration can be several months long. Blowouts from storage caverns of liquid or liquefied hydrocarbons have a different character, as the amount of products immediately released after wellhead failure is relatively small (Bérest and Brouard, 2003). Several examples of blowouts in gas storage caverns have been described in the literature, such as that in an ethane storage at Fort Saskatchewan, Canada (Alberta Energy and Utilities Board, 2002) or in a natural gas storage at Moss Bluff, Texas (Rittenhour and Heath, 2012). There were no casualties in these instances, as the gas rapidly ignited, although the entire inventory was lost. A somewhat similar accident occurred in a "compressed air storage" (in fact, an abandoned salt mine) at Kanopolis, Kansas; a complete description can be found in Van Sambeek (2009).

The most striking difference between a blowout in a well tapped in an oil or gas reservoir and a blowout in a gas-cavern well is that the blowout in a gas-cavern well is completed within a couple of days, as the gas inventory in a cavern is much smaller than in a reservoir. Another difference is that the modeling of the thermodynamic behavior of gas in the cavern is much simpler than in a permeable reservoir, allowing a complete computation of the blowout.

In this paper, a simple method for computing a blowout from a salt cavern is proposed. It involves relatively simple formulae and light numerical computations. Discussion of the numerical results focuses on blowout duration, gas rates at ground level and the evolutions of gas temperature in the cavern. This last issue is of special significance. The drop of gas temperature in a cavern during a blowout often is severe, and thermal tensile stresses are generated at the cavern wall (Bérest et al., 2013). It has been suspected that these stresses lead to fracturing at the cavern wall, spalling and loss of cavern tightness, and correct assessment of cavern temperature is important in this context. In fact, it will be proved that low gas temperatures in the cavern are experienced during a relatively short period of time, and that the depth of penetration of temperature changes at cavern walls is too small to generate deep tensile fractures.

# 1 EVOLUTION OF GAS TEMPERATURE AND PRESSURE IN THE CAVERN

## 1.1 Salt Caverns

Salt caverns currently are used for storing hydrocarbons, air or hydrogen. These caverns are created through solution mining. In this system, a well is drilled to the salt formation, and cased and cemented to the formation. A smaller tube is set inside the casing, as a straw in a bottle. Soft water is injected through the central tube. The water leaches out the salt, and the formed brine is withdrawn from the cavern through the annular space between the central string and the casing. After a year or so, a large cavern is created. Its depth ranges from 200 m to 2 000 m, and its volume ranges from 10 000 $m^3$ to several millions $m^3$. When solution mining is completed, the cavern is filled with saturated brine. Gas then is injected through the annular space, and brine is withdrawn through the central tubing ("debrining"). A small amount of brine is left at the cavern bottom, and the gas in the cavern is wet. A typical operation cycle includes withdrawal during winter and injection during summer. Minimum and maximum gas pressures typically range from, respectively, 15% to 90% of the geostatic pressure (*i.e.*, the weight of the overburden).

## 1.2 Energy Balance

Gas temperature, $T_c(t)$, and gas pressure, $P_c(t)$, can be considered almost uniform throughout the entire volume of a cavern (Bérest *et al.*, 2012). The stored gas is characterized by its state equation, which defines gas pressure ($P$) as a function of gas specific volume ($v = 1/\rho$, $\rho$ is the gas density) and of gas (absolute) temperature ($T$); by a thermodynamic potential, for instance, its enthalpy per unit of mass ($h$) or its internal energy per unit of mass ($e = h - Pv$):

$$P = P(v, T) \qquad (1)$$

$$h = h(T, P) \qquad (2)$$

The kinetic energy of gas in the cavern is neglected. The energy balance equation can be written:

$$m(\dot{e}_c + P_c \dot{v}_c) = Q + <\dot{m}> (h_{inj} - h_c) + L\dot{C} \qquad (3)$$

where $m$ is the mass of gas in the cavern, $e_c$ is the gas internal energy, and:

$$\dot{e}_c + P_c \dot{v}_c = C_v \dot{T}_c + T_c (\partial P / \partial T)|_{v_c} \dot{v}_c$$

$<\dot{m}> = \dot{m}$ when $\dot{m} > 0$ and $<\dot{m}> = 0$ when $\dot{m} < 0$. When gas is injected in the cavern ($<\dot{m}> = \dot{m} > 0$), the difference between the enthalpy of the injected gas ($h_{inj}$) and the enthalpy of the cavern gas ($h_c$) must be taken into account. $C$ is the amount of water vapor in the cavern, and $L$ is the phase-change heat of water (from liquid phase to vapor phase). $Q$ is the heat flux transferred from the rock mass to the cavern gas through the cavern wall, or:

$$Q = \int -K_{salt} \frac{\partial T_{salt}}{\partial n} da \qquad (4)$$

where $K_{salt}$ is the thermal conductivity of salt (typically, $K_{salt} = 6$ W/m.°C) and $T_{salt}$ is the temperature of the rock mass.

## 1.3 Simplifications

In the following, it is assumed that the gas is ideal, $Pv = rT$, $e = C_v T$, and $\dot{e} + P\dot{v} = C_v \dot{T} + rT\dot{v}/v$. During a blowout, $\dot{m} < 0$ and $<\dot{m}> = 0$. During gas withdrawal, cavern volume, or $V_0$, experiences only a small change and can be considered constant, or $V_0 = mv$ and $m\dot{v} = -\dot{m}v$. Water vapor condenses during gas depressurization ("raining", or even "snowing", in the cavern); however, from the perspective of energy balance, this term is small and can be neglected.

The heat flux from the cavern is much more significant; it can be computed as follows. The evolution of temperature in the rock mass is governed by thermal conduction. Generally speaking, penetration of temperature changes in the rock mass is slow. For instance, when a cold gas temperature, $T_0$, has been kept constant over a $t$-long period of time on the (flat) surface of a half-space ($x > 0$) whose initial temperature (at $t = 0$) was $T_\infty$, temperature evolution can be written:

$$T(x, t) = T_\infty + (T_0 - T_\infty) \text{erfc} \left( \frac{x}{2\sqrt{kt}} \right)$$

where $k \approx 3 \times 10^{-6} m^2/s$ is the rock thermal diffusivity. Heat flux per unit area at the surface is:

$$Q = -K_{salt}(T_\infty - T_0) / \sqrt{\pi kt}$$

Rock temperature changes significantly (by more than: $(T - T_\infty)/(T_0 - T_\infty) - \text{erfc}(1/2) \simeq 50\%$) in a domain with a thickness of $d = \sqrt{kt}$, or $d \approx 1$ m after $t = 4$ days. Blowout in a gas cavern is a rapid process: it is completed within a week or less. During such a short period of time, temperature changes are not given time enough to penetrate deep into the rock mass and, from the perspective of thermal conduction, cavern walls can be considered as a flat surface whose area equals the actual area of the cavern (in other words, for numerical computations, actual surface must be smoothed to eliminate shape irregularities whose radii of curvature are smaller than $d$), as was noted by Crotogino *et al.* (2001) and Krieter (2011).

When a varying temperature, $T_c = T_c(t)$, is applied on the surface, the heat flux per surface unit can be expressed as:

$$Q = \int_0^t -K_{salt} \frac{\dot{T}_c(\tau)}{\sqrt{\pi k(t-\tau)}} d\tau \qquad (5)$$

When these simplifications are accepted, the heat balance equation can be written:

$$\frac{\dot{T}_c(t)}{v_c(t)} + (\gamma - 1)\frac{\dot{v}_c(t)T_c(t)}{v_c^2(t)} = -\frac{\Sigma_c K}{C_v V_0 \sqrt{k}} \int_0^t \frac{\dot{T}_c(\tau)}{\sqrt{\pi(t-\tau)}} d\tau \qquad (6)$$

where $\Sigma_c$ is the (actual) surface of the cavern walls. When the heat flux is neglected ($K = 0$), the thermodynamic behavior of gas is isentropic and $Tv^{\gamma-1}$ is constant as expected.

## 1.4 The case of a van der Waals Gas

In Section 6, the case of a hydrogen storage is discussed, in which a van der Waals equation — instead of an ideal gas equation — is more appropriate:

$$P = -a/v^2 + RT/(v - b)$$

and

$$h = C_v T - 2a/v + rTv/(v - b)$$

The heat balance equation can be written:

$$\frac{\dot{T}_c(t)}{v_c(t)} + (\gamma - 1)\frac{\dot{v}_c(t)T_c(t)}{v_c^2 - bv_c} = -\frac{\Sigma_c K}{C_v V_0 \sqrt{k}} \int_0^t \frac{\dot{T}_c(\tau)}{\sqrt{\pi(t-\tau)}} d\tau \qquad (7)$$

## 1.5 An Example

Equation (6) was validated against the results of a withdrawal test performed in a gas-storage cavern at Melville (Canada), described by Crossley (1996). The measured flow rate, cavern pressure and temperature are drawn in Figure 1. The withdrawal period was 5 days

long. The following values were selected: $\gamma = 1.305$, and $C_p = 2\ 237$ J/kg.K. The cavern volume is $V_0 = 46\ 000$ m$^3$. Cavern shape was unknown, and the surface of the cavern walls was selected to be twice the surface of a sphere whose volume equals the actual cavern volume. Note that slightly before the end of the withdrawal phase (day 5), gas starts warming, as the heat flux from the rock mass becomes quite high.

## 2 EVOLUTION OF GAS TEMPERATURE AND PRESSURE IN THE WELLBORE

In this section, the flow of gas through the well is discussed. Here, gas temperature, pressure or specific volume are functions of $z$ ($z = 0$ at the cavern top or casing shoe and $z = H$ at the wellhead).

## 2.1 Main Assumptions

### Duct

Duct diameter, $D$, is assumed to be constant all along the well; hence, the cross-sectional area of the duct, $\Sigma$, is constant as well.

### Adiabatic Flow

Gas temperature decreases in the cavern and borehole. Casing steel, cement and rock at the vicinity of the well experience large temperature changes and thermal contraction. Conversely, the amount of heat transferred from the rock mass to the gas is not able to change gas temperature significantly, as the flow rate of gas is extremely fast. Heat transfer from the rock mass is neglected, and gas flow is considered adiabatic. This issue was discussed in Brouard Consulting and RESPEC (2013).

### Turbulent Flow

Except maybe at the end of the blowout, gas flow is turbulent. The effects of friction are confined to a thin boundary layer at the steel casing wall. The average gas velocity is uniform through any cross-sectional area (except, of course, in the boundary layer).

### Steady-State Flow

The gas rate in a borehole typically is a couple hundreds of meters per second (more, when hydrogen is considered). In other words, only a few seconds are needed for gas to travel from the cavern top to ground level. Such a short period of time is insufficient for

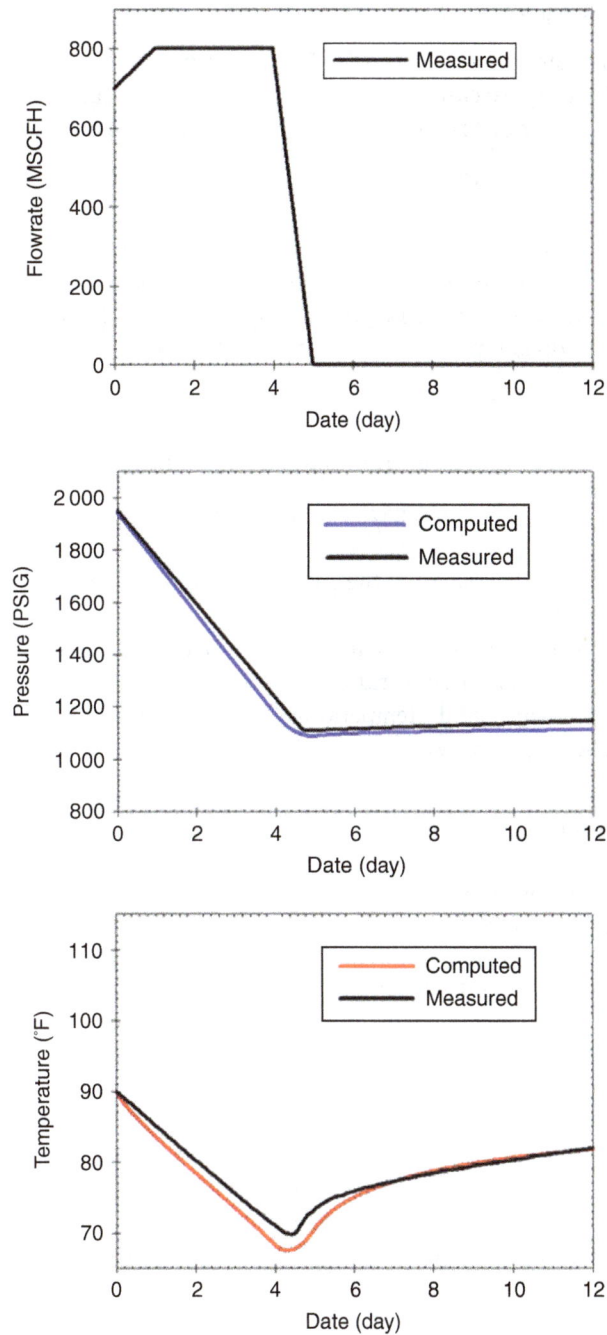

Figure 1

Melville Cavern: gas withdrawal rate, pressure and temperature evolutions, as observed (after Crossley, 1996) and computed.
1 MSCFH = 28 317 Nm³/h, 1 MPa = 145 psig and 20°C = 68°F.

cavern pressure and temperature change significantly. Steady-state flow is assumed and, for simplicity, gas temperature, velocity etc. will be noted $T_c = T_c(z)$,

$u = u(z)$, etc. (Obviously, when longer periods of time are considered, cavern pressure slowly decreases).

These assumptions, which are part of the so-called "Fanno-flow" model, are commonly accepted (Von Vogel and Marx, 1985), although Ma et al. (2011) consider an "isothermal" flow in the well.

## 2.2 Equations

In addition to Equations (1) and (2), gas flow can be described by the following set of equations:

$$\dot{\mu} = \frac{u(z)}{v(z)} = -\frac{V_0}{\Sigma} \frac{\dot{v}(z)}{v^2(z)} \qquad (8)$$

$$\frac{dh}{dz} + u\frac{du}{dz} + g = 0 \qquad (9)$$

$$v\frac{dP}{dz} + u\frac{du}{dz} + g = -f(u) \qquad (10)$$

$$u\frac{dS}{dz} \geq 0 \qquad (11)$$

Equation (8) is the mass conservation equation, where $u$ and $v$ define celerity and specific volume, respectively, and $\dot{\mu} = u(z)/v(z)$ is a constant in the well. Equation (9) is the energy equation, where $g$ is the gravity acceleration. Equation (10) is the momentum equation. Head losses per unit of length are described by $f(u) > 0$, where $f = f(u; D, \varepsilon, ...)$ is a function of gas velocity, duct diameter, wall roughness, etc. Statement (11) is the condition of positivity of entropy ($S$) change, which plays an important role in the context of the Fanno-flow model.

## 2.3 Boundary Conditions and Subsonic Flow

Equations (8) to (10) allow computation of gas pressure and gas temperature in the borehole. Boundary conditions are needed. Pressure and temperature in the cavern [$P_c = P(z = 0)$ and $T_c = T(z = 0)$] are assumed known at any instant. Then, the specific volume of gas, or $v_c$, can be computed through the state equation $P_c = P(v_c, T_c)$. In principle, gas pressure at the wellhead, $P_{wh}$, should be atmospheric: $P_{wh} = P_{atm}$.

However, this boundary condition cannot always be satisfied. It is known from thermodynamics textbooks that $dh(S, P) = TdS + vdP$; hence:

$$dh(S, v) = TdS - c^2 dv/v$$

where $c$ is the velocity of sound. (For an ideal gas, $c^2 = \gamma Pv$.) When gravity is disregarded in a Fanno flow, $dh + udu = 0$, and $TdS = (c^2 - u^2)dv/v$. However, the sign of $dS$ must not change; in other words, when the flow of gas is subsonic $(u_c < c_c)$ at cavern top $(z = 0)$, it must remain subsonic $[u(z) < c(z)]$ in the borehole (except perhaps at ground level, $z = H$).

When applying the boundary condition, $P_{wh} = P_{atm}$, leads to a solution such that gas flow is supersonic in a part of the well (this may occur when cavern pressure is sufficiently high), another solution must be selected (Landau and Lifchitz, 1971, Section 91). This solution is constructed such that $u_{wh} = c_{wh}$. (The flow, which is sonic at ground level, is said to be "choked".) In such a case, no constraint is applied to $P_{wh}$, which, in general, is larger than atmospheric. Conversely, when the cavern pressure is relatively small, the gas flow is said to be "normal". Even at ground level, the gas rate is significantly slower than the speed of sound, and the boundary condition $P_{wh} = P_{atm}$ applies.

## 2.4 Simplifications

The following simplified version of this set of equations allows a closed-form solution to be obtained.

### Body Forces

Body forces are disregarded, $g = 0$. (However, the case when $g \neq 0$ is discussed in Appendix A; it has been proven that, in practical terms, this case does not lead to significant differences.)

### Colebrook's Equation

Our main interest is in gas average velocities larger than $u = 1$ m/s (and up to several hundreds of m/s). Typically, air viscosity is $\eta = 1.3 \times 10^{-5}$ m$^2$/s, duct diameter is $D = 0.2$ m, and Reynolds number (Re $= uD/\eta$) is larger than $10^4$. In this context, head losses can be written $f(u) = Fu^2$, where $F = f/2D$ is the friction coefficient, and $f$ is the friction factor. Especially at the beginning of a blowout, when the velocity of gas is high and the Reynolds number is very large, the Colebrook's equation is written:

$$\frac{1}{\sqrt{f}} = -2\log_{10}\left(\frac{\varepsilon}{3.71D}\right) \quad (12)$$

where $\varepsilon$ is the well roughness. ($\varepsilon = 0.02$ mm is a typical value.)

### Gas State Equation

The gas state equation, $P = P(v, T)$, can be simplified as in the following two cases. In the first case, the gas (typically, natural gas or air) is ideal — i.e., its state equation is:

$$Pv = rT, r = C_p - C_v, \gamma = C_p/C_v$$

and its enthalpy can be written $h = C_p T$, where the heat capacity of the gas at constant pressure $C_p$ is constant. In the second case, the gas (typically, hydrogen) is of the van der Waals type — i.e., its state equation is:

$$P = -a/v^2 + RT/(v - b)$$

where $a, b$ are two constants, and its enthalpy is:

$$h = C_v T - 2a/v + rTv/(v - b)$$

where the heat capacity of the gas at constant volume $C_v$ is constant.

The properties of gases used in this paper are presented in Table 1 (from gas encyclopedia, *Air Liquide*, 2012; pressure and temperature are $10^5$ Pa and 298.15 K respectively).

## 2.5 Model Assessment

The aim of this paper is to provide a clear picture of the main phenomena that affect gas flow during a blowout, although results are indicative rather than exact. In fact, the model suffers from the following three flaws:
- head losses are roughly estimated by the simplified Colebrook's equation; in fact, actual coefficient $F$ is a function of the flow rate, especially when this rate is small;
- even the van der Waals state equation is a less than perfect description of the actual behavior of hydrogen, and heat capacity, $C_v$, is a function of temperature;
- at the end of a blowout, gas flow rates are low, and the simplifications considered in Section 2.4 no longer hold.

A more precise description of gas behavior can be taken into account in Equations (1) and (2) — for instance, $C_v = C_v(T, v)$ and $Pv = rTZ(T, P)$. However, when such a description is accepted, numerical computations are required to compute the flow of gas in the well.

## 3 AIR OR NATURAL-GAS FLOW

### 3.1 Momentum Equation

In this Section, Equations (8) to (11) are used to obtain a relation between gas temperature in the cavern, $T_c$, and the specific mass of the gas in the cavern, $v_c$. Taking into

TABLE 1

Gases constants

| Gases | $C_p$ (J/kg.K) | $C_v$ (J/kg.K) | $\gamma$(-) | M (g/mol) | $a$ (J.m$^3$/kg$^2$) | $b$ (m$^3$/kg) |
|-------|------|------|-------|--------|---------|---------|
| Air | 1 010 | 719 | 1.402 | 28.95 | – | – |
| CH$_4$ | 2 237 | 1 714 | 1.305 | 16.043 | – | – |
| H$_2$ | 14 831 | 10 714 | 1.384 | 2.016 | 6 092 | 0.013 |

account the simplifications noted in Section 2.4, energy Equation (6) can be re-written as:

$$C_pT(z) + \dot{\mu}^2 v^2(z)/2 = C_pT_c + \dot{\mu}^2 v_c^2 2 \qquad (13)$$

Or

$$P(z) = \left(P_c + \frac{\gamma-1}{2\gamma}\dot{\mu}^2 v_c\right)\frac{v_c}{v(z)} - \left(\frac{\gamma-1}{2\gamma}\right)\dot{\mu}^2 v(z) \qquad (14)$$

and momentum Equation (10) can be written as:

$$\left(\frac{rT_c}{\dot{\mu}^2} + \frac{\gamma-1}{2\gamma}v_c^2\right)\frac{1}{v^3(z)} - \frac{\gamma+1}{2\gamma v(z)} = F\frac{dz}{dv}(z) \qquad (15)$$

Note that this equation also can be written:

$$c^2(z) - u^2(z) = \gamma\dot{\mu}^2 v^3(z)F\frac{dz}{dv}(z)$$

Because only solutions resulting in $u^2 < c^2$ are considered, $v$ is an increasing function of $z$.

Integration of momentum Equation (15) from $z = 0$ (casing shoe) to $z = H$ (wellhead) leads to:

$$\frac{1}{2}\left[\left(\frac{rT_c}{\dot{\mu}^2} + \frac{\gamma-1}{2\gamma}v_c^2\right)\left(\frac{1}{v_c^2} - \frac{1}{v_{wh}^2}\right)\right] - \left(\frac{\gamma+1}{4\gamma}\right)Log\frac{v_{wh}^2}{v_c^2} = FH \qquad (16)$$

### 3.2 Normal Flow

Gas pressure at ground level is, in principle, atmospheric, $P_{wh} = P_{atm}$:

$$P_{wh} = \left(rT_c + \frac{\gamma-1}{2\gamma}\dot{\mu}^2 v_c^2\right)\frac{1}{v_{wh}} - \left(\frac{\gamma-1}{2\gamma}\right)\dot{\mu}^2 v_{wh} = P_{atm} \qquad (17)$$

The positive solution of this second-degree equation (with respect to $v_{wh}$) can be computed easily, and its combination with Equation (8) leads to the following differential equation:

$$\dot{v}_c = -\Sigma_c v_c \dot{\mu}(rT_c, v_c, \gamma, FH, P_{atm})/V_0 \qquad (18)$$

However, this solution is valid only when Equation (11) is true (normal flow) — i.e., when:

$$c^2 - u^2 = \gamma Pv - \dot{\mu}^2 v^2 =$$
$$\gamma\left(rT_c + \frac{\gamma-1}{2\gamma}\dot{\mu}^2 v_c^2\right) - \left(\frac{\gamma+1}{2}\right)\dot{\mu}^2 v^2 > 0 \qquad (19)$$

### 3.3 Choked Flow

When condition (19) is not met, the solution (normal flow) must be rejected. The boundary condition $P_{wh} = P_{atm}$ can no longer be satisfied. Instead of $P_{wh} = P_{atm}$ (Eq. 17), the choked-flow condition, $c_{wh} - u_{wh} = 0$, must be used:

$$rT_c + \left(\frac{\gamma-1}{2\gamma}\right)\dot{\mu}^2 v_c^2 - \left(\frac{\gamma+1}{2\gamma}\right)\dot{\mu}^2 v_{wh}^2 = 0 \qquad (20)$$

Eliminating $\dot{\mu}^2$ between (16) and (20) leads to:

$$\left(v_{wh}^2/v_c^2\right) - 1 - Log\left(v_{wh}^2/v_c^2\right) = [4\gamma/(\gamma+1)]FH$$

which proves that $v_{wh}/v_c$ is a function of $\gamma$ and $FH$ and that:

$$\dot{\mu}v_c = I(\gamma, FH)\sqrt{rT_c}$$

Combining again with Equation (8) leads to:

$$\dot{v}_c = \Sigma\sqrt{rT_c}I(\gamma, FH)/V_0 \qquad (21)$$

Equations (18) or (21), together with Equation (6), allow computation of gas temperature and specific-volume evolutions of gas during a blowout.

### 4 THE MOSS BLUFF BLOWOUT

In August 2004, Cavern #1 of the Moss Bluff natural gas storage in Texas experienced a major gas release and fire (Fig. 2). The cavern bottom was filled with saturated brine, and the volume of the gas-filled part of the cavern was $V_0 = 1\ 268\ 000$ m$^3$. The blowout initiated during de-brining of the cavern when gas entered the 8-⅝" brine string, causing the pipe to burst at ground level. The ensuing fire resulted, 21 hours (0.88 day) later, in

Figure 2

Moss Bluff blowout. From a) Rittenhour and Heath (2012); b) Cavern#1 profile, from Brouard Consulting and RESPEC (2013).

separation of the wellhead assembly and the uncontrolled loss of gas from the 20" production casing. The fire self-extinguished about 6-½ days later, when all the gas was burned off. More than 6 sbcf of gas had been released.

Several witnesses report that, after the natural-gas release had been completed, air at ground level was "sucked" into the cavern over several dozens of minutes. One possible explanation is that, during the blowout, the cavern gas was oversaturated with water vapor: partial pressure of the vapor dropped, but not enough time was left for the vapor to condense fully and to reach thermodynamic equilibrium with the brine sump at the bottom of the cavern. When the blowout was complete, additional condensation took place, leading to a decrease in the cavern gas pressure, and air was sucked into the cavern.

The surface of the cavern walls (not including the brine-gas interface) was computed to be $\Sigma_c = 84\,200\,\text{m}^2$. The friction factor was assumed to be $f = 0.012$ when diameter is $D = 8\text{-}\frac{5}{8}$ inches and $f = 0.010$ when the well diameter is $D = 20$ inches. The borehole length is $H = 765$ m. The gas initial pressure and temperature were assumed to be $P_c^0 = 13.89$ MPa and $T_c^0 = 51°C$ (324.15 K).

Main results are presented in Figure 3. It was said that after 21 hours (0.88 day), the well diameter increased from $D = 8\text{-}5/8"$ to $D = 20"$. As a result of this diameter increase, head losses are smaller and gas velocities are faster; the rate of pressure and temperature changes, and the heat flux from the rock mass abruptly increases. Total duration of the computed flow is slighly less than

6 days (the actual duration was 6.5 days). Gas flow is choked (gas velocity is sonic at the wellhead) during the first 3.5 days. Later, the cavern gas pressure becomes much smaller, resulting in slower velocities and normal flow.

Gas temperature in the cavern drops to $T_c = -5°C$ in two days before slowly warming. At this point, gas temperature at the wellhead is $T_{wh} = -40°C$. Cavern temperature reaches a minimum when the energy change rate generated by gas expansion exactly balances the heat-flux rate from the rock mass, as predicted by Equation (6). At the blowout climax, heat flux from the rock mass is approximately 50 MW.

The "end" of the blowout is a difficult notion to define. It can be seen in Figure 3, however, that, during day 5, the gas-flow velocity rapidly decreases and is almost nill after 5.8 days. However, thermal equilibrium between cavern gas and rock mass is not reached, and the cavern gas slowly warms, resulting in a gas outflow of approximately $u_c \simeq u_{wh} \approx 1$ m/s; the pressure difference between the cavern and ground level is no longer the driving force for gas flow. As was mentioned before, the actual validity of the mathematical solution at the end of the blowout is arguable, as water vapor condensation, for instance, may play a significant role. Note also that, at the end of the blowout, gas temperature increase is fast: gas density is small (pressure is atmospheric) and gas volumetric heat capacity is much smaller than it was before the blowout. For this reason, as explained in Section 1.3, significant temperature changes are not

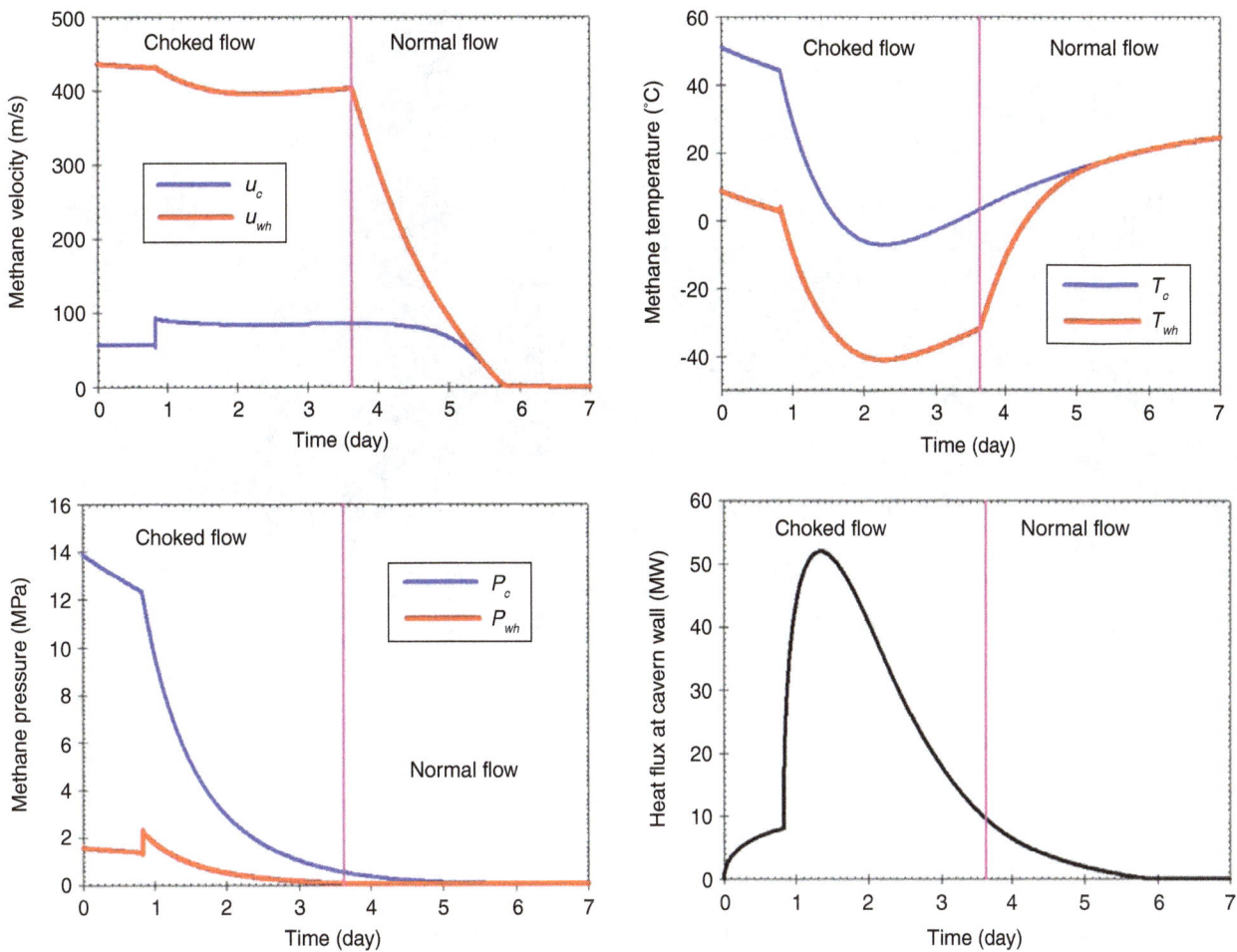

Figure 3

Computed evolutions of gas velocity, gas temperature, gas pressure and heat flux from the cavern wall as a function of time during the Moss Bluff blowout.

given time enough to penetrate deep into the rock mass. Significant tensile stresses are observed in a thin "skin" at cavern wall and deep fractures cannot develop (Brouard Consulting and RESPEC, 2013).

## 5 THE KANOPOLIS BLOWOUT

Compressed Air Energy Storage (CAES) is experiencing a rise in interest, as it can be used as buffer energy storage in support of intermittent sources of renewable energy, such as wind mills. We discuss here an air blowout that occurred in an abandoned mine that presents some similarities with the case of a blowout in an air storage cavern and provide some validation of the model. Obviously, in sharp contrast with natural gas or hydrogen, air cannot burn or explode, and less

severe consequences must be expected from an air blowout.

Van Sambeek (2009) gave a complete account of a remarkable accident in an abandoned salt mine, and provided a comprehensive and convincing explanation of that event. "*On October 26, 2000, a brick factory in Kanopolis, Kansas, was substantially destroyed by bricks, sand, and water falling from the sky. The jet of air blew from a previously sealed salt mine shaft and through a pile of bricks next to the brick factory*" (p. 620; *Fig. 4*). Bricks and sand were blown into the air more than 100 m "*for longer than 5 minutes but less than 20 minutes*" (p. 621). The hypothesis analyzed by Van Sambeek was that "*groundwater had entered the mine through the shafts over a long time and compressed the air within the mine*" until the shaft plug collapsed in October 2000 and air escaped from the mine.

Figure 4

Kanopolis brick factory after the blowout (Van Sambeek, 2009).

The Kanopolis Mine is not a salt cavern and, in the context of a blowout, several differences are noted. The cross-sectional area of the access well (a mine shaft) is larger than the cross-sectional area of a cavern borehole by two orders of magnitude, making blowout duration much shorter (10 or so minutes instead of several days).

The shaft, lined with wood timber, had a length of $H = 240$ m, and its inside dimensions were about $3.6$ m $\times 5.2$ m. An equivalent circular cross-section, $S = 18.72$ m$^2$, and a friction coefficient, $F = 0.225/$m, were selected, (this value was selected to match observed data, as Colebrook's equation hardly applies to gas flow in an old wood-lined mine shaft). For air, $\gamma = 1.4$. Van Sambeek (2009) suggests that the compressed-air (absolute) pressure might have been $P_c^0 = 0.272$ MPa (0.172 MPa relative) and that the air volume in the mine was $V_0 = 670\,000$ m$^3$. We assume that the initial air temperature was $T_c^0 = 15°$C (288 K).

It might have been expected that the drop in the mine's air temperature would be much more severe than in a conventional gas-storage cavern, as heat provided by the rock mass seems not to be given enough time to warm the air in the mine. In fact, heat was provided by the mine roof and by the dry surface of the pillars, whose overall surface is approximately $\Sigma_c = 1\,000\,000$ m$^2$. (No attempt was made to take into account the heat transferred from the brine that filled the lower part of the mine rooms.) The ratio between the wall area and the mine volume in Equation (9) (or $\Sigma_c/V_0 = 1.3$ m) is much larger in a mine than it is in a cavern, making the heat flux from the rock mass much faster.

Results are provided in Figure 5. The flow is normal (not choked). At ground level, air speed decreases from 180 m/s (650 km/h) to a few m/s in 11 minutes. This result is consistent with what was reported by Van Sambeek (2009) who proved that an air-stream speed of 180 m/s generates a drag force that is able to propel bricks perhaps as high as 100 m (*Fig. 6*). Figure 5 also displays air pressure and temperature during the blowout. Air temperature at ground level drops first to 1°C before increasing to 14°C at the end of the blowout. Air temperature in the mine does not experience changes larger than 1°C. Note (*Fig. 5*) that the heat flux from the rock mass reaches 330 MW after a few dozens of seconds.

## 6 HYDROGEN BLOWOUT

Salt caverns storing hydrogen (which have gas pressure in the 7-21 MPa range) are operated in the UK (Teesside,

Figure 5

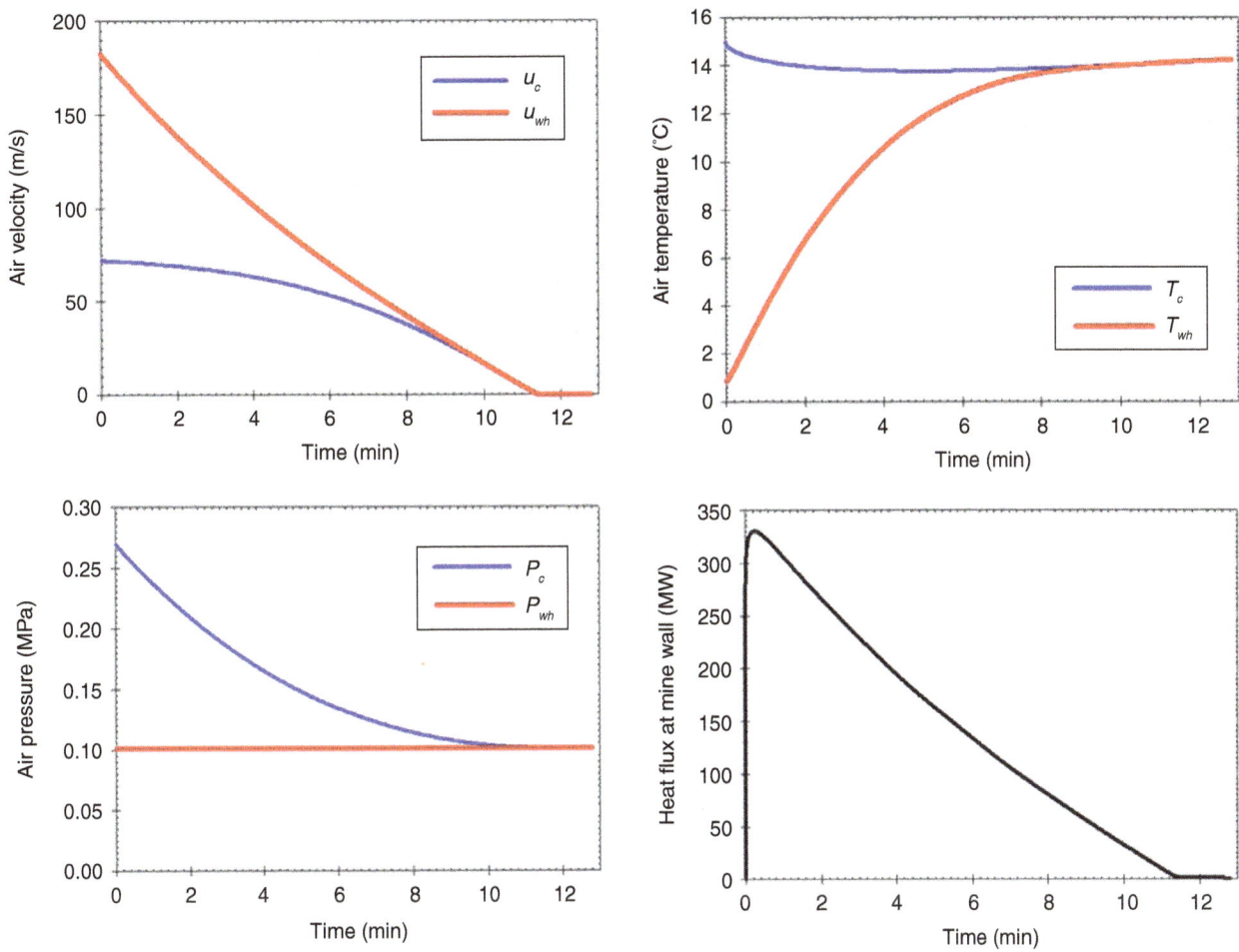

Computed evolution of air velocity, air temperature, air pressure and heat flux from cavern wall as a function of time during the Kanopolis blowout.

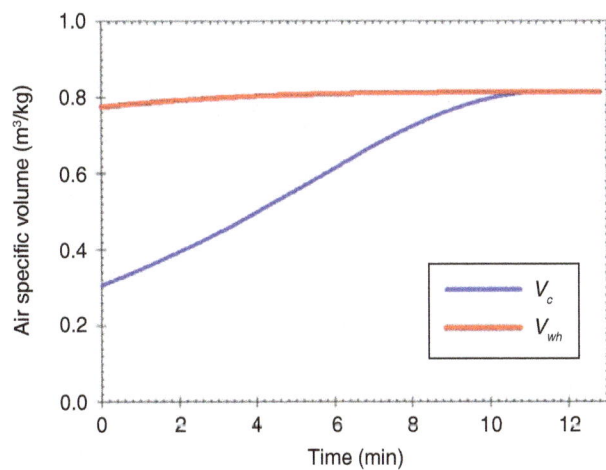

Figure 6

a) Air outflow during the blowout, videotaped from a distance of 2.5 km (Van Sambeek, 2009); b) calculated evolution of the specific volume of air as a function of time.

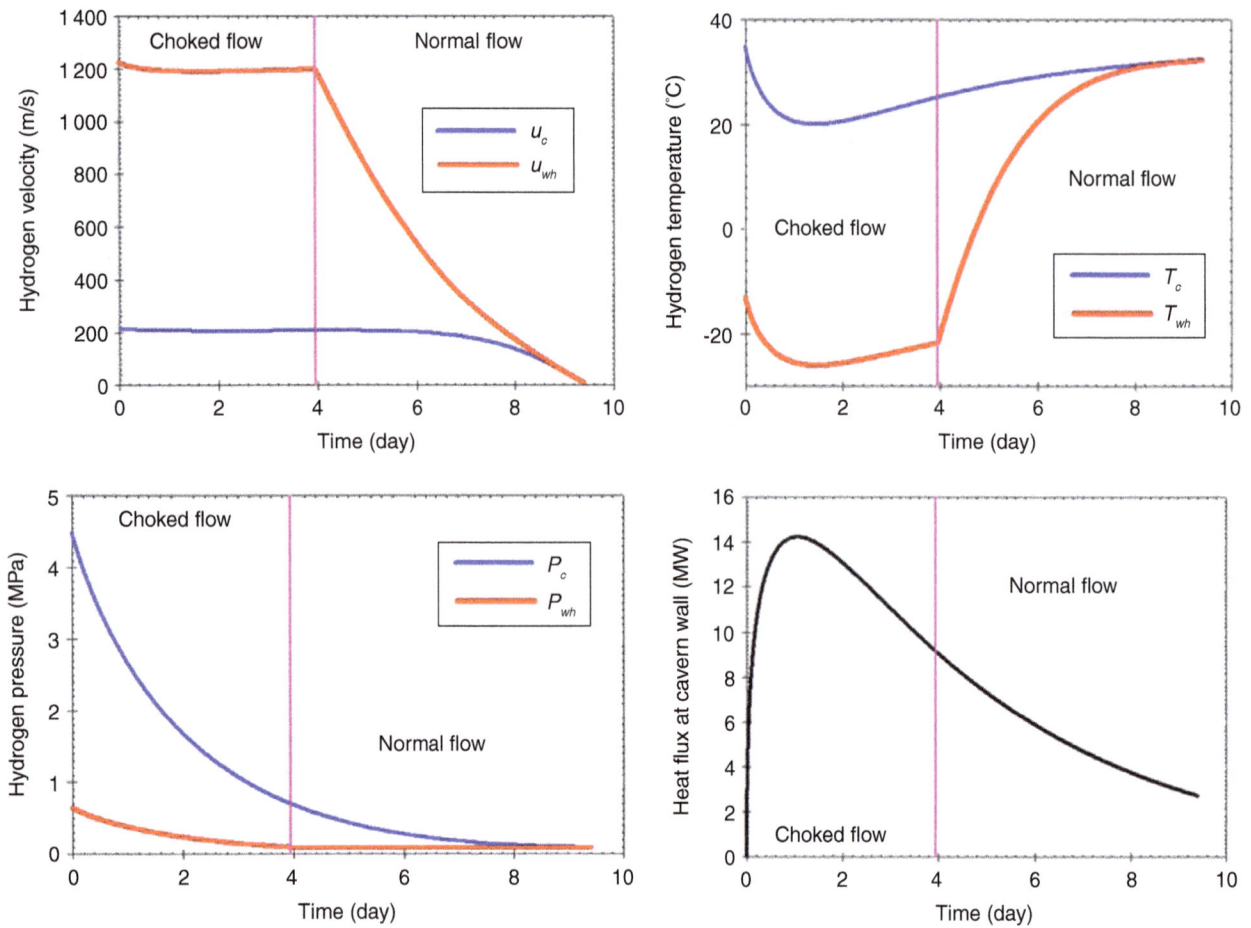

Figure 7

Computed evolution of hydrogen velocity, hydrogen temperature, hydrogen pressure and heat flux from the cavern as a function of time during a blowout.

three 70 000-m$^3$ caverns at a 370-m depth) and in Texas (Clemens Dome, Mont Belvieu, Moss Bluff). Hydrogen storage raises interesting problems, as its state equation and thermo-dynamic potentials differ significantly from those of an ideal gas.

## 6.1 Blowout from a Generic Hydrogen Storage

A discussion of a blowout from a generic hydrogen storage cavern is provided here. In this discussion, the cavern is cylindrical, with a volume $V_0 = 1\,000\,000\,\text{m}^3$, an overall surface $\Sigma_c = 60\,000\,\text{m}^2$, a casing-shoe depth $H = 370$ m, and tubing diameter $D = 7$". A friction factor $f = 0.01$ was selected, and the initial cavern pressure and temperature were $P_c^0 = 4.5$ MPa and $T_c^0 = 35\,°\text{C}$, respectively.

The thermodynamic behavior of hydrogen exhibits some specific features of interest (in particular, an isenthalpic depressurization leads to hydrogen warming); so, instead of the standard state equation of an ideal gas, a van der Waals state equation was selected to describe the gas behavior (*Appendix B*).

Main results are provided in Figure 7. The blowout is approximately 9 days long. The flow is choked during the first 4 days and is normal during the second half of the blowout. Gas velocities are high, as the speed of sound in hydrogen ($c \approx 1200\,\text{m/s}$) is much faster than in air or natural gas ($c^2 \approx \gamma r T$, $r = C_p - C_v$, and hydrogen heat capacities are large; *Tab. 1*). The same argument (large heat capacity) explains why the temperature drop in the cavern (the temperature plummets to 20°C) is not very large (*Eq. 13*). However, the temperature drop in the

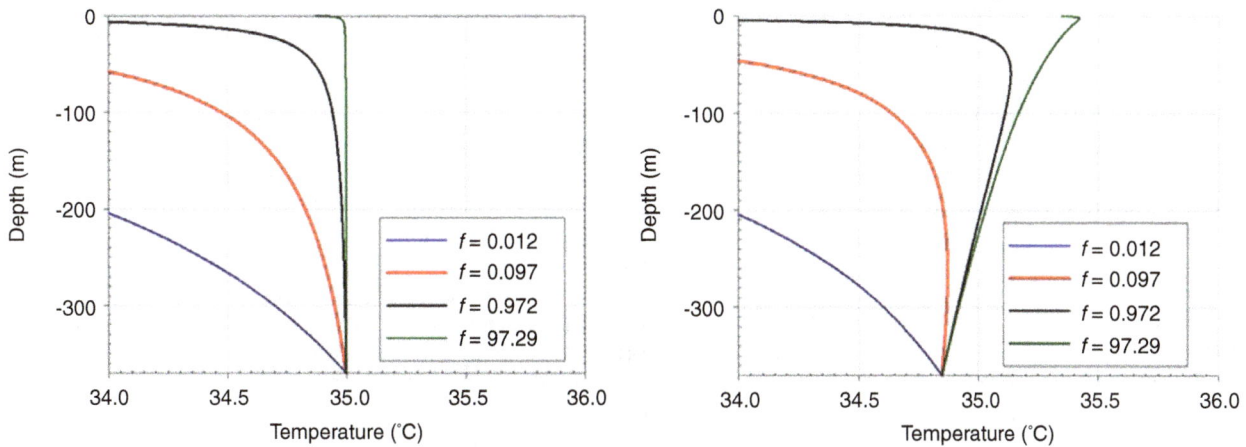

Figure 8

Temperature distribution in the well: when head losses are extremely large, hydrogen temperature increases in the well a), an effect that is not captured when the state equation of gas is ideal b).

well is not very different from what it is during a natural gas blowout, as the value:

$$T_c - T_{wh} = \left[u^2/2C_p\right]_c^{wh} \approx \gamma r T_c/2C_p \approx (\gamma - 1)T_c/2\gamma$$

is not very different from one gas to another. No hydrogen warming in the well was observed.

The same computations were performed when assuming that hydrogen is an ideal gas (instead of a van der Waals gas). Only tiny differences were observed; however, this conclusion may be wrong in some cases, as explained in the next section.

## 6.2 Joule-Thomson Effect

It can be observed in Figure 7 that, except at the end of the blowout, the wellhead temperature ($T_{wh}$) is much colder than the cavern temperature ($T_c$): the temperature of hydrogen decreases when it travels from the cavern top to the ground surface. However, it is known that when a real gas (as differentiated from an ideal gas) expands through a throttling device (the so-called Joule-Thomson expansion), its enthalpy remains constant, and gas temperature may either decrease or increase. Gases have a Joule-Thomson inversion temperature above which the gas temperature increases during an isenthalpic expansion. For hydrogen, this inversion temperature is $-71°C$ (much lower than the temperatures considered here).

However, the expansion of a gas during a blowout is not a Joule-Thomson expansion, as kinetic energy

cannot be neglected (Eq. 9). It is expected that hydrogen temperature increases during its expansion when kinetic energy can be neglected — i.e., when head losses are large.

An example of a hydrogen blowout involving gas-temperature increase is presented in Figure 8. Parameters are the same as presented in Section 6.1 except for the friction factor, which is $f = 97.3$ instead of $f = 0.01$. (Head losses are larger by a factor of 10 000.) The celerity of gas flow in the well is much smaller (it is divided by a factor of 10) than in the example described in Section 6.1, and the cavern is emptied in one year or so ("blowout" is somewhat of a misnomer). The hydrogen temperature is slightly warmer at ground level than it is in the cavern, an effect that is not captured when the state equation of gas is ideal. It can be concluded that, when realistic blowout scenarios are considered, the Joule-Thomson effect plays a minor role.

## CONCLUSIONS

A simplified solution was proposed to compute the evolution of gas pressures, temperatures and velocities during a blowout in a gas storage cavern. It was shown that, in general, the flow is choked when gas pressure in the cavern is high and is normal when the cavern pressure is low. Results must be considered as indicative rather than exact, as simple gas state equations and thermodynamic potentials were selected. Validation of the model is difficult: for obvious practical reasons, few parameters can be measured accurately during a blowout. However, the thermodynamic model of the cavern is able to explain correctly the evolution of cavern gas temperature during a (controlled) gas withdrawal; duration of the Moss Bluff blowout can be back-calculated correctly; and the computed air velocities are compatible with the ballistic flight of bricks observed during the Kanopolis blowout, as was proved by Van Sambeek (2009). It is believed that this model provides a good basis for computation of the thermomechanical behavior of cavern walls during a blowout, a concern of special significance for two reasons: it is important, before a blowout, to establish a credible scenario (gas rate, duration) and, after a blowout, to assess if the caverns can be operated again.

## ACKNOWLEDGMENTS

The authors thank Ron Benefield and his colleagues from Spectra Energy Transmission, and the Solution Mining Research Institute (SMRI), which granted permission that some results of the Moss Bluff Cavern#1 analysis performed for the SMRI (Research Report 2013-01) be published in this paper. They also are indebted to Leo van Sambeek, who provided them with additional comments on his remarkable analysis of the Kanopolis accident. Special thanks to Kathy Sikora. This study was funded partially by the French *Agence Nationale de la Recherche* (ANR) in the framework of the SACRE Project devoted to adiabatic CAES design. This project includes researchers from EDF, GEO-STOCK, PROMES (Perpignan), HEI (Lille) and École Polytechnique ParisTech (Palaiseau).

## REFERENCES

Air liquide (2012) *Gas Encyclopedia*, < http://encyclopedia.air-liquide.com > .

Alberta Energy and Utilities Board (2002) *BP* Canada Energy Company: Ethane Cavern well fires, Fort Saskatchewan, Alberta, Aug./Sept. 2001, *EUB post incident report*.

Bérest P., Brouard B. (2003) Safety of salt caverns used for underground storage, *Oil & Gas Science and Technology – Rev. IFP* **58**, 3, 361-384.

Bérest P., Djizanne H., Brouard B., Hévin G. (2012) Rapid Depressurizations: can they lead to irreversible damage? *Proc. SMRI Spring Technical Conference*, Regina, Canada, 24-23 April, 63-86.

Bérest P., Djizanne H., Brouard B., Frangi A. (2013) A Simplified Solution For Gas Flow During a Blow-out in an $H_2$ or Air Storage Cavern, *Proc. SMRI Spring Technical Conference*, Lafayette, Louisiana, 23-23 April, 86-94.

Brouard Consulting and RESPEC (2013) Analysis of Cavern MB#1 Moss Bluff Blowout data, *Research Report 2013-01*, *Solution Mining Research Institute*, 197 pages.

Crotogino F., Mohmeyer K.U., Scharf R. (2001) Huntorf CAES: More than 20 Years of Successful Operation, *Proc. SMRI Spring Technical Conference*, Orlando, Florida,15-18 April, 351-362.

Crossley N.G. (1996) Salt cavern Integrity Evaluation Using Downhole Probes. A Transgas Perspective, *Proc. SMRI Fall Meeting*, Cleveland, Ohio, 21-54.

Krieter M. (2011) Influence of gas cavern's surface area on thermodynamic behavior and operation, *Proc. SMRI Fall Technical Conference*, York, UK, 179-184.

Landau L., Lifchitz E. (1971) *Mécanique des Fluides*. Éditions MIR. (in French).

Ma L., Liu X., Xu H., Yang S., Wang Z. (2011) Stability analysis of salt rock gas storage cavern under uncontrolled blowout, *Rock and Soil Mechanics* **32**, 9, 2791-2798 (in Chinese).

Rittenhour T.P., Heath S.A. (2012) Moss Bluff Cavern 1 Blowout, *Proc. SMRI Fall Technical Conference*, Bremen, Germany, 119-130.

Van Sambeek L. (2009) Natural compressed air storage: a catastrophe at a Kansas salt mine, *Proc. 9th Int. Symp. on Salt*, Beijing, Zuoliang Sha ed., Vol. 1, 621-632.

Von Vogel P., Marx C. (1985) Berechnung von Blowoutraten in Erdgassonden, *Erdoel-Erdgas*, 101.Jg, Heft 10, Oktober 1985, 311-316 (in German).

## Appendix A – Gravity Forces

From Equations (9) and (10) in the main text, it is seen that gravity forces may play a significant role when $gH$ is not much smaller than $[u^2/2]_c^{wh}$. Because $gH$ typically is 10 000 m$^2$/s$^2$, gravity forces must be taken into account when $u_{wh} \approx 100$ m/s — i.e., when the flow is normal. When gravity forces are taken into account ($g \neq 0$), the following equations apply (instead of Eq. 13, 14 and 15).

$$C_P T + \dot{\mu}^2 v^2/2 + gz = C_P T_c + \dot{\mu}^2 v_c^2/2 \tag{A1}$$

$$P = \left( P_c + \frac{\gamma-1}{2\gamma} \dot{\mu}^2 v_c \right) \frac{v_c}{v} - \frac{\gamma-1}{2\gamma} \left( \dot{\mu}^2 v + \frac{2gz}{v} \right) \tag{A2}$$

$$\left( rT_c + \frac{\gamma-1}{2\gamma} \dot{\mu}^2 v_c^2 \right) \frac{1}{v} - \frac{\gamma+1}{2\gamma} \dot{\mu}^2 v - \frac{\gamma-1}{\gamma} \frac{gz}{v} = \left( \frac{g}{\gamma} + F\dot{\mu}^2 v^2 \right) \frac{dz}{dv} \tag{A3}$$

The solution of the differential equation (A3) can be written as:

$$F\dot{\mu}^2 H = v_{wh}^{1-\gamma} \left( \frac{g}{\gamma F \dot{\mu}^2} + v_{wh}^2 \right)^{\frac{\gamma-1}{2}}$$

$$\int_{v_c}^{v_{wh}} \frac{\left( rT_c + \frac{\gamma-1}{2\gamma} \dot{\mu}^2 v_c^2 \right) \frac{1}{w} - \frac{\gamma+1}{2\gamma} \dot{\mu}^2 w}{w^{1-\gamma} \left( \frac{g}{\gamma F \dot{\mu}^2} + w^2 \right)^{\frac{\gamma+1}{2}}} dw \tag{A4}$$

$$P_{wh} = \left( rT_c + \frac{\gamma-1}{2\gamma} \dot{\mu}^2 v_c^2 \right) \frac{1}{v_{wh}} - \frac{\gamma-1}{2\gamma} \left( \dot{\mu}^2 v_{wh} + \frac{2gH}{v_{wh}} \right) \tag{A5}$$

Equations (A4) and (A5) allow to compute $v_{wh}$, $\dot{\mu}^2$ etc. Main results are provided in Figure A1 ($g = 10$ m/s$^2$), which must be compared to the corresponding results provided in Figure 5 ($g = 0$ m/s$^2$) in the main text. Differences are

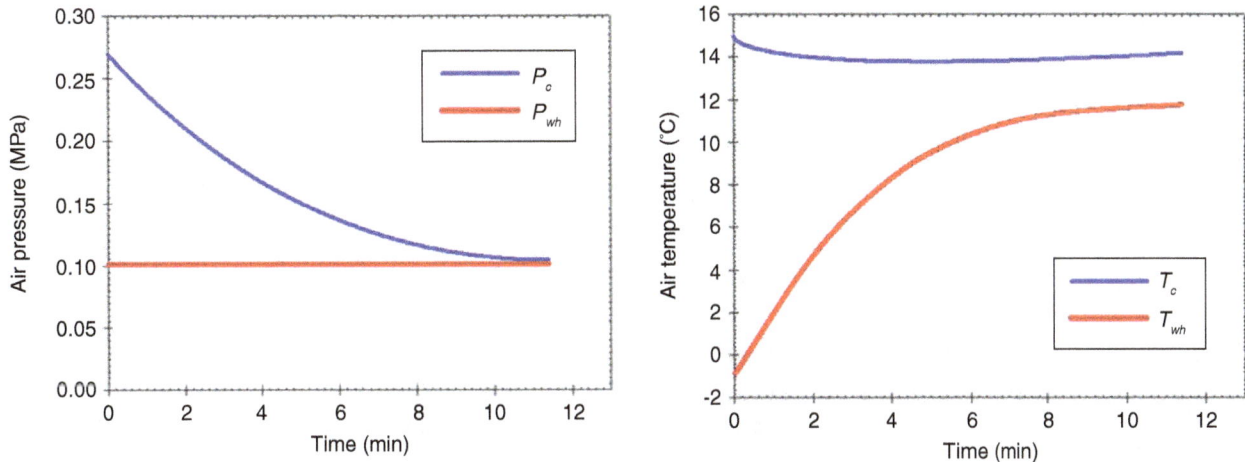

Figure A1

Kanopolis blowout ($g = 10$ m/s$^2$): computed evolutions of a) air pressure and b) air temperature as a function of time.

exceedingly small, except at the end of the blowout, as static pressure and temperature distributions in the well are modified by gravity forces. Equations (A1) and (A2) prove that, when thermodynamic equilibrium is reached, $T_{wh} = T_\infty - (gH/C_p) = 12.6°C$ (instead of: $T_{wh} = T_\infty = 15°C$) and that $P_c = P_{wh}(T_\infty/T_{wh})^{\frac{\gamma-1}{\gamma}} = 1.003\ P_{wh}$ (instead of: $P_c = P_{wh} = P_{atm}$). This solution is partly artificial, as it results from various assumptions that are not fully consistent. (Well walls are adiabatic; cavern walls are not.)

## Appendix B – Hydrogen Flow

It is known that hydrogen exhibits special behavioral characteristics — in particular, during depressurization such that hydrogen enthalpy is constant (for instance, when hydrogen leaks from a pressurized vessel through a pinhole), the temperature of hydrogen "increases". (For most gases in similar circumstances, the temperature decreases). For this reason, instead of the ideal gas equation, the more precise van der Waals equation is used:

$$P = -\frac{a}{v^2} + \frac{rT}{v-b} \tag{B1}$$

where $a$ and $b$ are constants. It can be assumed that the temperature is above the critical temperature, $T_{cr} = 33.2$ K (which means that (B1) allows computation of $v$ when $P$ and $T$ are known). The internal energy and enthalpy of hydrogen are:

$$\begin{cases} e(T,v) = C_v T - \frac{a}{v} \\ h(T,v) = e + Pv = C_v T - \frac{2a}{v} + \frac{rTv}{v-b} \end{cases} \tag{B2}$$

The same method as used for air can be used here: energy equation allows computation of pressure and temperature. Let $\bar{C}_p = C_v + r$ (for a van der Waals gas, $\bar{C}_p$ is not the heat capacity at constant pressure, $C_p$) and $\bar{\gamma} = \bar{C}_p/C_v$:

$$T = (\bar{\gamma} - 1)(v-b)\frac{h_c + \frac{1}{2}\mu^2 v_c^2 + \frac{2a}{v} - \frac{1}{2}\mu^2 v^2}{r(v\bar{\gamma} - b)} \tag{B3}$$

$$P = -\frac{a}{v^2} + (\bar{\gamma} - 1)\frac{h_c + \frac{1}{2}\mu^2 v_c^2 + \frac{2a}{v} - \frac{1}{2}\mu^2 v^2}{\bar{\gamma}v - b} \tag{B4}$$

When one sets:

$$\psi(v,\mu) = \frac{\bar{\gamma}}{b^2}\left(h_c + \frac{1}{2}\mu^2 v_c^2 + \frac{2a\bar{\gamma}}{b}\right)\left(\frac{b}{(\bar{\gamma}v - b)} + Log\left(\frac{v\bar{\gamma} - b}{v}\right)\right)$$

it can be inferred that:

$$\left[-\frac{2a}{3v^3} + (\bar{\gamma} - 1)\left[-\frac{a}{bv^2} + \psi(v,\mu) - \frac{1}{2\bar{\gamma}}\mu^2 Log(v\bar{\gamma} - b) - \frac{\mu^2 v}{2(\bar{\gamma}v - b)}\right]\right] + \mu^2 Log(v) = -F\mu^2\frac{dz}{dv} \tag{B5}$$

Instead of Equations (16, 17) and (19) when $g = 0$, we now have:

$$\left[-\frac{2a}{3v^3} + (\bar{\gamma} - 1)\left[-\frac{a}{bv^2} + \psi(v,\mu) - \frac{1}{2\bar{\gamma}}\mu^2 Log(v\bar{\gamma} - b) - \frac{\mu^2 v}{2(\bar{\gamma}v - b)}\right] + \mu^2 Log(v)\right]_{v_c}^{v_{wh}} = -FH\mu^2 \tag{B6}$$

$$P_{wh} = -\frac{a}{v_{wh}^2} + (\bar{\gamma} - 1)\frac{h_c + \frac{1}{2}\mu^2 v_c^2 + \frac{2a}{v_{wh}} - \frac{1}{2}\mu^2 v_{wh}^2}{\bar{\gamma}v_{wh} - b} \tag{B7}$$

$$c^2 - u^2 = -\frac{2a}{v} + \frac{v^2}{(v-b)} \bar{\gamma}(\bar{\gamma}-1) \frac{h_c + \frac{1}{2}\dot{\mu}^2 v_c^2 + \frac{2a}{v} - \frac{1}{2}\dot{\mu}^2 v^2}{v\bar{\gamma} - b} - \dot{\mu}^2 v^2 > 0 \qquad (B8)$$

Here, again, when inequality (B8) is met (normal flow), Equations (B6) and (B7) allow elimination of $\dot{\mu}^2$ and computation of $v_{wh}$. When inequality (B8) is not met (choked flow), Equation (B6) together with the condition $c_{wh}^2 - u_{wh}^2 = 0$, as given in (B9):

$$-\frac{2a}{v_{wh}^3} + \frac{\bar{\gamma}(\bar{\gamma}-1)\left(h_c + \frac{2a}{v_{wh}}\right)}{(v_{wh}-b)(v_{wh}\bar{\gamma}-b)} = \dot{\mu}^2\left[1 + \frac{\bar{\gamma}(\bar{\gamma}-1)(v_{wh}^2 - v_c^2)}{2(v_{wh}-b)(v_{wh}\bar{\gamma}-b)}\right] \qquad (B9)$$

allow for computation of $v_{wh}$ This is shown in Figure B1.

An example is provided here. Hydrogen properties are provided in Table 1 in the main text. In Figure B1a, cavern temperature and pressure are $T_c = 313.15$ K and $P_c = 0.5$ MPa, respectively, resulting in an inlet hydrogen specific volume $v_c = 2.6$ m$^3$/kg. The well is 1000-m deep, the well diameter is 0.5 m and the friction factor is $f = 0.01$. Equations (B6, B7) and (B9) allow three curves to be drawn: $\dot{\mu}_{FH}^2 = \dot{\mu}_{FH}^2(v_{wh}), \dot{\mu}_{atm}^2 = \dot{\mu}_{atm}^2(v_{wh})$ and $\dot{\mu}_{son}^2 = \dot{\mu}_{son}^2(v_{wh})$, respectively. The grey zone is the supersonic zone, in which $c_{wh} < u_{wh}$ (not acceptable). The intersection of the curves described by Equations (B6) and (B7), obtained when the specific volume at the exit is $v_{wh} = 11.4$ m$^3$/kg, can be accepted, as it lays outside the supersonic zone: the flow is normal.

In Figure B1b, cavern temperature and pressure are $T_c = 313.15$ K and $P_c = 13$ MPa, respectively, resulting in an inlet gas specific volume $v_c = 0.10$ m$^3$/kg. Here, again, $FH = 10$. The intersection of the curves representing (B6) and (B7) belongs to the gray supersonic zone and does not provide an acceptable solution: the flow cannot be normal. The flow is choked, and the specific volume at the exit, $v_{wh} = 0.55$ m$^3$/kg, is given by the intersection of the curves described by (B6) and (B9): $\dot{\mu}_{FH}^2 = \dot{\mu}_{FH}^2(v_{wh})$ and $\dot{\mu}_{son}^2 = \dot{\mu}_{son}^2(v_{wh})$. Note that, at this intersection, the curve $\dot{\mu}_{FH}^2 = \dot{\mu}_{FH}^2(v_{wh})$ reaches a maximum, as $v_{wh} - c_{wh} = 0$.

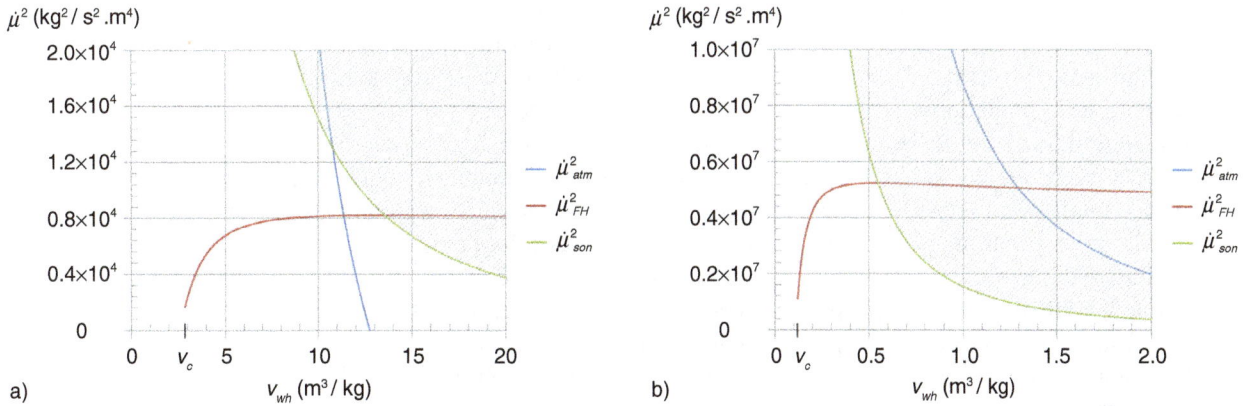

Figure B1

Determination of the specific volume of hydrogen at the exit, $v_{wh}$, in the case of a) normal flow and b) choked flow.

# 8

# Hollow Fiber Membrane Contactors for $CO_2$ Capture: Modeling and Up-Scaling to $CO_2$ Capture for an 800 MW$_e$ Coal Power Station

Erin Kimball[1]*, Adam Al-Azki[2], Adrien Gomez[3], Earl Goetheer[1], Nick Booth[2], Dick Adams[2] and Daniel Ferre[3]

[1] TNO, Leeghwaterstraat 46, 2628 CA Delft - The Netherlands
[2] E-ON New Build & Technology Ltd, Ratcliffe on Soar Nottingham, NG11 0EE - United Kingdom
[3] IFP Energies nouvelles, Rond-point de l'échangeur de Solaize, BP 3, 69360 Solaize - France
e-mail: erin.kimball@tno.nl

* Corresponding author

**Résumé — Contacteurs à membrane à fibres creuses pour la capture de $CO_2$ : modélisation et mise à l'échelle de la capture du $CO_2$ d'une centrale électrique au charbon de 800 MWe** — Une analyse technico-économique a été effectuée pour comparer les modules de membrane à fibres creuses (HFMM, *Hollow Fiber Membrane Modules*) avec les colonnes à garnissage structuré, plus conventionnelles, pour une utilisation comme absorbeur dans les systèmes de capture de $CO_2$ à base d'amine pour centrale électrique. Afin de simuler le fonctionnement d'un système HFMM de taille industrielle, un modèle bidimensionnel a été développé et validé en se basant sur les données issues d'un HFMM de laboratoire. Après diverses expériences réussies et une validation du modèle, un HFMM à l'échelle du pilote a été construit et simulé avec le même modèle. Les résultats des simulations, pour les deux tailles de HFMM, ont été utilisés pour évaluer la faisabilité d'une mise à l'échelle vers un système HFMM à même de capturer le $CO_2$ d'une centrale électrique de 800 MWe. Les exigences du système – longueur totale de la membrane, surface totale de contact et volume du module – ont été déterminées à partir des simulations et utilisées pour établir une comparaison économique avec les colonnes à garnissage structuré. Les résultats indiquent qu'une réduction significative des coûts d'au moins 50% est nécessaire pour rendre les HFMM compétitifs par rapport aux colonnes à garnissage structuré. Des études approfondies restent nécessaires pour optimiser plusieurs paramètres de conception des HFMM de taille industrielle, tels que le rapport d'aspect (longueur/diamètre du module), la durée de vie de la membrane, le matériau et la forme du module, tout en abaissant le coût global. Cependant, les HFMM présentent l'avantage d'une surface de contact par unité de volume et d'une capacité de mise à l'échelle du module plus importantes, ces paramètres étant clés pour les applications nécessitant une empreinte limitée ou une configuration flexible.

*Abstract — Hollow Fiber Membrane Contactors for $CO_2$ Capture: Modeling and Up-Scaling to $CO_2$ Capture for an 800 MW$_e$ Coal Power Station — A techno-economic analysis was completed to compare the use of Hollow Fiber Membrane Modules (HFMM) with the more conventional structured packing columns as the absorber in amine-based $CO_2$ capture systems for power plants. In order to simulate the operation of industrial scale HFMM systems, a two-dimensional model was developed* *and validated based on results of a laboratory scale HFMM. After successful experiments and valida-tion of the model, a pilot scale HFMM was constructed and simulated with the same model. The results of the simulations, from both sizes of HFMM, were used to assess the feasibility of further up-scaling to a HFMM system to capture the $CO_2$ from an 800 $MW_e$ power plant. The system requirements – membrane fiber length, total contact surface area, and module volume – were determined from simu-lations and used for an economic comparison with structured packing columns. Results showed that a significant cost reduction of at least 50% is required to make HFMM competitive with structured packing columns. Several factors for the design of industrial scale HFMM require further investiga-tion, such as the optimal aspect ratio (module length/diameter), membrane lifetime, and casing mate-rial and shape, in addition to the need to reduce the overall cost. However, HFMM were also shown to have the advantages of having a higher contact surface area per unit volume and modular scale-up, key factors for applications requiring limited footprints or flexibility in configuration.*

## INTRODUCTION

The need to reduce the amount of $CO_2$ that is emitted into the atmosphere is an issue that is being addressed glob-ally. Targets for $CO_2$ emissions reduction have been set in several countries, such as a reduction to 17% below 2005 levels by 2020 in the United States, a reduction per unit of GDP (Gross Domestic Product) of 40-45% below 2005 levels by 2020 in China, and a reduction of 20% below 1990 levels by 2020 in the European Union [1]. In order to meet these ambitions, a multi-headed approach will be needed where less fossil-based energy is consumed, by increasing energy efficiency and the share of renewable energy, and the fossil based-energy that is consumed becomes cleaner. For the latter, Carbon Capture, Utiliza-tion, and Storage (CCUS) will be key in reducing the emissions from the $1.2 \times 10^8$ GWh of energy that is gen-erated from fossil fuels every year [2].

This work is concerned with carbon capture, specifi-cally post-combustion carbon capture, as it will be the most important solution for existing power plants in the near-term [3]. As an example, the flue gas from a coal-fired power plant, a dilute stream of $CO_2$ of about 14 vol% in $N_2$, was considered. The conditions of the flue gas are mild – atmospheric pressure and about 50°C – but the volumes are immense, around 600 m$^3$/s for an 800 $MW_e$ coal-fired power plant. In order to separate the relatively small amount of $CO_2$ (10-15 vol%) from the flue gas, the gas is usually brought into contact with a chemical solvent, typically an amine, such that the $CO_2$ reacts with the solvent and can be later released from the solvent upon heating of the liquid or a reduction in pressure. The systems involved in such a separation are huge, upwards of 16 000 m$^3$ in volume and 2 million m$^2$ of contact area for the absorption columns alone, for the 800 $MW_e$ case[1].

In order to provide the high surface area needed, mul-tiple large towers, more than 10 m in diameter and 40 m in height, are filled with a structured packing material. The flue gas is fed into the bottom of the tower and flows upwards while reacting with the liquid solvent that is fed at the top. The liquid flows over the packing material, which thus provides the high amount of contact surface needed. The large dimensions can make the retrofit of such systems to existing power plants difficult, due to space limitations, while also requiring a high initial investment. Furthermore, operational issues, such as foaming, entrainment, and channeling, can occur and are difficult to mitigate for structured packing columns [4].

An alternative option to the standard configuration, which helps to solve some of these key issues is the use of membrane contactor modules. With membrane mod-ules, non-selective, porous, hydrophobic membranes are most often used, as investigated in this work, but com-posite membranes consisting of a porous support and selective dense top layer have also been considered [5]. The purpose of the membrane is to provide the contact surface area required between the gas and liquid phases. The gas flows on one side of the membrane and is trans-ported through the membrane pores in order to come into contact with the liquid, flowing on the other side of the membrane. Two membrane configurations are commonly employed for such modules: flat sheet and hollow fiber. With flat sheet membrane modules, the absorption liquid and flue gas are passed along either side of a membrane sheet, with many membranes in par-allel separated by spacers. These systems are advanta-geous in that they are robust against particulates in the flows and are more flexible in the membrane materials that can be used, but have the disadvantages of having a relatively low specific surface area due to the inclusion of spacer material and having relatively high pressure drops [6]. The other configuration, Hollow Fiber Mem-brane Modules (HFMM), uses hollow fiber membranes,

---

[1] Values based on internal TNO calculations completed as part of the EU FP7 project CESAR.

which have specific surface areas of more than $1\,000\,m^2/m^3$ and are already used commercially in water treatment systems. In HFMM, hollow membrane fibers, usually a few millimeters in diameter, are used to provide the contact surface area between the gas and liquid phases. The gas usually flows usually around the outside of the fibers (shell side) and the liquid usually flows on the inside of the fiber tubes (tube side). A module is made up of thousands of fibers tightly bundled together, with several modules connected in parallel in order to achieve the required capture rate of $CO_2$. The flows are still counter current, as with the packed columns, but the flow rates can be controlled independently such that foaming, flooding, and entrainment are no longer problems. As will be shown here, the system volume is also much smaller, making HFMM well suited for retrofit applications.

While much of the work with HFMM has been focused on predicting the mass transfer of the various components within the different phases (gas, membrane, liquid) [7-10] the work presented here will discuss the impact that these systems can have on large scale post-combustion $CO_2$ capture. A techno-economic analysis was completed in order to compare the use of HFMM with more conventional structured packing columns based on the design of a $CO_2$ capture plant for an 800 $MW_e$ coal fired power plant.

The approach taken in this work was a combination of experimental, modeling, and techno-economic analysis,

as shown in Figure 1. Experimental results from a pilot scale HFMM, constructed by *Polymem* (Toulouse, France), with 10 $m^2$ of gas-liquid contact surface area were obtained to investigate the operation of larger scale modules. The results were then compared with a 2D HFMM mass transfer model, similar to those described in literature, which was previously validated as described in Chabanon *et al.* (this same journal issue) [11]. Fitting the model results to the experimental data allowed for the calculation of the membrane mass transfer coefficient, $k_m$, and other correlations for the mass transfer of components within the liquid, membrane, and gas phases. The $k_m$ value and correlations were then used as input for a 1D HFMM model implemented in Aspen. The actual process in developing the 1D Aspen model, including the derivation of correction factors to account for the change from 2D to 1D, was not straightforward. Therefore, the work with the 1D Aspen model is outside of the scope of this paper and will be addressed in future work (the authors may be contacted for more information).

The results presented here for the industrial design are based on the calculations with the 2D mass transfer model. Two cases will be discussed – a more ideal case, assuming uniform flows of the liquid and gas for each of the fibers, and a worse case, assuming maldistribution of flows as seen with the pilot scale HFMM. The industrial 800 $MW_e$ design for each case is given followed by

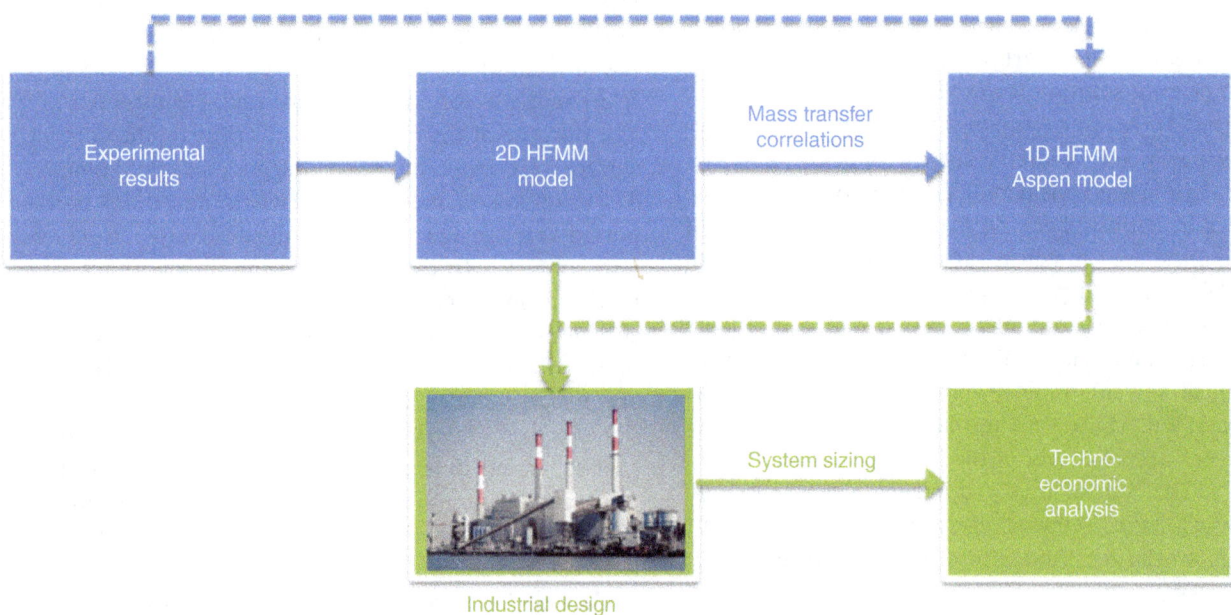

Figure 1

Schematic of the approach taken to complete the techno-economic analysis of a large scale HFMM system.

an analysis of the economics, as compared to a conventional structured packing system for the 800 $MW_e$ plant.

# 1 EXPERIMENTAL

The experimental work was conducted in two phases, as described in detail in Chabanon *et al.* [11] and briefly described here. The first phase concerned the operation of a small laboratory scale HFMM, made up of only 119 PTFE fibers for a total contact surface area of 0.2 $m^2$, with the other parameters given in Table 1. The experimental results fit very well with the 2D mass transfer model, which assumed uniform flows of both the gas and liquid, indicating that the small module represented nearly ideal conditions.

The second phase concerned the construction and operation of a pilot scale module with a total contact surface area of 10 $m^2$ and 8 521 fibers. The parameters of the module are also given in Table 1, where length is the actual length of the fibers and effective length is the length that is available for gas liquid contacting due to the sealant material. It should be noted that the membrane fibers were the same as those used for the laboratory scale module. The other details of the design of this module are described in [11]. In short, the gas flowed from top to bottom, counter-current to the liquid, which flowed from the top. An inline transparent section followed the liquid outlet to allow for observation of any gas bubbles and an inline filter followed the gas outlet to catch any liquid that could have percolated through

the membranes. These features, along with two pressure sensors each for the gas and liquid phases, ensured that the operational parameters were set to allow for good gas-liquid separation without bubbling of the gas into the liquid phase (avoidance of bubbling was necessary for accurate measurements).

With both modules, the feed gas consisted of a mixture of 14-15% $CO_2$ in $N_2$ in each experiment while the total flow rate was on the order of liters per minute for the lab scale 100's of liters per minute for the pilot scale (10 $m^3/h$). The flow rates of the liquid, 30% MEA in $H_2O$ in each case, was varied between $1 \times 10^{-2}$ and 1 liters per minute for the lab and pilot scale modules, respectively. All measurements were taken after steady-state was reached.

Table 2 shows a set of the experimental results from the laboratory scale HFMM for two different liquid flow rates and several gas flow rates. The results show a smooth trend (*Fig. 2*) of increasing $CO_2$ at the outlet with increasing gas flow rate. This is as expected as a lower liquid to gas ($Q_l/Q_g$) ratio has been shown to result in a lower removal of $CO_2$ from the gas stream, resulting in a higher concentration of $CO_2$ at the outlet [12]. With an increase in liquid flow rate, the $CO_2$ concentration at the outlet decreases for a given gas flow rate. This is again as expected since with higher liquid flow rates, $Q_l/Q_g$ is increased and the driving force for mass transfer of the $CO_2$ into the MEA solution is higher. These results were then used to validate the 2D mass transfer model.

For the pilot scale module, the results of several of the experiments are given in Table 3, for a range of gas and liquid flow rates. The inlet $CO_2$ concentration in

TABLE 1

Details of the laboratory scale and pilot scale HFMM

|  |  | Lab scale | Pilot scale |
|---|---|---|---|
| Module | Inner diameter (m) | $1.24 \times 10^{-2}$ | 0.105 |
|  | Length (m) | 0.35 | 1 |
|  | Effective length (m) | 0.30 | 0.88 |
|  | Number of fibers (-) | 119 | 8521 |
|  | Packing ratio (-) | 0.59 | 0.648 |
|  | Specific interfacial area ($m^2/m^3$) | 1 331 | 1 329 |
| Fiber | Inner diameter (m) | $4.3 \times 10^{-4}$ | $4.3 \times 10^{-4}$ |
|  | Outer diameter (m) | $8.7 \times 10^{-4}$ | $8.7 \times 10^{-4}$ |
|  | Porosity (-) | 0.336 | 0.336 |
|  | Material (-) | PTFE | PTFE |

TABLE 2

Experimental data from the laboratory scale HFMM

| No. | Experimental data | | | | | $Q_l/Q_g$ (kg/kg) | CO$_2$ removal | |
| | Liquid | | | Gas | | | | |
| | $Q_l$ (mL/min) | Temp. at inlet (°C) | CO$_2$ lean loading (mol/mol) | $Q_g$ (L/min) | CO$_2$ conc. at inlet (%) | | CO$_2$ conc. at outlet (%) | Experiment efficiency (%) |
|---|---|---|---|---|---|---|---|---|
| 1 | 10 | Room temp. | 0 | 0.1 | 15.2 | 3.29 | 0.1 | 99.3 |
| 3 | 10 | Room temp. | 0 | 0.2 | 15.2 | 1.64 | 0.1 | 99.3 |
| 39 | 10 | Room temp. | 0 | 1.5 | 15.7 | 0.22 | 4.8 | 69.4 |
| 41 | 10 | Room temp. | 0 | 3.0 | 14.9 | 0.11 | 9.0 | 39.6 |
| 43 | 10 | Room temp. | 0 | 4.5 | 14.9 | 0.07 | 9.2 | 38.3 |
| 38 | 10 | Room temp. | 0 | 6.0 | 15.0 | 0.05 | 10.7 | 28.7 |
| 25 | 50 | Room temp. | 0 | 0.6 | 14.5 | 2.75 | < 0.05 | 99.9 |
| 13 | 50 | Room temp. | 0 | 0.2 | 14.9 | 8.23 | 0.1 | 99.5 |
| 27 | 50 | Room temp. | 0 | 1.5 | 15.0 | 1.10 | 2.0 | 86.7 |
| 57 | 50 | Room temp. | 0 | 3.0 | 15.6 | 0.55 | 4.4 | 71.8 |
| 30 | 50 | Room temp. | 0 | 4.5 | 14.7 | 0.37 | 6.9 | 53.1 |
| 58 | 50 | Room temp. | 0 | 6.0 | 14.8 | 0.27 | 8.6 | 41.9 |

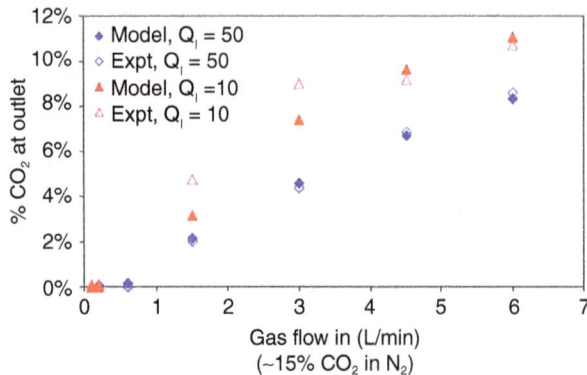

Figure 2

Comparison of simulation results and experimental data for the laboratory scale HFMM of the CO$_2$ outlet concentration for gas inlet flow rates of $50 \times 10^{-3} - 6.00$ L/min and two liquid flow rates, $Q_l = 10 \times 10^{-3}$ and $50 \times 10^{-3}$ L/min. Taken from [11].

the MEA can be seen to increase slightly with each experiment, but this was not thought to influence the results significantly. Similar trends were observed as with the lab scale module and, again, the major influence on

the CO$_2$ capture efficiency was $Q_l/Q_g$ (7th column in *Tab. 3*). For larger ratios of 12 or more, the capture efficiency was above 70%, while for smaller ratios of less than 2, the capture efficiency was usually less than 50% (except for experiment No. 64). These experiments were then simulated with the 2D model, as discussed in the following section.

## 2 MODELING

The 2D mass transfer model is described in detail in [11]. Briefly, the model is based on mass and energy balances written for each phase – liquid, membrane, and gas – and each relevant component – CO$_2$, MEA species, and water. Between the gas and membrane phases, continuity in the concentrations and temperatures are stipulated. Between the membrane and liquid phases, equilibrium relations are used for the concentrations and continuity is used for the temperature. The model was validated based on the data from the laboratory scale HFMM, as given in Table 2. The fit between the model and experimental data, shown in Figure 2, was very good and allowed for the calculation of the membrane mass transfer coefficient. The best fit resulted in

TABLE 3

Experimental data from the pilot scale HFMM

| No. | Experimental data | | | | | | CO$_2$ removal | |
| | Liquid | | | Gas | | | | |
| | $Q_l$ (L/h) | Temp. at inlet (°C) | CO$_2$ lean loading (mol/mol) | $Q_g$ (Nm$^3$/h) | CO$_2$ conc. at inlet (%) | $Q_l/Q_g$ (kg/kg) | CO$_2$ conc. at outlet (%) | Experiment efficiency (%) |
|---|---|---|---|---|---|---|---|---|
| 64 | 30.7 | 16.0 | 0.046 | 10.2 | 14.9 | 1.88 | 6.5 | 56.5 |
| 67 | 30.6 | 16.0 | 0.084 | 30.2 | 15.0 | 0.53 | 11.8 | 21.5 |
| 68 | 100.8 | 16.0 | 0.088 | 30.1 | 14.8 | 1.77 | 8.8 | 40.8 |
| 69 | 151.0 | 16.0 | 0.107 | 30.3 | 14.4 | 2.67 | 8.2 | 42.8 |
| 70 | 50.9 | 16.0 | 0.133 | 10.2 | 12.8 | 3.27 | 4.7 | 63.4 |
| 72 | 200.9 | 16.0 | 0.157 | 10.1 | 13.8 | 12.94 | 3.8 | 70.5 |
| 75 | 200.8 | 16.0 | 0.183 | 30.2 | 14.5 | 3.66 | 9.6 | 33.6 |
| 76 | 15.7 | 16.0 | 0.179 | 5.5 | 14.6 | 1.90 | 8.4 | 42.5 |
| 80 | 100.9 | 16.0 | 0.195 | 5.5 | 15.5 | 12.10 | 4.1 | 73.7 |

a value for $k_m$ of $2.58 \times 10^{-4}$ m/s, as also reported in [11]. Given the assumptions of ideality in the model and the good fit, this $k_m$ value was taken as the "ideal case" $k_m$.

The 2D model was then used to simulate the operation of the pilot scale module. Due to non-uniformities in the liquid and, especially, the gas flows, the same $k_m$ value could not be used to fit the data. The $k_m$ that was used to fit the pilot scale module data was $5.31 \times 10^{-5}$ m/s, an order of magnitude smaller than the ideal case $k_m$ (for details, see [11]). As the membrane fibers were the same in both of the modules, the decrease in the $k_m$ value needed to fit the pilot scale data was not caused by a change in the actual $k_m$, but instead in the overall mass transfer coefficient, $K_{ov}$, incorporating the gas, membrane, and liquid phase mass transfer. Most of the decrease in the mass transfer was thought to be caused by maldistribution of the fibers, as observed by X-ray imaging of the fiber bundle, and bypassing of the gas flow around the fibers, but inconsistent liquid flow could have also been a factor. Still, since $k_m$ was used as the fitting parameter, these non-idealities were lumped into the calculation of a modified membrane mass transfer coefficient, $k_m'$. As this value represented a case with much lower CO$_2$ capture efficiency than the ideal case, it was taken as the "worse case" $k_m'$.

Given $k_m$ and $k_m'$, the model was then used to study how an HFMM system could be designed for the capture of 90% of the CO$_2$ from an 800 MW$_e$ coal-fired power plant. The results are described in the following section.

## 3 EVALUATION AND DISCUSSION OF THE 800 MW$_e$ STUDY

### 3.1 Case Descriptions

Two different cases were defined for this part of the study for an 800 MW$_e$ coal-fired power plant, a base case with a gas flow temperature of 50°C, and a cooled case, with a gas flow temperature of 30°C after a Direct Contact Cooler (DCC). The cooled case was also analyzed as it resulted in a lower flue gas volume and a slightly higher CO$_2$ concentration, and thus required less membrane surface area and a smaller (cheaper) design. The characteristics of flue gases are detailed in Table 4. Also shown are the specifications for the absorption solvent conditions. These were chosen to be equivalent to those for the packed column reference case and were adjusted accordingly in the model.

For the CO$_2$ absorption, the solvent used was based on MEA, 30 wt%, with a CO$_2$ lean loading of 0.271 mol CO$_2$/mol MEA, and a rich loading of 0.468 mol CO$_2$/mol MEA. This implies that the total amine molar flow rate was calculated to be equal to 504 000 kmol/h (11 500 m$^3$/h) to reach 90% CO$_2$ capture.

### 3.2 Results for Module Requirements

The number of membrane modules was fixed at 100 and, for technical and feasibility considerations. The diameter

TABLE 4

Specifications of the flue gas conditions, solvent conditions, and HFMM designs for the two cases considered: with and without a direct contact cooler

|  | Base case (no DCC) | Cooled case (w/ DCC) |
|---|---|---|
| *Flue gas* | | |
| Temperature (°C) | 50 | 30 |
| Pressure (atm) | 1 | 1 |
| Total gas flow rate ($m^3/s$) | 680 | 586.5 |
| Molar composition: | | |
| $CO_2$ (%) | 13.4 | 14.6 |
| $N_2$ (%) | 71.1 | 77.4 |
| $O_2$ (%) | 3.6 | 3.9 |
| $H_2O$ (%) | 11.9 | 4.2 |
| *Solvent* | | |
| Temperature (°C) | 40 | 40 |
| Total liquid flow rate ($m^3/s$) | 3.48 | 3.48 |
| MEA (wt%) | 30 | 30 |
| Inlet loading (mol/mol) | 0.271 | 0.271 |

of the modules was chosen to allow for a superficial gas velocity of 1.8 m/s, which resulted in a pressure drop of about 100 mbar along the length of a 2 m fiber, given the same packing density as in the pilot scale HFMM. The number of fibers in each module was based on the calculated module diameter and a value of 1 250 $m^2/m^3$ as a feasible specific membrane module contact surface area, following the manufacturing recommendation of *Polymem*, the module provider involved in this work.

The membranes for this part of the study were considered to be the same as those used in the laboratory and pilot scale modules described above with the parameters as given in Table 1. Regarding the fibers themselves, details of their characterization are given in [11]. Even for longer membranes, the structure of the membranes and loss in length due to the sealant around the fibers were assumed to not change given discussions with the membrane manufacturer, *Polymem*. The construction of such modules and support of the membrane bundle for modules longer than 1 meter is a subject of ongoing work at *Polymem*. With longer fibers, the bundle must be

supported somehow and the liquid pressure at the inlet of the fibers must not be too high as to cause wetting of the pores. In the worst case, for a large scale system, a few modules in series would be required to achieve the required $CO_2$ capture.

The length of the fibers in each case was adjusted in the model in order to achieve a simulation result that met the 90% $CO_2$ capture criteria. This then allowed for the calculation of the total required membrane contact surface area and module volume. The results are shown below in Table 5 for both cases of the flue gas conditions and for both values of $k_m$ – the low worse case value, $k_m$' and the higher ideal case value, $k_m$. Also included in Table 5 are the comparisons of the contact surface area and volume required for each $k_m$ value relative to that required for a structured packing column (the numbers for the structured packed column were developed within a different part of the same EU FP7 project, CESAR; surface area = $2.01 \times 10^6$ $m^2$, volume = 16 500 $m^3$).

The key points to be noted from Table 5 concern the dramatic impact of the $k_m$ value on the sizing of the system. A decrease in $k_m$ by a factor of 5 results in an increase in required membrane area and module volume of 325% for the base case and 335% for the cooled case. When these values are compared to a structured packing column (designed for the base case), the influence of the $k_m$ is again apparent. With the higher $k_m$ value from the ideal case, the required surface area for the HFMM and the packed column are essentially the same (row 10 in *Tab. 5*), indicating that the mass transfer is not limited by the membrane in the HFMM, but instead is dominated by the transfer of the gas into the liquid phase. With the lower $k_m$' from the worse case, the extra mass transfer resistance from the non-idealities in the flows result in the situation where the surface area required for gas-liquid contact is three times higher with the HFMM than with the packed column. Still, in both cases there is a substantial reduction in volume with the HFMM relative to that required for the packed column – about 70% with the worse case $k_m$' value and as much as 90% with the ideal case $k_m$.

### 3.3 Configuration Design Options

Several factors must be considered in the design of HFMM for $CO_2$ capture from power plants, for example the casing material, the shape (round or square), the total number of modules, the number of fiber bundles within each module, and the aspect ratio (length/ diameter). Large membrane modules would be difficult to install and maintain; furthermore, the aspect ratio

TABLE 5

Results for the HFMM requirements for each case and for the two $k_m$ values discussed previously

| | Base case (no DCC) | Cooled case (w/DCC) |
|---|---|---|
| Number of modules (-) | 100 | 100 |
| Packing density ($m^3$ fibers/$m^3$ module) | 0.709 | 0.709 |
| Specific surface area ($m^2/m^3$) | 1 250 | 1 250 |
| Module diameter (m) | 4.07 | 3.78 |
| # fibers/module (-) | $1.20 \times 10^7$ | $1.04 \times 10^7$ |
| Ineffective membrane length (m) | 0.10 | 0.10 |
| *$k_m$ from ideal case* | | |
| Length of fibers (m) | 1.23 | 1.24 |
| Total surface area ($m^2$) | $2.00 \times 10^6$ | $1.74 \times 10^6$ |
| Total volume ($m^3$) | 1 597 | 1 389 |
| *Comparison with structured packing (ratio HFMM/packed column)* | | |
| Surface area (-) | 0.99 | 0.86 |
| Volume (-) | 0.097 | 0.084 |
| *$k_m'$ from worse case* | | |
| Length of fibers (m) | 4.00 | 4.16 |
| Total surface area ($m^2$) | $6.49 \times 10^6$ | $5.82 \times 10^6$ |
| Total volume ($m^3$) | 5 194 | 4 659 |
| *Comparison with structured packing (ratio HFMM/packed column)* | | |
| Surface area (-) | 3.23 | 2.90 |
| Volume (-) | 0.315 | 0.283 |

would make it difficult to ensure good gas flow penetration to each element. Hence, a smaller diameter module design should be considered for efficient absorption. The implications of smaller modules on capital cost would be a matter best addressed by manufacturers; however, the other factors must be considered at the same time to achieve the optimal design. To keep the absorber footprint (floor size) to a minimum, square modules would provide a better solution. The difficulty in this case would be the design of the liquid system header box which operates above atmospheric pressure, for which purpose a circular design would be preferable.

From Figure 3, it can be seen that the overall module design could be very much like a 'shell-and-tube' heat exchanger, except that each tube is a bundle of membrane fibers. If required, the whole module or individual bundles could be lifted and removed from the gas stream, for which purpose the header plate would have a top and bottom clamp for each bundle. Each module may also contain several loose-fitting baffle plates, providing further gas distribution. However, the benefit of baffle plates may not be substantial enough to justify the incurred pressure drop. This would be the case for a module design with a good aspect ratio (small diameter, large length). Another consideration of this module design may also be the possibility to design the membrane potting in such a way as to direct the gas path more effectively at the inlet and outlet of the module (shown as "sloped potting" in *Fig. 3*). This may also help overcome the pressure drop incurred by any baffles.

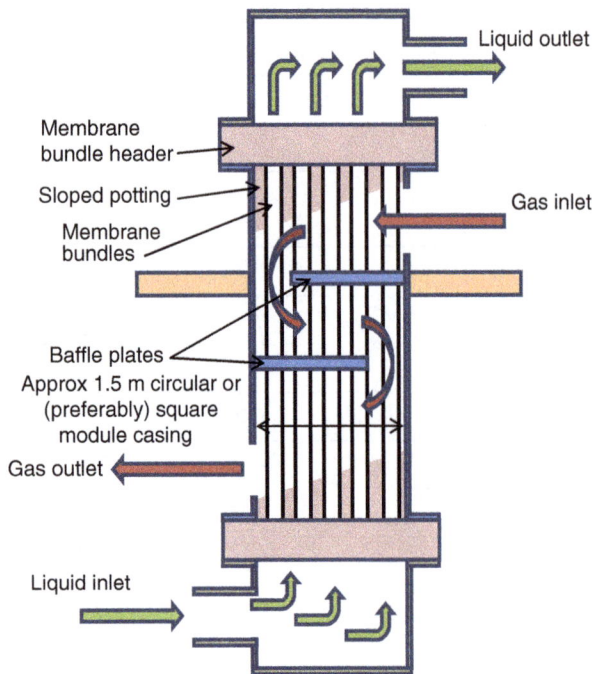

Figure 3

Detailed schematic of a possible design of a HFMM and internals.

The footprint of the gas absorption plant will depend mostly upon the total number of modules required for the desired $CO_2$ capture rate, and further work should be carried out with regards to available options for ductwork sizing. Another consideration will be the very low gas velocities in the HFMM. This may result in ash drop out, calling for an improved absorber design which can collect and remove the ash before entering the modules. Lower gas velocities will provide the advantage of less erosion; however, the impact of increased pressure drop must be further studied.

The design considerations of industrial scale-up, as well as the large differences in the two experimental systems discussed earlier suggest that an accurate prediction of the overall mass transfer coefficient is required in order to provide a more appropriate fiber length estimation. Accurately approximating the fiber length would provide a solid basis for design and industrial scale-up; however, ensuring a good distribution of gas and liquid flows (through baffles, for example), minimizing bending of fibers, and minimizing pressure drop will all be key aspects in the final design of an industrial scale system. Still, with the configurations as calculated with the model (*Tab. 5*), it can be concluded that scale-up to some degree should be feasible.

## 4 ECONOMIC COMPARISON WITH STRUCTURED PACKING COLUMNS

The economic feasibility of the $CO_2$ capture system incorporating the HFMM was also studied based on the ideal case $k_m$ and worse case $k_m'$ values and the system parameters, as given in Table 5. The values for the conventional structured packing columns capture system were based on calculations completed as part of the same EU FP7 CESAR project and based on the EBTF guidelines [13]. For $CO_2$ capture with HFMM, an equivalent system was assumed with only the absorption columns replaced by the membrane contactor modules. The capital and operating costs of the other components – stripper, pumps, blowers, etc. – remained the same (in future work, this will need to be investigated in more detail as the membrane modules cause more gas-side pressure drop and will result in higher operating costs for the blowers or even the need for additional blowers). The calculations for the structured packing absorption column were based on state-of-the-art designs with the total values given in Table 6, along with the other assumptions used in the economic calculations. The results for the packed column configuration were only available when no DCC was added to the system. Also given in Table 6 are the cost of the DCC, as calculated for the specified flue gas inlet and outlet conditions, and the cost of the membrane modules, based on past experience of the authors.

The total capital costs (not including installation costs) of the absorber for each of the cases considered are compared in Figure 4. With the lower $k_m'$ value from the worse case, the absorber system is exceedingly more expensive than the conventional structured packing

TABLE 6

Values used in the economic analysis. Costs do not include installation costs

| Description | Value |
|---|---|
| Capital costs exclusive of absorber (M€) | 155.9 |
| Total operating costs (M€/yr) | 30 |
| Annuity factor (1/yr) | 0.08 |
| Project life (yr) | 40 |
| Working hours (h/yr) | 7 500 |
| Amount $CO_2$ produced (ton/yr) | $3.90 \times 10^6$ |
| Cost of DCC (M€) | 5.39 |
| Cost of membrane modules (€/m$^2$) | 20 |

column and even with the $k_m$ value from the ideal case, requiring the same contact surface area as the packed column, the absorber capital investment is still more than twice as high. A benefit of adding a DCC to the system is seen for the worse case $k_m$' and results in a capital savings of 8 M€. With the ideal case $k_m$, however, there is almost no difference between the two configurations.

A more detailed analysis was completed to compare the base case (no DCC) with the analysis already completed for the structured packing column. The operating

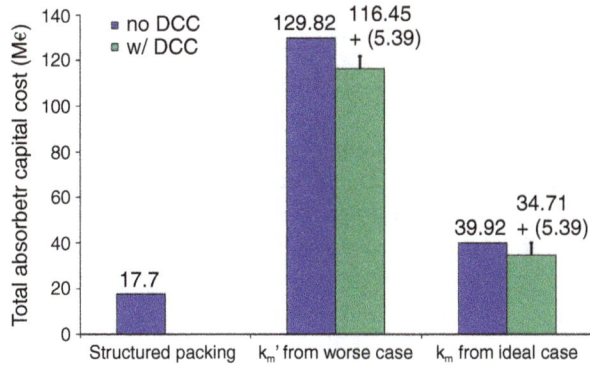

Figure 4

Comparison of total absorber capital costs for the structured packing column and each of the HFMM cases. The error bars shown for the case with the DCC indicate the cost of the DCC itself.

costs and capital costs were accounted for annually with the formula:

$$Annual\ cost = Annuity\ factor\ \times\ capital\ costs + operating\ costs$$

The total yearly costs and cost per ton $CO_2$ avoided are shown in Table 7. The results show that the process is dominated by the operating costs, set constant at 30 M€/yr (*Tab. 6*). Given this and the magnitude of the amount of $CO_2$ that would be captured each year, the cost per ton of $CO_2$ avoided does not vary greatly – an increase of only 2.05 €/ton $CO_2$ with the ideal $k_m$ value relative to the packed column. With the worse case $k_m$' value, the difference is more significant and results in a rise of 9.13 €/ton $CO_2$. With the assumption that the operating costs were equal between the packed column and HFMM systems and given the large influence of the operating costs, a more detailed analysis may result in significantly different results. However, when looking at the initial investment, the HFMM system is always much more costly. Furthermore, back calculating to determine the price of the membrane modules required to make HFMM competitive with packed columns shows a dramatic cost reduction is necessary –56% for the ideal case with the higher $k_m$ value and up to 86% for the worse case with the lower $k_m$' value, as shown in Table 8.

TABLE 7

Results of economic analysis for the structured packing case and two cases with HFMM (no DCC)

| | Structured packing | Membrane contactor (no DCC) | |
| --- | --- | --- | --- |
| | | $k_m$', worse case | $k_m$, ideal case |
| Absorber/module capital (M€) | 17.7 | 130 | 40 |
| Total capital costs (M€) | 173 | 285 | 195 |
| Total annual cost (M€/yr) | 43.24 | 52.81 | 45.62 |
| Total cost per ton $CO_2$ avoided (€/ton) | 51.62 | 60.75 | 53.67 |

TABLE 8

Required values to make HFMM cases competitive with the structured packing case

| | Membrane contactor (no DCC) | |
| --- | --- | --- |
| | $k_m$', worse case | $k_m$, ideal case |
| % Reduction in module price required (%) | 86 | 56 |
| Required specific module price (€/m²) | 2.75 | 8.85 |

TABLE 9

Results of economic analysis assuming a membrane lifetime of 5 years and a replacement cost of 50%

|  | Structured packing | $k_m$', worse case | $k_m$, ideal case |
|---|---|---|---|
| Total annual cost (M€/yr) | 43.24 | 65.79 | 49.61 |
| Additional annual cost (M€/yr) | n/a | 12.98 | 3.99 |
| Total cost per ton $CO_2$ avoided (€/ton) | 51.62 | 65.53 | 55.15 |
| Required module price for equivalent cost per ton $CO_2$ captured (€/m$^2$) | n/a | 1.23 | 3.95 |

Another key assumption of the analysis above is that no degradation of the membranes occurs. This is not very realistic, so the analysis was repeated assuming a membrane lifetime of 5 years and a replacement cost of 50% of the module cost, or 10 €/m$^2$ for the base case, to account for replacement of only the membrane bundles and not the entire module. The economic feasibility is much worse, as shown in Table 9, and the price of the modules must be reduced by 80% to 3.95 €/m$^2$ (assuming a constant replacement cost of 50%) even in the ideal case.

## CONCLUSIONS

A case study for an industrial scale $CO_2$ capture system was completed using the 2D mass transfer model described in Chabanon et al. [11]. Based on data from two experimental systems – a laboratory scale and a pilot scale – two membrane mass transfer coefficients resulted from the fit of the simulation results and the experimental data – the "ideal case" $k_m$ and the "worse case" $k_m$', respectively. Four cases were simulated for comparison utilizing the two different membrane mass transfer coefficient values ($k_m$ and $k_m$'), and two different system configurations, with and without a direct contact cooler. The results showed an increase in membrane area (and thus module volume) of about 330% for the worse case relative to the ideal case.

Further economic calculations based on the design calculations showed that with the worse case $k_m$' value, the system is exceedingly more expensive than the conventional structured packing column, and even with the ideal case $k_m$ value, requiring the same contact surface area as the packed column, the total capital investment is still more than twice as high. In this case, a minimum reduction of 56% of the cost of the membrane modules is required to be competitive. In the more realistic case,

including degradation of the membranes, the required reduction in capital cost was a minimum of 80%.

The design considerations of industrial scale-up, as well as the large differences in the studied cases suggest that an accurate overall mass transfer coefficient value is required in order to provide a more appropriate fiber length estimation. Accurately approximating the fiber length would provide a solid basis for design and industrial scale-up; however, with the current configuration, it can be concluded that scale-up to some degree should be feasible.

This preliminary study shows enormous challenges for HFMM technology to become feasible and that it may only be attractive where substantial benefits are gained by the much smaller volume and high flexibility of hollow fiber membrane module systems relative to packed columns. Several key parameters – overall mass transfer coefficient, membrane lifetime, and aspect ratio – must be carefully determined to ensure accurate calculations of the economic feasibility. Improvements should be made not only in the membrane materials, with increased transport across the membrane, but also with the module construction and ensuring a uniform distribution of the gas and liquid flows. Substantial decreases in the costs are required, especially in those of the membranes and module casings, in order for hollow fiber membrane module contactors to be competitive with structured packing absorption columns. Still, if trends such as those observed with reverse osmosis membrane-based system can be realized, including an 86% reduction in membrane cost between 1990 and 2005 [14], hollow fiber membrane modules for $CO_2$ capture may one day become competitive. Furthermore, it is important to note that any comparison between HFMM and packed columns must consider the fact that a new technology undeveloped for power plant applications is being compared with a technology that is relatively well established.

## ACKNOWLEDGMENTS

This work has been financially supported by the European Commission through the CESAR Project (Grant agreement No. 213 569) as part of the FP7 program.

## REFERENCES

1 From Copenhagen Accord to Climate Action [Online]. Available: http://www.nrdc.org/international/copenhagenaccords/. [Accessed 5 November 2012].

2 [Online]. Available: http://www.iea.org/stats/balancetable.asp?COUNTRY_CODE=29. [Accessed 5 November 2012].

3 International Energy Agency (2009) Technology Roadmap: Carbon Capture and Storage.

4 IPCC (2005) *Carbon Dioxide Capture and Storage*, Cambridge University Press, Cambridge.

5 Simons K., Nijmeijer K., Wessling M. (2009) Gas-liquid membrane contactors for $CO_2$ removal, *Journal Membrane Science* **340**, 214-220.

6 Cui Z.F., Muralidhara H.S. (2010) *Membrane Technology: A Practical Guide to Membrane Technology and Applications in Food and Bioprocessing*, Elsevier, Oxford.

7 Al-Marzouqi M., El-Naas M., Marzouk S., Abdullatif N. (2008) Modeling of chemical absorption of $CO_2$ in membrane contactors, *Separation Purification Technology* **62**, 499-506.

8 Boucif N., Favre E., Roizard D. (2008) Capture in HFMM contactor with typical amine solutions: A numerical analysis, *Chemical Engineering Science* **63**, 5375-5385.

9 Keshavarz P., Ayatollahi S., Fathikalajahi J. (2008) Mathematical modeling of gas-liquid membrane contactors using random distribution of fibers, *Journal Membrane Science* **325**, 98-108.

10 Porcheron F., Drozdz S. (2009) Hollow fiber membrane contactor transient experiments for the characterization of gas/liquid thermodynamics and mass transfer properties, *Chemical Engineering Science* **64**, 265-275.

11 Chabanon E., Kimball E., Favre E., Lorain O., Goetheer E., Ferre D., Gomez A., Broutin P. (2012) Hollow fiber membrane contactors for post-combustion $CO_2$ capture: A scale-up study from laboratory to pilot plant, *Oil Gas Science Technology*. same issue. DOI: 10.2516/ogst/2012046.

12 Freguia S., Rochelle G.T. (2003) Modeling of $CO_2$ capture by aqueous monoethanolamine, *AIChE Journal* **49**, 1676-1686.

13 European Best Practice Guidelines for Assessment of $CO_2$ Capture Technologies, *European Commission 7th Framework Program CESAR* Deliverable 2.4.3 (2011).

14 Reddy K., Ghaffour N. (2007) Overview of the cost of desalinated water and costing methodologies, *Desalination* **205**, 340-353.

# 9

# Enhanced Selectivity of the Separation of $CO_2$ from $N_2$ during Crystallization of Semi-Clathrates from Quaternary Ammonium Solutions

J.-M. Herri*, A. Bouchemoua, M. Kwaterski, P. Brântuas, A. Galfré, B. Bouillot, J. Douzet, Y. Ouabbas and A. Cameirao

Centre SPIN, Département PROPICE, UMR CNRS 5307, École Nationale Supérieure des Mines de Saint-Étienne,
158 Cours Fauriel, 42023 Saint-Étienne - France
e-mail: herri@emse.fr

* Corresponding author

**Résumé** — **Amélioration de la sélectivité du captage du $CO_2$ dans les semi-clathrates hydrates en utilisant les ammoniums quaternaires comme promoteurs thermodynamiques** — La réduction des émissions de $CO_2$ est très probablement l'un des enjeux importants de ce siècle. La capture puis le stockage géologique de ce gaz, à partir de sources industrielles ponctuelles et massives, est une voie d'importance. L'une des voies technologiques consiste à utiliser les clathrates hydrates, ou semi-clathrates hydrates, qui nécessitent de pressuriser le gaz en amont du procédé. Sous pression, l'eau et les gaz forment un solide qui encapsule préférentiellement le $CO_2$, puis le gaz peut-être ensuite récupéré sous pression après la dissociation du solide. L'abaissement de la pression opératoire est un objectif en soi afin de faire baisser les coûts opératoires. Cet abaissement peut être obtenu par l'utilisation de promoteurs thermodynamiques, dont les sels d'ammonium quaternaires constituent une famille intéressante puisqu'ils forment des solides naturellement anti-agglomérants, et plus facilement manipulables. Dans ce travail, nous présentons de nouveaux résultats expérimentaux sur les équilibres des semi-clathrates de ($CO_2$, $N_2$) en présence de Tetra-$n$-Butyl Ammonium Bromide (TBAB). Nous donnons des mesures expérimentales de pression et température en fonction de la concentration en TBAB. La pression opératoire peut être abaissée jusqu'à la pression atmosphérique. Nous donnons aussi une information supplémentaire portant sur la composition de l'hydrate. Nous observons que la sélectivité du $CO_2$ dans les semi-clathrates hydrates est bien meilleure que pour les clathrates hydrates traditionnels, sans promoteur thermodynamique.

*Abstract* — ***Enhanced Selectivity of the Separation of $CO_2$ from $N_2$ during Crystallization of Semi-Clathrates from Quaternary Ammonium Solutions*** — *$CO_2$ mitigation is crucial environmental problem and a societal challenge for this century. $CO_2$ capture and sequestration is a route to solve a part of the problem, especially for the industries in which the gases to be treated are well localized. $CO_2$ capture by using hydrate is a process in which the cost of the separation is due to compression of gases to reach the gas hydrate formation conditions. Under pressure, the water and gas form a solid that encapsulates preferentially $CO_2$. The gas hydrate formation requires high pressures and low* *temperatures, which explains the use of thermodynamic promoters to decrease the operative pressure. Quaternary ammoniums salts represent an interesting family of components because of their thermodynamic effect, but also because they can generate crystals that are easily handled. In this work, we have made experiments concerning the equilibrium of $(CO_2, N_2)$ in presence of Tetra-n-Butyl Ammonium Bromide (TBAB) which forms a semi-clathrate hydrate. We propose equilibrium data (pressure, temperature) in presence of TBAB at different concentrations and we compare them to the literature. We have also measured the composition of the hydrate phase in equilibrium with the gas phase at different $CO_2$ concentrations. We observe that the selectivity of the separation is dramatically increased in comparison to the selectivity of the pure water gas clathrate hydrate. We observe also a benefice on the operative pressure which can be dropped down to the atmospheric pressure.*

## LIST OF SYMBOLS

A, B    A particular type of semiclathrate hydrate

$C$    Langmuir constant of a guest molecule in a given cavity, $[C]$ depends on corresponding concentration/concentration dependent variable in relation to which it is defined, for example $[C_f] = Pa^{-1}$, whereas $[C_x]$ is dimensionless, or heat capacity $[C] = J.K^{-1}$

$\Delta$    Finite difference between two values of a quantity

$\Delta_\alpha^\beta$    Finite difference between two values of a quantity for a process from a given initial state $\alpha$ to a final state $\beta$

EOS    Equation of State

$f$    Fugacity, $[f] = Pa$

$G$    Growth rate, $[G] = m.s^{-1}$, or Gibbs energy, $[G] = J$

$\gamma$    Activity coefficient, dimensionless, or specific energy of surface $[\gamma] = J.m^{-2}$

$k$    Rate (kinetic) constant $[k] = s^{-1}$

$k_B$    Boltzmann's constant
$k_B = (1.380\,648\,8 \pm 0.000\,001\,3) \times 10^{-23}\,J.K^{-1}$

$k_H^\infty$    Henry's constant at saturation pressure of the pure solvent, *i.e.*, at infinite dilution of the dissolved species, $[k_H^\infty] = Pa$

$M$    Molar mass, $[M] = g.mol^{-1}$

$n$    Amount of substance, *i.e.* mole number, $[n] = mol$

R    Avogadro's number,
$N_{Av} = (6.022\,141\,29 \pm 0.000\,000\,27) \times 10^{23}\,mol^{-1}$

$\omega$    Intermolecular interaction potential, $[\omega] = J$

$v$    Stoichiometric coefficient, or number of water molecules per number of guest molecules in a cage of of a given type $I$ (hydration number), dimensionless

$P$    Pressure, $[P] = Pa$

$r$    Distance between the centre of the cavity and the guest molecule $[r] = nm$

$\rho$    (Mass) density, $[\rho] = kg.m^{-3}$

$R$    Universal gas constant, $R = (8.314472 \pm 0.000015)\,J.K^{-1}.mol^{-1}$, or radius of a cavity, assumed to be of spherical geometry, $[R] = nm$

$\sigma$    Core distance at which attraction and repulsion between a guest host-pair balance each other, $[\sigma] = pm$

$T$    Absolute temperature, $[T] = K$

TBAB    Tetra-*n*-Butyl Ammonium Bromide

$\theta$    Fraction of sites occupied (by a particular species and for a specific type of cavity as indicated by additional subscripts), dimensionless

$V$    Volume, $[V] = m^3$

$w$    Mass fraction (dimensionless)

$x$    Mole fraction of a chemical species, dimensionless; here mainly used to designate the mole fraction of guest species dissolved in the liquid phase in the immediate vicinity of the hydrate surface

$y$    Mole fraction of a chemical species, dimensionless; here mainly used to designate the mole fraction of guest species in the gas phase

$z$    Mole fraction of a chemical species, dimensionless; here mainly used to designate the mole fraction of guest species in the cavities of the clathrate hydrate structure

Z    Compressibility factor

## SUBSCRIPTS

A    Anionic species $A^{|z_A|-}$

C    Cationic species $C^{z_C+}$

cell    Refering to the cell inside the experimental reactor

eq    Referring to a state of equilibrium

$f$    Indicating reference fugacity used as concentration dependent quantity

int        Interface between the integration layer and the diffusion layer

$i$          Index identifying a particular type of cavity

$j, j'$       Index characterising chemical species or chemical component (depending on the context), or guest specie

TBAB   Refering to Tetra-$n$-Butyl Ammonium Bromide

w         Water

$x$         Indicating the reference to the mole fraction as reference frame for the composition variable

R         Reactor

0          Indicating initial conditions for temperature and pressure

## SUPERSCRIPTS

\*          Indicating the unsymmetric convention for normalisation of activity coefficients, i.e., the pure component reference frame for the solvent component and the infinitely dilution reference frame for the solute components and the solute species, respectively, or corresponding to the critical nucleï

⊖          Standard state; well defined state at reference conditions, where besides the state of aggregation or dilution, particularly the value for the pressure is fixed to its standard state value of 0.1 MPa. In this study used in the context of the standard value of the molality of $m^{\ominus} = 1\,mol.kg^{-1}$

○          Pure component state

∞          State of infinite dilution

G         Gas/vapour phase

H         Hydrate phase

L          Liquid phase

$L_w$        Liquid aqueous phase (depending on the context either an aqueous phase consisting of pure water) or a liquid mixed aqueous phase composed of an aqueous solution of a single binary electrolyte

ref        Reference state/frame in general

V         Vapour phase

## INTRODUCTION

The $CO_2$ capture by using clathrates is a concept which takes profit of the physical interaction of gas molecules with water to be adsorbed selectively in a water-based solid structure being crystallized simultaneously, here called gas hydrates, or clathrate hydrates, or semi-clathrates hydrates. The process consists in mixing water and the gas mixture containing $CO_2$ under hydrate-forming conditions of pressure and temperature. A thermodynamic promoter can be added in the system. It enters the hydrate structure and/or cavities to stabilize the hydrates. It enables the formation of hydrates at mild conditions of temperature and pressure and increases the selectivity of $CO_2$ in the hydrate phase.

The hydrate phase is made of clathrate hydrates and enclathred gas. According to the operative conditions of pressure and temperature, favored encapsulation of carbon dioxide in the hydrate phase is made while the other gases remain preferably in the gas, or liquid, phase. Pure carbon dioxide can then be recovered by the depressurisation and/or heating of the hydrate phase.

A concept of $CO_2$ capture by using clathrate hydrates has been patented by Dwain F. Spencer, firstly in 1997 and extended in 2000 (Spencer, 1997, 2000). This research (supported by the US Department of Energy) has been completed by additional works on the costing and design of single-stage or multi-stage processes for hydrate crystallisation. Deppe et al. (2002) have compared a hydrate-based technology for capturing $CO_2$ from $CO_2/H_2$ mixtures to conventional technologies (by amine and Selexol). The industrial context is a nominal 500 MW IGCC (Integrated Gasification Combined Cycle) coal gaseifier: the gas to be treated is the so-called syngas. The added capital cost for implementing the hydrate, amine and Selexol technologies is estimated to be respectively 23.9, 56.9 and 85.1 $millions, and the operational cost is respectively 8, 21 and 14 $/ton of $CO_2$. As a result, gas hydrate crystallization seems to offer the best economical potential.

An energy consumption estimation of a $CO_2$ separation from flue gases of natural gas-fired thermal power plant has also been made by Tajima et al. (2004). They showed that clathrate hydrate process consumes a considerable amount of energy, mainly due to the extremely high pressure conditions required for hydrate formation. By using thermodynamic additives, Duc et al. (2007) showed that the operative pressure can be reduced, improving the competitiveness of the process. They performed a complete sizing and costing for a hydrate-based treatment of flue gases (usually $CO_2/CO/N_2$ mixture) emitted by the steelmaking industry. The work has been done in the framework of the ULCOS European project during 6th EEC program (ULCOS, Ultra Low $CO_2$ emission for Steelmaking industry). The preliminary costing shows that the process is viable with a total cost (capital costing plus operative costing) of 22 € and 40 € per ton of captured $CO_2$, corresponding respectively to nitrogen free black furnace (with shaft injection and plasma) and conventional black furnace (top gas and flue gas). The main part of the cost is due to compressors.

The concept of $CO_2$ capture by using hydrates implies the crystallization and the handling of slurries. From 15 years, the handling of high concentrated slurries has also questioned the community of refrigeration because of the application to thermal storage and transportation, by using Phase Change Materials (PCM). For example, PCM based on ice slurries has been developed by several industries because of their potential in refrigeration (negative temperature). To that aim, the fluid is composed of water that can crystallize to form ice and the melting point is adjusted by adding appropriate additives, such as alcohols. Different kinds of ice slurry generators have been developed (Meunier et al., 2007). The more accessible technology is the scraped surface heat exchanger patented by Sunwell system (Gibert, 2006) or the Heatcraft generator patente by *Lennox* company (Compingt et al., 2009).

However, for applications above 273.15 K, i.e. air conditioning systems, the use of ice slurry is no longer applicable because the ice melting point cannot be adjusted to positive temperature. Several authors (Lipkowski et al., 2002; Obata et al., 2003; Oyama et al., 2005) have shown that some kind of quaternary ammonium can be used as PCM for air conditioning applications. For example, Tetra-Butyl Ammonium Bromide (TBAB) forms hydrate slurries very similar to ice slurries, but with a melting point that can be adjusted from 273.15 K to 285.15 K depending on the TBAB concentration (*Fig. 1*). A previous study has demonstrated the possibility to handle such suspensions (Darbouret, 2005) up to solid concentration of 30% vol.

The *JFE Engineering Corporation* has designed and commercialized a technology which uses TBAB hydrates slurry also (Takao et al., 2001, 2002, 2004; Mizukami, 2010; Ogoshi et al., 2010). Their first prototype was built in 2005 and today, approximately 10 systems have been sold and have shown the feasibility of this process to cooling power up of 2 MW. In general, this company transforms an old system of air conditioning which is using cold water. Water circulating in the secondary loop is replaced by a TBAB slurry.

So, there is a clear convergence on the technologies that can be used to capture $CO_2$, or to store energy. For low scale applications (air conditioning), the technology begins to be mature. However, for $CO_2$ capture, at a bigger scale, there is no prototype to validate the concept. Only laboratory pilot scale units exist: bubble reactors for $CO_2/N_2$ separation (Douzet et al., 2011, 2013) or $CO_2/H_2$; (Xu et al., 2012), combination of a bubble and stirred reactor (Linga et al., 2010), or spray reactor (Brinchi et al., 2011) for $CO_2/CH_4$ separation.

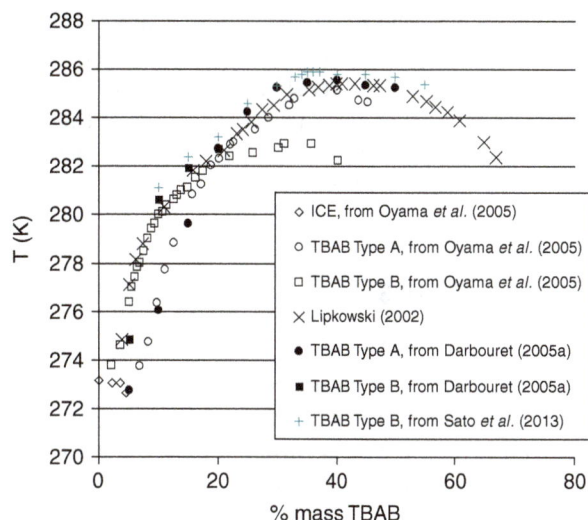

Figure 1

*T,x*– phase diagram of TBAB-water binary system in the region of crystallization of clathrates polyhydrates (Oyama et al., 2005; Lipkowski et al., 2002; Darbouret, 2005).

Also, and because the operative cost is mainly driven by the operative pressure, studies are carrying on to understand the benefices of additives which can deplete the pressure (thermodynamic promotors) and/or orientate the selectivity (kinetic promotors).

## 1 STATE OF THE ART

The clathrates hydrates, and semi-clathrate hydrates, are ice-like compounds in the sense that they correspond to a re-organisation of the water molecules to form a solid. The crystallographic structure is based on H-bonds. The clathrates of water are also designated improperly as "porous ice" because the water molecules build a solid network of cavities in which gases, volatile liquids or other small molecules could be captured.

The clathrates hydrates of gases, called gas hydrates, have been studied intensively due to their occurrence in deep sea pipelines where they cause serious problems of flow assurance.

Each structure is a combination of different types of polyhedra sharing faces between them. Jeffrey (1984) suggested the nomenclature $e^f$ to describe each polyhedra: $e$ is the number of edges of the face, and $f$ is the number of faces with $e$ edges.

The clathrates of gas can be stabilized by thermodynamic promotors. Two classes of thermodynamic promoters can be distinguished. The first ones are species which are enclathred in the cavities without modifying the structure of the hydrates. This kind of hydrate is still called clathrate hydrate and its structure belongs to the classical sI, sII or sH structure. The second class of thermodynamic promoters modifies the structure. The new structure is called semi-clathrate hydrates.

## 2 CLATHRATE HYDRATES

The classical sI, sII or sH structure (*Tab. 1*) can be stabilized by the presence of promoters. Two promoters, tetrahydrofuran and cyclopentane, have been well described in the literature for their remarkable properties to drop the operative pressure to form gas hydrates.

## 2.1 Tetrahydrofuran: An Example of Water Soluble Additive

TetraHydroFuran (THF) forms structure II hydrates in which THF occupies the large cavity $5^{12}6^4$ and gas competes with THF for the occupation of the large cavity and/or occupies the small cavity. THF is a water-soluble additive. With water, they are completely miscible in the liquid state over the whole composition range in the pressure and temperature domain of hydrate formation (Riesco and Trusler, 2005). The equilibrium pressure reduction effect to form hydrates is dependent on the relative concentration of THF. It is important to notice that in presence of pressurized carbon dioxide (temperature above 290 K and a pressure above 2.0 MPa), Sabil *et al.* (2010) observed that water and THF phase split in two liquids phases (in presence of a solution of 5% mole of THF in water). Over the years, many authors reported hydrate dissociation data of the system

{THF + water + gas}. Only works where nitrogen and carbon dioxide were used are presented in the following paragraph and summed up in Table 2.

### 2.1.1 Benefits of THF on the Equilibrium of Pure Gases

Seo *et al.* (2001, 2008) studied the hydrate dissociation pressure – temperature data {$H_2O$ + THF + pure gas ($CO_2$ or $N_2$)} for several compositions of THF (1 to 5%mol of THF in water) and the hydrate dissociation pressure – temperature data {$H_2O$ + THF + $CH_4$} for 3%mol of THF.

Firstly, their studies focused on showing the stabilization effect of THF on hydrate formation compared to the effect of others water miscible promoters (propylene oxide, 1,4-dioxane, acetone). For a concentration of promoter of 3%mol, THF was found to be the most interesting for each gas ($CO_2$, $N_2$ and $CH_4$).

Secondly, the effect of THF concentration from 1 to 5%mol on the equilibrium hydrate dissociation pressure-temperature data has been studied. The pressure decreased rapidly up to 1%mol of THF, but very slowly above 1%mol for each case study ($CO_2$ and $N_2$).

Delahaye *et al.* (2006) also investigated the hydrate dissociation pressure – temperature data {$H_2O$ + THF + $CO_2$} for THF concentration in the range of 1.6 to 3.0%mol, and the latent heat of dissociation of the mixed hydrates ($CO_2$/THF/water). Like Seo *et al.* (2008), the authors observed the drastic reduction of the hydrate pressure formation with a few mole of THF in the system.

Sabil *et al.* (2010) also examined the complete hydrates dissociation lines for the system {$H_2O$ + THF + $CO_2$} at a THF concentration of 5%mol for different carbon dioxide concentrations.

TABLE 1

Structure of gas hydrates (Sloan and Koh, 2008)

| | SI | | SII | | SH | | |
|---|---|---|---|---|---|---|---|
| | | | | | | | |
| Cavity (*) | $5^{12}$ | $5^{12}6^2$ | $5^{12}$ | $5^{12}6^4$ | $5^{12}$ | $4^3 5^6 6^3$ | $5^{12}6^8$ |
| Type of cavity | 1 | 2 | 1 | 3 | 1 | 5 | 4 |
| Number of cavity | 2 | 6 | 16 | 8 | 3 | 2 | 1 |
| Average cavity radius (nm) | 0.395 | 0.433 | 0.391 | 0.473 | 0.391 | 0.406 | 0.571 |
| Variation in radius (%) | 3.4 | 14.4 | 5.5 | 1.73 | | | |

TABLE 2

Experimental studies on gas hydrate in presence of tetrahydrofuran

| Authors | Gas(es) | Promoters | Pressure and temperature area |
|---|---|---|---|
| Seo et al. (2001) | $N_2$, $CH_4$ | THF (3%mol) > acetone (3%mol) > 1,4-dioxane (3%mol) > propylene oxide (3%mol)<br><br>THF (5%mol) > THF (3%mol) > THF (1%mol) | For $N_2$/THF/water system:<br>$T$ = [280.85-293.75] K<br>$P$ = [3.12-10.872] MPa<br><br>For $CH_4$/THF (3%mol)/water system:<br>$T$ = [292.16-302.46] K<br>$P$ = [2.02-8.91] MPa |
| Delahaye et al. (2006) | $CO_2$ | THF (1.6-3%mol) | $P$ = [0.2-3.5] MPa<br>$T$ = [286-291] K |
| Seo et al. (2008) | $CO_2$ | THF (3%mol) > propylene oxide (3%mol) > > 1,4 dioxane<br><br>THF (5%mol) > THF (3%mol) > THF (1%mol) | For $CO_2$/THF/water system:<br>$T$ = [270-290] K<br>$P$ = [0.2-4.7] MPa |
| Sabil et al. (2010) | $CO_2$ | THF (5%mol) | For $CO_2$/water system:<br>$T$ = [275.12-290] K<br>$P$ = [1.51-5.28] MPa<br><br>For $CO_2$/THF/water system and depending on $CO_2$ concentration:<br>$T$ = [285.39-301.46] K<br>$P$ = [0.9-7.05] MPa |
| Yang et al. (2011) | $N_2$, $O_2$, air | THF (5-5.13%mol) | For $N_2$/THF/water system:<br>$T$ = [295.75 -303.60] K<br>$P$ = [10.95- 29.527] MPa<br><br>For $O_2$/THF/water system:<br>$T$ = [286.10 -303.63] K<br>$P$ = [2.130-25.812] MPa<br><br>For air/THF/water system:<br>$T$ = [281.84 -299.41] K<br>$P$ = [0.981-16.685] MPa |
| Kang and Lee (2000)<br>Kang et al. (2001) | $CO2$/$N_2$ | THF (1, 3%mol) | For $CO_2$/$N_2$/water system:<br>$T$ = [274, 277, 280] K<br>$P$ = [1.394-32.308] MPa<br><br>For $CO_2$/$N_2$/THF/water system:<br>$T$ = [274.85-295.45] K<br>$P$ = [0.2-12.905] MPa |
| Linga et al. (2007a,b) | (16.9%/83.1%) $CO_2$/$N_2$ | THF (1%mol) | For $CO_2$/$N_2$/water system ($y CO_2$, initial mole fraction = 0.169):<br>$T$ = 273.7 K<br>$P$ = [9, 10, 11] MPa<br><br>For $CO_2$/$N_2$/THF/water system:<br>$T$ = 273.7 K<br>$P$ = 1.5 MPa |
| Linga et al. (2010) | (16.9%/83.1%) $CO_2$/$N_2$ | THF (1%mol) | $P$ = [1.5, 2.3, 2.5] MPa<br>$T$ = 273.7 K |
| Herslund et al. (2013) | $CO_2$/$N_2$ | THF 5%mol | $T$ = 283.3 to 285.5 K |

The authors discovered a four-phase equilibrium region with three fluid phases at temperatures above 290 K and pressures above 2.0 MPa: a water-rich with a constant amount of carbon dioxide, a carbon dioxide in an organic-rich, the vapor phase and the hydrate phase.

Finally, Yang *et al.* (2011) studied the hydrate dissociation pressure – temperature data {$H_2O$ + THF + pure gas ($N_2$)} for THF concentration of 5%mol. The authors also provided equilibrium phase hydrate data for {$O_2$ + THF + water}, {air + THF + water}. Their research focused on finding valuable information on the air separation by hydrate crystallization.

### 2.1.2 Benefits of THF on Gas Mixtures

Recovering $CO_2$ from a gas mixture of $CO_2/N_2$ in presence of THF was the purpose of the works of (Kang and Lee, 2000; Kang *et al.*, 2001; Linga *et al.*, 2007a,b). The authors measured the hydrate dissociation data for several mixtures of $CO_2$ and $N_2$ without any promoter and in presence of THF. As for pure gases, with aqueous solutions containing 1 and 3%mol of THF, a drastic drop of equilibrium dissociation pressure has been observed from the moment the THF concentration reached 1 mole percent (Kang and Lee, 2000; Kang *et al.*, 2001). Moreover, a benefit of THF is observed on the selectivity of the separation. Pressure-composition diagrams of the {$H_2O$ + THF + $CO_2/N_2$} system for 1 mole percent of THF and the {$H_2O$ + $CO_2/N_2$} system have been drawn at three temperatures (*Tab. 2*) for several compositions of the gas mixture $CO_2/N_2$. The respective hydrate compositions of the mixed hydrates with and without THF have been added on the diagrams. Kang and Lee (2000) showed that the $CO_2$ selectivity in the mixed hydrate phase has been lowered in the mixed hydrate when the THF has been used as a hydrate promoter.

Linga *et al.* (2007a,b) tested a THF concentration of 1%mol and studied the gas uptake and the rate of the crystallization. THF reduced the induction time but also the growth rate of the crystallization.

The authors did not estimate the carbon dioxide selectivity and did not calculate the gas storage capacity.

Finally, Linga *et al.* (2010) investigated the formation of gas hydrate with one composition of gas mixture at THF concentrations of 1 and 1.5%mol. The authors estimated the gas uptake and the $CO_2$ recovery and compared them to the results reported in the literature with Tetra-*n*-Butyl Ammonium Bromide and Tetra-*n*-Butyl Ammonium Fluoride (Fan *et al.*, 2001; Li S. *et al.*, 2009).

### 2.2 Cyclopentane: An Immiscible Organic Additive

CycloPentane (CP) is described in the literature as an excellent thermodynamic promoter. It forms structure II hydrate without any gas (Nakajima *et al.*, 2008) but competes with $CO_2$ to occupy the large cavity $5^{12}6^4$. Cyclopentane is a hydrophobic compound and needs to be dispersed in water. So, the main difference with THF is the low solubility of CP in water.

Fan *et al.* (2001) were the first to report the quadruple equilibrium point (CP hydrate-liquid water-organic liquid-vapor) at a temperature of 280.22 K and a pressure of 0.0198 MPa (abs).

As for THF, many authors reported the hydrate dissociation data of the system (THF + water + pure $CO_2$, pure $N_2$, or $CO_2$ + $N_2$): their works are summed up in Table 3 and detailed in the following paragraph.

Zhang and Lee (2009a) and Zhang *et al.* (2009) determined hydrate dissociation data for {$H_2O$ + ($CO_2$ + CP)} system. Dissociation conditions for {$H_2O$ + ($CO_2$ + CP)} hydrate have been compared with the dissociation data for {$H_2O$ + ($CO_2$ + TBAB)} with a TBAB weight fraction of 0.427 (Arjmandi *et al.*, 2007) and for {$H_2O$ + ($CO_2$ + THF)} with a THF molar fraction of 3 percent (Delahaye *et al.*, 2006). CP appears to be a better additive than TBAB to decrease the equilibrium pressure.

Zhang and Lee (2009b) also studied the potential of using CP as a kinetic promoter in a static autoclave at a low temperature.

Mohammadi and Richon (2010) compared the stabilization effect of CP on hydrate formation of {$H_2O$ + ($CO_2$ + CP)} gas hydrate to the effect of several organic promoters (example: methyl-cyclopentane, methyl-cyclohexane, and cyclohexane). For volume fraction of promoter of 10% in water, among the promoters, CP promotion effect was the highest. Experimental dissociation data for clathrate hydrates of cyclopentane and carbon dioxide have been also reported. Data were in good agreement with the study of Zhang and Lee (2009a) and Zhang *et al.* (2009).

Experimental hydrate dissociation data for {$H_2O$ + ($N_2$ + CP)} system were first published by Tohidi *et al.* (1997). They tested the potential of using cyclopentane to decrease the equilibrium pressure when gas hydrates of nitrogen and promoter (cyclopentane or neopentane) are formed. CP promotion effect has also been compared in that case to several organic promoters of the literature (cyclohexane and benzene). It was found to be the strongest promoter, just above neopentane.

Mohammandi and Richon (2011) and Du *et al.* (2010) completed the experimental hydrate dissociation data for {$H_2O$ + ($N_2$ + CP)} system and there results were in good agreement.

For $CO_2/N_2$ gas mixture, Li S. *et al.* (2010) preliminary showed that gas hydrates could be enriched in $CO_2$ in presence of cyclopentane. The authors studied two different situations: CP dispersed in an emulsion or a buoyant CP phase on the top of the liquid water phase. Differences in the separation efficiency have been

TABLE 3

Experimental studies on gas hydrate in presence of cyclopentane

| Authors | Gas(es) | Promoters | Pressure and temperature area |
|---|---|---|---|
| Tohidi *et al.* (1997) | $N_2$ | Promoter volume fraction: no indicated<br><br>Cyclopentane > neopentane > cyclohexane > benzene | For $N_2$/CP/water system:<br>$T$ = [282.9 -289.1] K<br>$P$ = [0.641-3.496] MPa |
| Mohammadi and Richon (2009, 2010, 2011) | $CO_2$, $N_2$ | For $CO_2$ (promoter volume fraction in water: 10%):<br><br>Cyclopentane > > cyclohexane > methylcyclohexane, isopentane, methylcyclopentane<br><br>Cycloheptane, cyclooctane = no promotion effect<br><br>For $N_2$ (promoter volume fraction in water: 10%):<br><br>Cyclopentane > > cyclohexane > methylcyclohexane = methylcyclopentane | For $CO_2$/CP/water system:<br>$T$ = [284.3-291.8] K<br>$P$ = [1.82-2.52] MPa<br><br>For $N_2$/CP/water system:<br>$T$ = [281.7-290.2] MPa<br>$P$ = [0.25-4.06] MPa |
| Zhang and Lee (2009a)<br>Zhang *et al.* (2009) | $CO_2$ | Cyclopentane (1-60%vol. in water) | $P$ = [0.89-3.51] MPa<br>$T$ = [286.65-292.61] K |
| Zhang and Lee (2009) | $CO_2$ | Cyclopentane (1-10%vol. in water) | $P$ = [1.90-3.42] MPa<br>$T$ = [273.41-273.76] K |
| Du *et al.* (2010) | $N_2$ | Cyclopentane (30.5%vol. in water) | $P$ = [2.27-30.40] MPa<br>$T$ = [281.3-303.1] K |
| Li S. *et al.* (2010) | (16.6%/83.4%) $CO_2/N_2$ | Cyclopentane (20% weight fraction of water) | $P$ = [2.49-3.95] MPa<br>$T$ = 281.25 K |
| Herslund *et al.* (2013) | $CO_2/N_2$ | Cyclopentane | $T$ = [275.25-285.2] K |

reported. This difference implies that the authors have not reached the thermodynamic equilibrium in their work. In fact, we will comment this point in our discussion, but it appears that the formed hydrate from a gas mixture are not at equilibrium, and that its composition is directly dependent on kinetic considerations, and so is indirectly dependent on the geometry of the system.

## 3 SEMI-CLATHRATE HYDRATES

Another class of clathrates, called semi-clathrates, can be formed in presence of electrolytes, such as alkyls Ammonium salts (this work) or alkyls Phosphonium salts (Sato *et al.*, 2013). It forms, in presence of water, and without any gas, a semi-clathrates hydrate crystal, even at atmospheric pressure (McMullan and Jeffrey, 1959). They are qualified as peralkylonium polyhydrates and have been the research project of the Russian team of Dyadin *et al.* over decades and were only published in english in 1984, 1985 and 1995 (Dyadin and Udachin, 1984, 1987; Dyadin *et al.* 1995). In contrast to gas hydrates, the cation is the guest situated in the frame-

work cavities and separated from the host- molecules by the distance not less than the sum of the van der Waals radii (hydrophobic inclusion). "*A simple anion of a halogenide type displaces the water molecule in the framework, forming H-bonds together with the neighbouring molecules (hydrophilic inclusion), causing the framework to become of water-anion type. The anion with a hydrophobic part includes in a hydrophilic way with the polar group, forming the framework knot or edge, the hydrocarbon part being situated in one of the framework cages*" (Dyadin and Udachin, 1984).

So, they are called semi-clathrates due to the fact that the crystalline water network is broken in order to incorporate the cation of the compound. For instance, in the case of TBAB hydrate, the nitrogen atom at the center of the four butyl radicals takes the place of a water molecule effectively, "breaking" the four surrounding cages and creating a larger cavity made from smaller ones. Therefore the hydration number will change regarding gas hydrates because less water molecules in a similar structure, due to their replacement by the cation of the semi-clathrate, will be present. Bromide atoms and water molecules form

TABLE 4

Anions of semi-clathrates (Dyadin and Udachin, 1987; Dyadin et al., 1995)

| Anion | Examples |
|---|---|
| Halogens | $Cl^-$, $Br^-$ |
| Hydroxide | $OH^-$ |
| Carboxylates | $HCO_2^-$, $C_3H_7CO_2^-$ |
| Branched carboxylates | iso-$C_3H_7CO_2^-$, meta-$ClC_4H_4CO_2^-$ |
| Linear dicarboxylates | $CH_2(CO_2^-)_2$, $(CH_2)_3(CO_2^-)_2$ |
| Branched dicarboxylates | iso-$C_3H_7(CO_2^-)_2$ |
| Aromatic dicarboxylates | o-$C_6H_4C_2O_4^{2-}$ |
| Oxalates | $C_2O_4^{2-}$, $(CH_2)_3C_2O_4^{2-}$ |
| Oxides | $O^-$ |
| Phosphates | $HPO_4^{2-}$, $PO_4^{3-}$ |
| Chromates | $CrO_4^{2-}$ |
| Nitrates | $NO_3^-$ |
| Bicarbonates | $HCO_3^-$, $CO_3^{2-}$ |
| Chlorates | $ClO_3^-$ |
| Alkane sulfonates | $HSO_3^-$ |
| Tungstates | $WO_4^{2-}$ |

TABLE 5

Cations of semi-clathrates (Dyadin and Udachin, 1987; Dyadin et al., 1995)

| Cation | Examples |
|---|---|
| Linear alkyls ammoniums | $(C_4H_9)_4N^+$, $(C_3H_7)_4N^+$ |
| Branched alkyl ammoniums | (iso-Am)$_4N^+$ |
| Linear alkyl phosphoniums | $(C_4H_9)_4P^+$ |
| Branched alkyl ammoniums | (iso-Am)$_4P^+$ |
| Linear alkyl phosphines | $(C_4H_9)_3P^+$ |
| Linear alkyl ternary ammoniums | $(C_4H_9)_3N^+$ |
| Linear alkyls arsines | $(C_4H_9)_3As^+$ |
| Linear amine phosphoniums | $(C_4H_9)_3(C_3H_5NCH_2)P^+$ |
| Linear alkyl diammoniums | $(CH_2)[(C_4H_9)_4N]^{2+}$ |

the cage structure. Tetra-n-Butyl Ammonium is located at the centre of four cages (Shimada et al., 2005b) and the butyl groups occupy the cavities. There is a large number of semi-clathrate compounds due to the different cation and anion they possess. Tables 4 and 5 shows the anions and cations that can be found in a semiclathrate. Furthermore, it is possible that some semi-clathrates have different cations.

With a wide range of anions and cations that can compose a semi-clathrate, it is not a surprise that many different crystalline structures can be formed. A detailed review has been made by Dyadin and Udachin in 1987 which continues to be the reference on the subject (Dyadin and Udachin, 1987).

TBAB has been extensively studied in reason of a direct application for air conditioning as a phase change material (Lipkowski et al., 2002; Oyama et al., 2005; Kamata et al., 2004; Darbouret et al., 2005; Lin et al., 2008; Arjmandi et al., 2007; Li S. et al., 2009). Tetra butyl ammonium bromide salt forms at least four different structures with hydration number of 24, 26, 32 and 36 (Lipkowski et al., 2002). Only one of the crystallographic

structures of semi-clathrate hydrates has been determined precisely by McMullan and Jeffrey (1959) and completed by Shimada (2005b).

Oyama et al. (2005) reported some thermal properties of TBAB semi-clathrates. They determined the phase diagram of semi-clathrates hydrates under the conditions of atmospheric pressure, and also measured latent and specific heats capacity. From the phase diagram, the congruent melting points of two different TBAB semi-clathrates structures were determined (285.15 K and 282.55 K for semi-clathrates with respectively a hydration numbers of 26 and 38). It is interesting to notice that Dyadin et al. (1995) did not find the structure with 38 as hydration number. It shows again that semiclathrate formation is a difficult process to predict. Figure 1 plots the data of Oyama et al. (2005) competed with some data from literature. A complete reviewing off all the equilibrium data can be in Sato et al. (2013).

Davidson (1973) suggested that these semi-clathrates crystals do not encage gas molecules, but Shimada et al. (2005a) and Duc et al. (2007) have given opposite results or opinions. Shimada et al. (2003, 2005a) has supposed that TBAB semi-clathrates could encage small gas molecules. In fact, based on the structure analyze of pure TBAB semi-clathrates (formula $(n-C_4H_9)_4N^+Br$, $38H_2O$), they supposed that the structure could encapsulate the gas in the free cavities (Fig. 2). Duc et al. (2007) have experimentally confirmed that semi-clathrates can encapsulate up to 40 $M^3$ STP of gas per cubic meter of solid.

Since this period, a huge quantity of experimental results has been produced which gives evidence of the

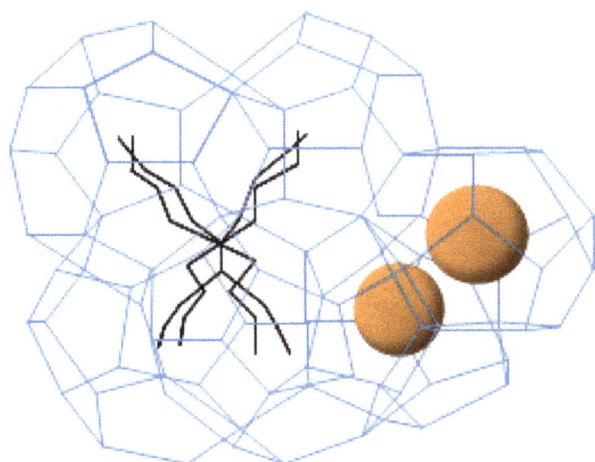

Figure 2

Structure of gas TBAB semi-clathrates (Shimada *et al.*, 2005b).

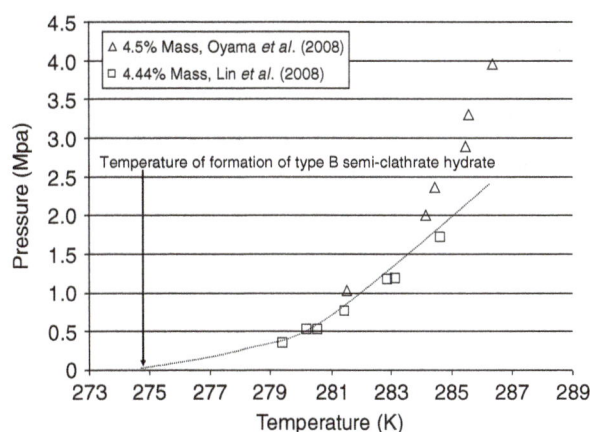

Figure 3

CO$_2$-TBAB semi-clathrate equilibrium curve at TBAB mass fraction of 4.44-4.5%, from Oyama *et al.* (2008) and Lin *et al.* (2008).

enclathration of gases in the TBAB structure. The question that remains is which structure is formed. In fact, facing the variety of possibilities, and except the work of Shimada *et al.* (2005b) who clearly identified one structure in presence of gas, we do not have evidence of the crystallographic structure formed.

### 3.1 (Pressure-Temperature) Equilibrium with TBAB, CO$_2$ and N$_2$

In the recent years, many experimental data have been produced about the equilibrium of TBAB semi-clathrate in presence of pure CO$_2$ and pure N$_2$ gas. Also, a model has been produced by Paricaud (2011) to predict the pressure and temperature of equilibrium as a function of the TBAB mass fraction for only one type of hydrate structure. This model will not be described in detail in this publication, but some of its fundamental equations will be used to explain the observations of the experimental results.

Arjmandi *et al.* (2007) have given CO$_2$-TBAB semi-clathrate equilibrium data at TBAB mass fraction of 0.1 and 0.4. Duc *et al.* (2007) have determined few points at TBAB mass fraction of 0.05, 0.09, 0.5 and 0.65. Lin *et al.* (2008) did a complete study for TBAB mass fraction of 0.044, 0.07 and 0.05. Oyama *et al.* (2008) gave equilibrium data of CO$_2$ semi-clathrate at TBAB mass fraction of 0.1. But also, they produced data at very low mass fraction of 0.01, 0.02, 0.03 and 0.045 without formation of the semi-clathrate hydrate but only formation of the clathrate hydrate. Lastly, Deschamps and

Dalmazzone (2009) have given few points at TBAB mass fraction of 0.4.

Experiments about N$_2$-TBAB semi-clathrate equilibrium are less numerous. Arjmandi *et al.* (2007) have given N$_2$-TBAB semi-clathrate equilibrium data at TBAB mass fraction of 0.1. Duc *et al.* (2007) have determined few points at TBAB mass fraction of 0.05, 0.09, 0.5 and 0.65. Deschamps and Dalmazzone (2009) have given few points at TBAB mass fraction of 0.4.

Figures 3, 4 and 5 show some of the experimental data for CO$_2$-TBAB semi-clathrate. All the data are not coherent. In Figure 3, for two similar TBAB mass fractions, we can distinguish two different equilibrium curves (Oyama *et al.* in 2008, and Lin *et al.* in 2008). At TBAB mass fraction of 9-10 percent (*Fig. 4*), the equilibrium data of Arjmandi *et al.* (2007), Duc *et al.* (2007), Oyama *et al.* (2008) and Lin *et al.* (2008) do not seem in coherence together. But, at TBAB mass fraction of 0.4 (*Fig. 5*), the data from Arjmandi *et al.* (2007), Duc *et al.* (2007) and Deschamps and Dalmazzone (2009) appear to be coherent together. The difference between the authors can be explained by the difficulty to crystallize the same structure. In fact, as we underlined before, the TBAB semi-clathrate can crystallize under 4 structures, and at least two of them can capture gas (Paricaud, 2011).

Independently of the difference between the experiments, it can be stated that the equilibrium pressure decreases as the temperature decreases. The different authors reported equilibrium pressure down to 0.5 MPa, but never close to the atmospheric pressure.

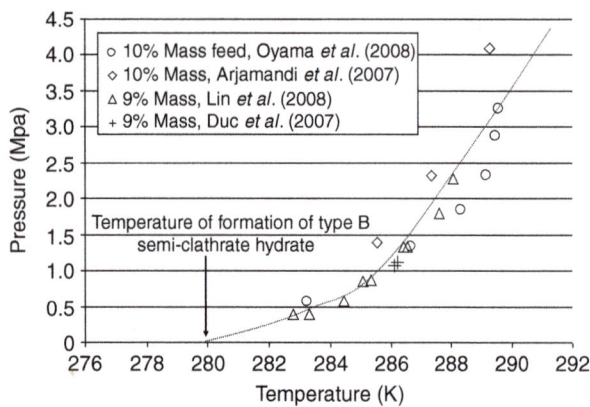

Figure 4

CO$_2$-TBAB semi-clathrate equilibrium curve at TBAB mass fraction of 9-10 percent, from Arjmandi et al. (2007), Duc et al. (2007), Oyama et al. (2008) and Lin et al. (2008).

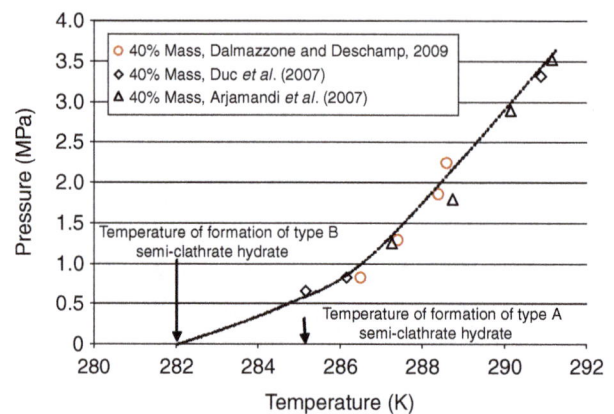

Figure 5

CO$_2$-TBAB semi-clathrate equilibrium curve at TBAB mass fraction of 40 percent, from Arjmandi et al. (2007), Duc et al. (2007), Deschamps and Dalmazzone (2009).

In fact, at low pressure, the kinetics slow down and the crystallization time becomes too long to be observed. But, from a theoretical point of view, we will demonstrate that the equilibrium pressure tends to 0 as the temperature tends to the temperature of equilibrium of the pure TBAB clathrate hydrate.

For example, at TBAB mass fraction of 4.5 percent, Figure 1 shows that type B semi-clathrate hydrate is stable at temperature down to 274-275 K. In Figure 3, this temperature can be understood as the limit temperature at which the extrapolated equilibrium pressure becomes zero. At TBAB mass fraction of 10 percent, Figure 1 shows that pure type A semi-clathrate hydrate (without gas) is stable at temperature down to 276-277 K, and that pure type B semi-clathrate hydrate (without gas) is stable at temperature down to 279.5-280.5 K. In Figure 4, we can observe that the temperature of 280 K is the limit temperature at which the extrapolated equilibrium pressure becomes zero. The same analysis can be done in Figure 5 by considering that the Type B still continues to be the crystallized structure in solution at TBAB mass fraction of 0.4. From Figure 1, we can observe that pure type A semi-clathrate hydrate (without gas) is stable at a temperature down to 285 K, and that pure type B semi-clathrate hydrate (without gas) is stable at a temperature down to 280 K. Figure 5 shows that the value of 280 K seems to be the limit temperature at which the extrapolated equilibrium pressure becomes zero.

Remark: Mixed promoter systems can also be an alternative system to enhance the carbon dioxide selectivity and to drastically decrease the operative pressure. In presence of carbon dioxide/hydrogen (38.6/61.4 mole percent mixture), Li X.-S. et al. (2011, 2012) tested hydrate formation with the simultaneous presence of TBAB and cyclopentane. They compared the gas uptake, the CO$_2$ selectivity in the hydrate phase to the single promoter systems ({CP + gas + water} and {TBAB + gas + water}). The system with the CP/TBAB solution volume ratio of 5 percent and a concentration of 0.29 mole percent of TBAB was the optimum to obtain the largest gas uptake and the highest CO$_2$ selectivity at 274.65 K and 4.0 MPa. The selectivity of CO$_2$ over hydrogen was 91.6 mole percent in the mixed hydrate phase when the carbon dioxide concentration in the residual gas was 13.5 mole percent. In comparison, in presence of the TBAB/(CO$_2$ + H$_2$)/water system, the selectivity of CO$_2$ was 93.6 mole percent when the carbon dioxide concentration in the residual gas was 18 mole percent in the same conditions of temperature and pressure. The authors observed that a synergetic effect may occur: cyclopentane does not only form sII hydrates but also takes part in the semi-clathrate hydrate structure (Li X.-S. et al., 2012).

## 4 EXPERIMENTAL SET-UP

An experimental apparatus (Fig. 6) has been built to investigate the gas hydrates formation and dissociation with and without the presence of TBAB. The

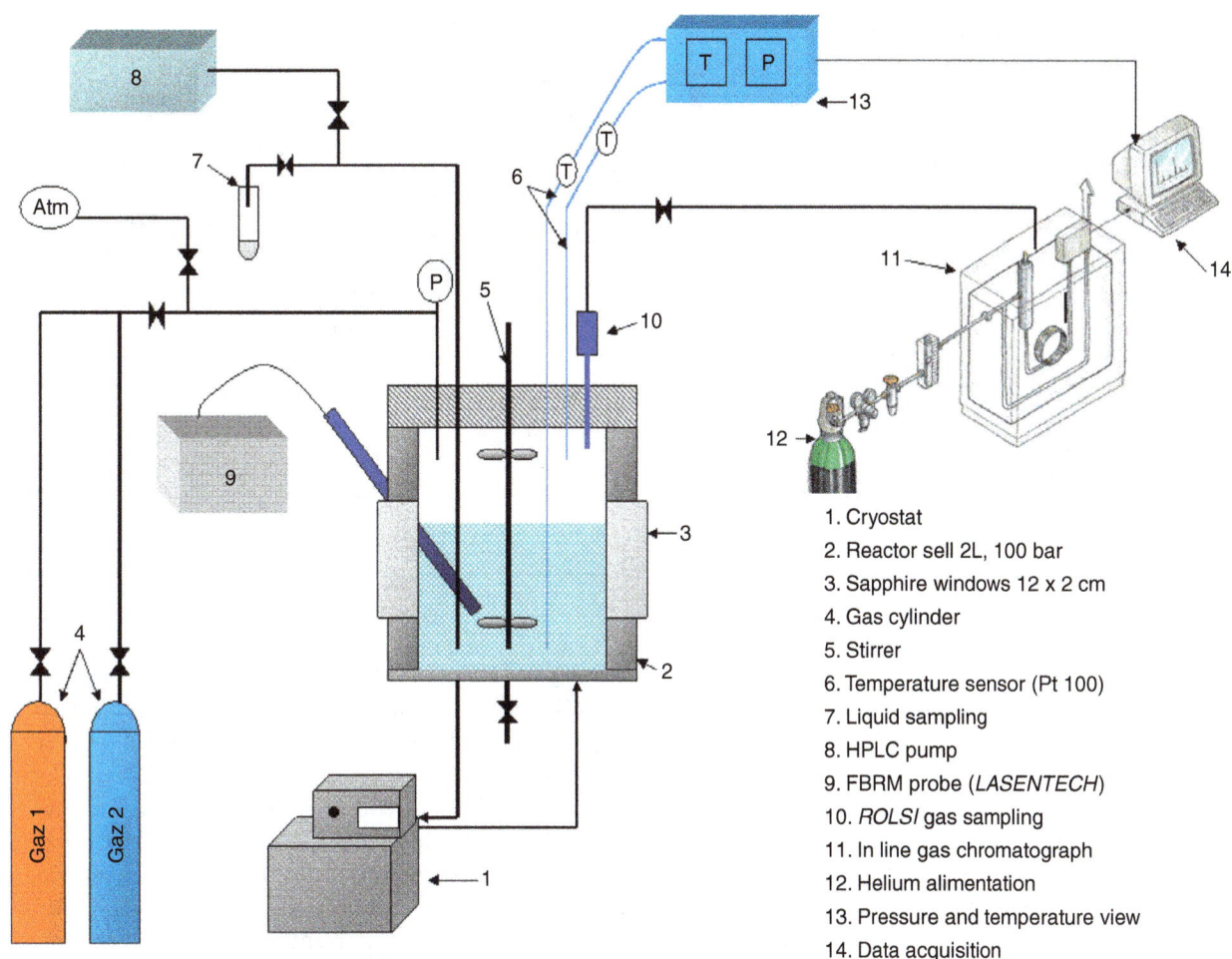

Figure 6

Experimental set-up.

1. Cryostat
2. Reactor sell 2L, 100 bar
3. Sapphire windows 12 x 2 cm
4. Gas cylinder
5. Stirrer
6. Temperature sensor (Pt 100)
7. Liquid sampling
8. HPLC pump
9. FBRM probe (*LASENTECH*)
10. *ROLSI* gas sampling
11. In line gas chromatograph
12. Helium alimentation
13. Pressure and temperature view
14. Data acquisition

thermodynamic equilibrium conditions have been obtained by determining the pressure, the temperature and the compositions of all gas, liquid and hydrate phases. The experimental set-up is a stainless steel high pressure batch reactor (autoclave) with a measured volume of $V_R = 2.51 \pm 0.03$ dm$^3$. It is surrounded with a double jacket connected to an externally cooler (*Huber* CC-505) with a controller CC3 of 0.02 K precision. Two polycarbonate windows (12 $\times$ 2 cm) are mounted on both sides of the reactor, allowing the visual observation of the hydrates. A *Pyrex* cell of volume $V_{cell}$ is located in the stainless steel autoclave in which the pressure can be raised up to 10 MPa. The *Pyrex* cylinder is filled with a volume of $V^{liq,0} = 1$ dm$^3$ of water containing TBAB at concentration between 5 wt% and 40 wt%. The liquid is injected in the pressurized reactor by using a HPLC (High Performance Liquid Chromatography)

pump (*JASCO*-PU-1587). The initial gas mixture is prepared by injecting each of the two gases directly into the reactor and is performed separately in two successive steps by means of two different valves. The composition of the gas phase is determined on-line by using a gas chromatograph, after sampling, by a *ROLSI* instrument. This device collects a small gas sample of a defined volume, which is then directly injected into the loop of the gas chromatograph. The volume of the sample is in between 1 μm$^3$ and 5 μm$^3$ and is considered as negligibly small relative to the total volume of the gas phase in the reactor which is of the order of magnitude of 1.5 dm$^3$.

A four vertical-blade turbine impeller ensures stirring of the suspension during crystallization. Temperature is monitored by two Pt100 probes in the bulk and the gas phases. Pressure is measured by mean of a pressure transducer (range: 0-10 MPa) with a precision of

$\pm 0.05$ MPa. The data acquisition unit ($P$, $T$) is connected to a personal computer.

A classical valve is used to take a sample of 1 cm$^3$ of liquid. It is analyzed by refractometry to determine the TBAB concentration. The sample is also analysed in a *DIONEX* ionic exchange chromatograph (off-line) to measure the concentration of the electrolyte tracer LiNO$_3$ (see experimental procedure for details, Herri *et al.*, 2011).

## 5 EXPERIMENTAL PROCEDURE

The hydrate is obtained by crystallization of pure gases (CO$_2$, N$_2$) or gas mixture with a liquid phase containing water and the salt.

At the beginning, the reactor is closed and vacuumed with a vacuum pump. Then, the cell is flushed three times with the gas to be studied. It erases any trace of other gases used in a previous experiment. Then, the reactor is vacuumed again.

After, the reactor is pressurized with the pure gas (CO$_2$ or N$_2$) at initial pressure $P_0$. The gas phase is sampled with the *ROLSI* instrument and on-line analysed by gas chromatography to check the gas purity. The gas phase is stirred, cooled down, and maintained at the operative temperature (here called $T_0$, typically in the temperature range of 278 to 286 K).

When a gas mixture of CO$_2$/N$_2$ is studied, the following experimental procedure is used. The reactor is pressurised with the first gas (CO$_2$ or N$_2$) to the desired pressure and analysed by gas chromatography to check the gas purity. After stabilisation of temperature and pressure, the second gas is injected until the desired operative pressure ($P_0$) is reached. The gas mixture is stirred, cooled down, and stabilized at the working conditions of temperature ($T_0$, typically in the temperature range of 278 to 286 K) and pressure. The gas phase is analysed again by gas chromatography to monitor the initial gas composition.

The temperature needs to be higher than the pure TBAB semi-clathrates equilibrium temperature (without gas), and lower than pure water gas hydrate equilibrium temperature to be sure that gas/TBAB semi-clathrates are formed, and not pure water gas hydrates or pure TBAB semi-clathrates. For example, at a TBAB mass fraction of 0.21, temperature cannot be inferior to 283 K to avoid the formation of pure Type A semi-clathrate (*Fig. 1*). At a TBAB mass fraction of 0.11, the temperature cannot be inferior to 290 K to avoid the formation of the Type B semi-clathrate.

Once ($P_0$, $T_0$) is stabilized, the stirrer is stopped and the liquid solution ($V_0^L = 1$ dm$^3$) is injected in the reactor

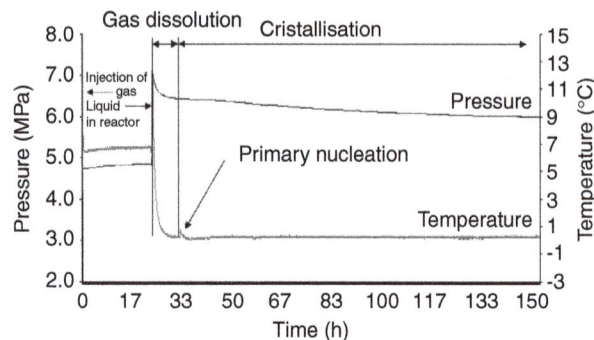

Figure 7

Typical example of the evolution of pressure and temperature during crystallization of pure gas -TBAB semi-clathrates.

by using the HPLC pump. The liquid is composed of a controlled quantity of water ($n_{w,0}^L$ moles) and TBAB ($n_{TBAB,0}^L$ moles). The liquid solution contains also a tracer at a low mole fraction $x_0^{tracer} = 10$ ppm (here LiNO$_3$). The tracer is not consumed during the crystallization and remains in the liquid phase. Once the liquid phase has been injected in the autoclave, an increase of temperature and pressure is simultaneously observed, firstly because the liquid is at ambient temperature. The temperature increase is also due to the gas compression resulting from the reduction of the gas volume by the liquid. The stirring is started and the pressure decreases due to gas dissolution in the liquid.

After a while (from some minutes to several hours, nucleation being a stochastic phenomenon), crystallization begins. It is accompanied by a sudden increase of temperature (*Fig. 7*). During crystallization (exothermic process), the pressure decreases due to gas consumption by hydrates and the temperature returns to the operative temperature. After a while, the system reaches the equilibrium (end of crystallization), the pressure and the temperature reach constant values.

Then, we can proceed to gas hydrate dissociation. The temperature of the reactor is heated by steps of 1 K (*Fig. 8*). During each step, the pressure increases due to gas hydrate dissociation and reaches a constant value which represents the thermodynamic equilibrium.

During the dissociation, at each stage ($i$) the liquid is sampled and analyzed by refractometry to evaluate the quaternary ammonium concentration, and by ionic chromatography to evaluate the concentration $C^{tracer,i}$.

The gas phase is also sampled and analyzed by gas chromatography.

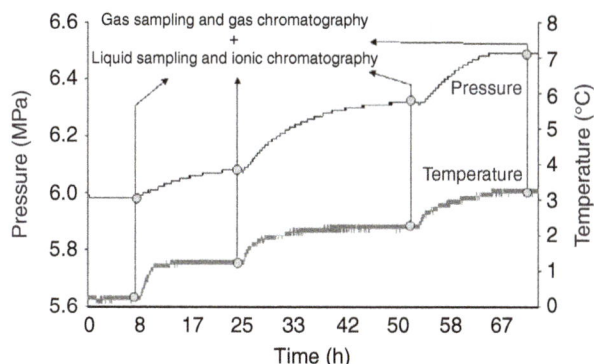

Figure 8

Typical example of the evolution of pressure and temperature during dissociation of pure gas -TBAB semi-clathrates.

## 6 MASS BALANCES

### 6.1 Mass Balance for the Gaseous Components

When at given values of the state variables (pressure, temperature, composition), a gaseous, a liquid and a solid hydrate phases are present in the system, the initial quantity of the molecules in the reactor is distributed between these three phases. Thus, in equilibrium, the quantity of gas in the hydrate phase can be determined from a mass balance according to:

$$n_{j,0} = n_j^H + n_j^L + n_j^G \qquad (1)$$

In Equation (1), $n_{j,0}$ stands for the initial (total) mole number and $n_j^H$, $n_j^L$ and $n_j^G$ are the mole numbers of component $j$ ($j = CO_2$, $N_2$) in the hydrate, the liquid and the gas phase, respectively.

The amount of substance in the liquid phase is estimated by means of a corresponding gas solubility data. The mole number in the gas phase is calculated by using an equation of state approach as outlined in the next sections.

The total mole number $n^G$ in the gas phase in any equilibrium state can be calculated by means of the classical Equation (2):

$$Z(T, P, \vec{y}) = \frac{PV}{nRT} \qquad (2)$$

For $n \equiv n^G$. In Equation (2), $T$, $P$, and $V$ are the temperature, pressure and total volume of the gas phase, respectively, while $\vec{y} = (y_{CO_2}, y_{N_2})$, $n$ and $R$ represent the vector of the mole fractions of the components in the mixture,

the total mole number in the gas mixture, and the universal gas constant, respectively. $Z$ is the compressibility factor that can be calculated by means of a suitable Equation of State (EOS), $e.g.$, a classical cubic EOS. In this study, the Soave-Redlich and Kwong (SRK) EOS has been used (parameters from Danesh, 1998).

Each time, the composition of the gas phase is determined by using gas chromatography analysis.

We recall here that the reactor of total inner volume $V_R = 2.51$ dm$^3$ is initially filled with the gaseous components at the initial temperature $T_0$ and under the initial total pressure $P_0$. Therefore, at this stage, the system only consists of a gas phase, being composed of the two gaseous components $CO_2$ and $N_2$. Knowing the temperature, the pressure and the gas composition, the initial mole number, $n_0^G$, can be calculated as follows:

$$n_0^G = \frac{P_0 V_R}{Z(T, P, \vec{y}_0) R T_0} \qquad (3)$$

Equation (2) has also been used to determine the total amount of substance of the gas phase in a state corresponding to the three phase hydrate-liquid-vapour equilibrium. In the latter case, the initial values of the variables are to be replaced by the corresponding measured values in that equilibrium state. Moreover, the volume of the reactor, $V_R$, has to be replaced by the actual value of the gas volume $V^G$:

$$V^G = V_R - V^{L+H} \qquad (4)$$

where $V^{L+H}$ stands for the volume of the liquid phase and hydrate phase. This volume is assumed to remain equal to the initial liquid volume, $(V_0^L)$, the density of the liquid and hydrate phase being close (1 080 kg.m$^{-3}$ for type A hydrate and 1 070 kg.m$^{-3}$ for type B hydrate according to Oyama et al. (2005), compared to the aqueous densities, measured between 1 021 kg.m$^{-3}$ and 1 039 kg.m$^{-3}$ according to Darbouret, 2005; Obata et al., 2003; and Belandria et al., 2009). At last, the mole numbers of the respective gaseous component $j$ ($j = CO_2$, $CH_4$, $N_2$) in the gas phase, $n_{j,0}^G$ and $n_j^G$, are respectively given by:

$$n_{j,0}^G = n_0^G y_{j,0} \quad \text{and} \quad n_j^G = n^G y_j \qquad (5)$$

### 6.2 Liquid Phase Volume

The liquid phase contains $LiNO_3$ as a tracer. Initially the concentration of lithium $[Li^+]_0$ and the initial volume of liquid $V_0^L$ are known. During the crystallization and dissociation steps, the concentration of lithium is measured by ion-exchange chromatography after sampling. So, we can calculate the volume of liquid water from a mass balance for the Li$^+$ ions:

$$V_0^L [Li^+]_0 = V^L [Li^+] \Rightarrow V^L = \frac{V_0^L [Li^+]_0}{[Li^+]} \qquad (6)$$

where $V^L$ and $[Li^+]$ are the volume of the liquid aqueous phase and the molar concentration of lithium in this phase.

## 6.3 Composition of the Liquid Phase

The mass fraction of TBAB in the liquid phase ($w_{TBAB}^L$) is determined experimentally after on-line sampling of a small amount of the liquid phase and measurement of the index of refraction. In fact, the Index of Refraction (IR) at temperature of 295.15 K is linearly dependent of the mass fraction $w_{TBAB}^L$ in the range of 0 to 0.4, following a correlation given by Darbouret (2005):

$$IR = 1.333 + 0.178 \, w_{TBAB}^L \qquad (7)$$

The mole number $n_j^L$ in the liquid phase (Eq. 9) is calculated in a good approximation by using solubility data of the gas in water (Holder et al., 1988) under the assumption that neither LiNO$_3$ (due to its low concentration, about 10 ppm), nor TBAB does not affect this solubility. The second hypothesis is done because of the lack of data about the solubility of CO$_2$ or N$_2$ in H$_2$O-TBAB liquid solutions. This is hardly defendable from a fundamental point of view. Thiam et al. (2008) showed that the solubility of CO$_2$ is decreased by 10 to 15% as the TBAB mass fraction is increased by 10%. But, from a practical point of view, this approximation can be done because the mole number $n_j^L$ is one order of magnitude lower than $n_j^H$ and $n_j^G$. It does not affect significantly the calculation of $n_j^H$.

In equilibrium, the equality of the fugacities of the gases in the liquid and the gas phase holds according to:

$$f_j^L(T, P, x_j) = f_j^G(T, P, y_j) \qquad (8)$$

Substituting the fugacity in the liquid phase for an extended form of Henry's law (Eq. 10) and expressing the gas phase fugacity in terms of fugacity coefficient $n_j^L$ can be expressed as:

$$n_j^L = \frac{V^L \rho_w^\circ}{M_w} \frac{y_j \varphi_j^G P}{K_{H,j}^\infty \exp\left(P v_j^\infty / RT\right)} \qquad (9)$$

where $V^L$ stands for the volume of the liquid phase in equilibrium, $\rho_w^\circ$ is the density, and $M_w$ is the molar mass of pure water. $v_j^\infty = 32 \, cm^3 . mol^{-1}$ is the partial molar volume of compound $j$ in water (an average value from Holder et al., 1988). In establishing Equation (9), the activity coefficient of CO$_2$ in water was in a good approximation neglected and the very good approximations $n_j^L \ll n_w^L$, was applied. $K_{H,j}^\infty$ (Pa$^{-1}$) represents Henry's constant at saturation pressure of the pure solvent, i.e., at infinite dilution of the gaseous component, which as function of temperature is calculated from the following correlation (Holder et al., 1988):

$$K_{H,j}^\infty(T)[Pa] = \exp\left(-A - \frac{B}{T}\right) \qquad (10)$$

A and B are constants listed in Holder et al. (1988).

## 6.4 Composition of the Hydrate Phase

After the amounts of substance of compound $j$ in the gas phase $n_j^G$ and in the liquid phase $n_j^L$ have been estimated, the mole number of the gas $j$ in the hydrate phase $n_j^H$ can be derived from Equations (1, 5, 9): the gas encapsulated in the hydrate equals the initial quantity in the feed $n_0^G$ minus the quantities in the gas at step $n_j^G$ and the quantity in the liquid $n_j^L$.

The amount of water and TBAB in the hydrate phase is assumed to be the part which has been consumed in the liquid phase.

$$\begin{aligned} n_w^H = n_{w,0}^L - n_w^L &= \left(1 - w_{TBAB,0}^L\right) \frac{V_0^L \cdot \rho_0^L}{M_w} \\ &- \left(1 - w_{TBAB}^L\right) \frac{V^L \cdot \rho^L}{M_w} \end{aligned} \qquad (11)$$

$$\begin{aligned} n_{TBAB}^H = n_{TBAB,0}^L - n_{TBAB}^L &= w_{TBAB,0}^L \frac{V_0^L \cdot \rho_0^L}{M_{TBAB}} \\ &- w_{TBAB}^L \frac{V^L \cdot \rho^L}{M_{TBAB}} \end{aligned} \qquad (12)$$

$V^L$ is calculated from Equation (6) and $w_{TBAB}^L$ from Equation (7). $M_w$ and $M_{TBAB}$ are respectively the molar mass of water and TBAB. The density $\rho^L$ of the solution has been measured experimentally by Darbouret (2005) in the temperature range [273.15-297.15] K and the weight fraction of $w_{TBAB}^L$ in the range of [0-0.4]. From these data, Douzet (2011) has proposed a correlation with a precision of 0.1%:

$$\rho^L = 1000 + 99.7 \, w_{TBAB}^L \qquad (13)$$

## 7 EXPERIMENTAL RESULTS

We present the results of two experiences (*Tab. 6*): one starting with a TBAB mass fraction of 0.21 and an initial $CO_2$ mole fraction of 0.317, and the other one starting with a TBAB mass fraction of 0.11, and an initial $CO_2$ mole fraction of 0.665. In both cases, we succeeded in measuring three equilibrium points following the procedure described before. We complete the data with the results of a previous work (*Tab. 7*) at a TBAB mass fraction of 0.40.

Results of this work are compared to equilibrium points at the same initial compositions and similar temperature, but without TBAB (only pure water). With pure water, the equilibrium points are calculated from the GasHyDyn software described in Herri *et al.* (2011).

Table 8 compares the experimental equilibrium pressure in presence of TBAB to the calculated ones, at the same temperature and gas composition, but for pure water gas hydrate. We can say that the pressure of formation of the semi-clathrate hydrate is considerably decreased, by a factor from 5.4 to 11 for these experiments. It confirms the pressure drop observed on pure gas-TBAB semi-clathrates (*Fig. 3-Fig. 5*).

Figure 9 presents the selectivity of the separation of $CO_2$ from $N_2$ during crystallization of semi-clathrate hydrates from TBAB solution. The data from this work are compared to the results from Duc *et al.* (2007). All the experimental points of this study and from Duc *et al.* (2007) are in the range of temperature [283.4-288.6] K. We have also plotted a reference case

TABLE 6

Experimental results from this study

| $w^L_{TBAB,0}$ | $w^L_{TBAB}$ | $P_{eq}$ (MPa) | $T_{eq}$ (K) | $n_{CO_2}/(n_{CO_2}+n_{N_2})$ | | | |
|---|---|---|---|---|---|---|---|
| | | | | Gas $y_{CO_2}$ | Liquid | Hydrate $z_{CO_2}$ | $\frac{z_{CO_2}y_{N_2}}{z_{N_2}y_{CO_2}}$ |
| 0.112 | 0.073 | 2.5 | 283.4 | 0.431 | 0.243 | 0.989 | 118.70 |
| | 0.079 | 2.6 | 284.4 | 0.456 | 0.249 | 0.984 | 73.37 |
| | 0.091 | 2.7 | 285.5 | 0.50 | 0.259 | 0.955 | 21.22 |
| 0.218 | 0.175 | 4.9 | 283.9 | 0.239 | 0.556 | 0.794 | 12.27 |
| | 0.183 | 5.0 | 287.5 | 0.245 | 0.581 | 0.764 | 9.98 |
| | 0.189 | 5.2 | 288.6 | 0.254 | 0.596 | 0.909 | 29.34 |

TABLE 7

Experimental results from Duc *et al.* (2007)

| $w^L_{TBAB,0}$ | $w^L_{TBAB}$ | $P_{eq}$ (MPa) | $T_{eq}$ (K) | $n_{CO_2}(n_{CO_2}+n_{N_2})$ | | | |
|---|---|---|---|---|---|---|---|
| | | | | Gas $y_{CO_2}$ | Liquid | Hydrate $z_{CO_2}$ | $\frac{z_{CO_2}y_{N_2}}{z_{N_2}y_{CO_2}}$ |
| 0.40 | – | 0.66 | 285.15 | 1 | – | 1 | – |
| | – | 2.01 | 285.15 | 0.175 | – | 0.789 | 17.63 |
| | – | 2.05 | 285.15 | 0.174 | – | 0.783 | 17.13 |
| | – | 3.28 | 285.15 | 0.061 | – | 0.275 | 5.84 |
| | – | 4.44 | 285.15 | 0 | – | 0 | – |
| 0.40 | – | 0.83 | 286.15 | 1 | – | 1 | – |
| | – | 2.217 | 286.15 | 0.205 | – | 0.927 | 49.25 |
| | – | 2.257 | 286.15 | 0.233 | – | 0.9182 | 36.95 |
| | – | 3.4 | 286.15 | 0.0656 | – | 0.2851 | 5.68 |
| | – | 4.681 | 286.15 | 0 | – | 0 | – |

TABLE 8

Comparison of equilibrium pressure $P_{eq}$ at given $(T_{eq}, x_{CO_2})$ between the semi-clathrate hydrate and the clathrate hydrate

| $w_{TBAB}^L$ | $T_{eq}$ (K) | $P_{eq}\left(w_{TBAB}^L\right)$ (MPa) | $P_{eq}(w_{TBAB}=0)$ (MPa) | $\dfrac{P_{eq}\left(w_{TBAB}^L\right)}{P_{eq}(w_{TBAB}=0)}$ (-) |
|---|---|---|---|---|
| 0.073 | 283.4 | 2.5 | 13.6 | 5.4 |
| 0.079 | 284.4 | 2.6 | 18.6 | 7.2 |
| 0.091 | 285.5 | 2.7 | 20.8 | 7.6 |
| 0.175 | 283.9 | 4.9 | 24.8 | 5.0 |
| 0.183 | 287.5 | 5.0 | 48.4 | 9.7 |
| 0.189 | 288.6 | 5.2 | 57.2 | 11.0 |

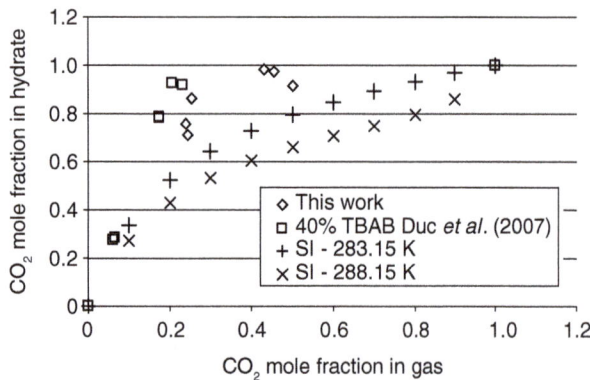

Figure 9

Selectivity of the separation of $CO_2$ from $N_2$ during crystallization of semi-clathrate hydrates from TBAB solution, and comparison to the selectivity of clathrate hydrate of structure SI.

consisting in the molar fraction of $CO_2$ in the SI structure (pure water clathrate hydrate), at two temperatures of 283.15 K and 288.15 K. We can observe an enhancement of the content in $CO_2$ in the case of the semi-clathrate hydrate of TBAB, even at low molar fraction of $CO_2$. However, it is observed a difference between the data of this work and the results of Duc et al. (2007). The data from Duc et al. (2007) has been determined with a liquid solution at TBAB mass fraction of 0.4. In our work, the TBAB mass fraction is in the range [0.07-0.18] and could explain the difference. But independently of the differences, it can be underlined that the semi-clathrate hydrate of TBAB has a better affinity to $CO_2$ in comparison to pure water clathrate hydrate.

For example, at a gas molar fraction of $CO_2$ around 0.2, the $CO_2$ molar fraction in the clathrate hydrate is in the range of 0.43 ($T$ = 288.15 K) to 0.42 ($T$ = 283.15 K) whereas the $CO_2$ molar fraction in the TBAB semi-clathrate hydrate can be up to 0.93 (from Duc et al., 2007, at a temperature of 286.15 K).

Similarly, at gas molar fraction of $CO_2$ around the value of 0.5, the $CO_2$ molar fraction in the clathrate hydrate is in the range 0.66 ($T$ = 288.15 K) to 0.79 ($T$ = 283.15 K) whereas the $CO_2$ molar fraction in the TBAB semi-clathrate hydrate can be up to 0.96 (this work) at a temperature of 285.5 K.

## 8 MODELING

The experimental results of this work show that TBAB decreases drastically the operative pressure, possibly down to the atmospheric pressure. It can be done by adjusting the operative temperature of formation of the Gas-TBAB semi-clathrate hydrate at a temperature close to the temperature of formation of the pure TBAB semi-clathrate hydrate.

Paricaud (2011) proposed a new expression to describe the Gas-TBAB semi-hydrate equilibrium from a model derived from the van der Waals and Platteeuw using $\theta_j$ (van der Waals and Platteeuw, 1959)

$$
\begin{aligned}
0 = {} & \frac{\Delta_{dis} G_m^{\ominus}\left(T_{cgr}^{HL_w}\right)}{R T_{cgr}^{HL_w}} + \frac{\Delta_{dis} H_m^{\ominus}\left(T_{cgr}^{HL_w}\right)}{RT}\left(1 - \frac{T}{T_{cgr}^{HL_w}}\right) \\
& + \frac{\Delta_{dis} C_{p,m}^{\ominus}\left(T_{cgr}^{HL_w}\right)}{R}\left(1 - \frac{T_{cgr}^{HL_w}}{T} + \ln\frac{T_{cgr}^{HL_w}}{T}\right) \\
& + \frac{\Delta_{dis} V_m^{ref}\left(T_{cgr}^{HL_w}, p^{\ominus}\right)}{RT}(p - p^{\ominus}) \\
& + \ln\ x_C^{L_w}\gamma_{x,C}^{*,L_w} + \ln x_A^{L_w}\gamma_{x,A}^{*,L_w} + v_w \ln x_w^{L_w}\gamma_w^{L_w} \\
& - v_i \ln(1 - \theta)
\end{aligned}
$$

(14)

$\gamma$ are the activity coefficients. In this work, a modification of the model approach of Paricaud (2011) is used in which in contrast to the use of the SAFT-VRE equation of state (Galindo et al., 1999) the electrolyte NRTL model (eNRTL-model) (Chen et al., 1982; Chen and Evans, 1986; Bollas et al., 2008) is incorporated into the model to describe the liquid phase non-idealities in presence of TBAB (Kwaterski and Herri, 2011)

In Equation (14), $T_{cgr}^{HL_w} = T_{cgr}^{HL_w}(p^{\ominus})$ stands for the temperature of the congruent melting point of the semi-clathrate hydrate, i.e., the temperature of the phase transition $\beta \rightarrow L_w \equiv H \rightarrow L_w$ at $p = p^{\ominus}$.

TABLE 9

Reference properties of type A and B TBAB semi-clathrate hydrates

| | Type A | Type B |
|---|---|---|
| $v_w$ (mole of water/mole of TBAB) | 26 | 38 |
| $T_{cgr}^{HL_w}$ (K) | 285.15 | 283.5 |
| $\Delta_{dis} G_m^{\ominus} \left( T_{cgr}^{HL_w} \right)$ (J/mol TBAB) | 23 804 | 24 867 |
| $\Delta_{dis} H_m^{\ominus}(T_{cgr}^{HL_w})$ (J/mole TBAB) | 146 350 | 193 060 |
| $\Delta_{dis} C_{p,m}^{\ominus}(T_{cgr}^{HL_w})$ | 0 | 0 |
| $\Delta_{dis} V_m^{ref}$ (m³/mole TBAB) | −0.00003 | −0.00003 |
| Type of free cavity to enclahtrate gas | ? | $5^{12}$ |
| $v_i$ (mole of free cavity/mole of TBAB) | ? | 2 |

Figure 10

Minimum occupancy factor of cavities $5^{12}$ in the type B semi-clathrate hydrate of TBAB.

Numerical values (*Tab. 9*) for the standard molar quantities, $\Delta_{dis} G_m^{\ominus} \left( T_{cgr}^{HL_w} \right)$ and $\Delta_{dis} H_m^{\ominus}(T_{cgr}^{HL_w})$ are gained through adjustments of the $HL_w$-coexistence curves given in Figure 1 (or to the dissociation enthalpies directly if allowable).

If values of the standard molar isobaric heat capacity upon hydrate dissociation are measured directly, they can additionally be used to adjust $\Delta_{dis} C_{p,m}^{\ominus}(T_{cgr}^{HL_w})$. $\Delta_{dis} V_m^{ref}$, evaluated at $T_{cgr}^{HL_w}$ and $p^{\ominus}$, accounts for the effect of pressure on the melting points for the water + salt binary systems (Paricaud, 2011). Paricaud points out that *"different values for $\Delta_{dis} V_m^{ref}(T_{cgr}^{HL_w}, p^{\ominus})$, $\Delta_{dis} H_m^{\ominus}(T_{cgr}^{HL_w})$, $T_{cgr}^{HL_w}$ and $\Delta_{dis} C_{p,m}^{\ominus}(T_{cgr}^{HL_w})$ should be used for different types of hydrates"*. He further reports that $v_w$ should be fixed to its experimental value. $v_w$ is the stoichiometric coefficient of water in the equation of dissociation of a semiclathrate hydrate $TBAB \cdot v_w H_2O$ regarded as a combined chemical reaction and phase equilibrium and can be written as:

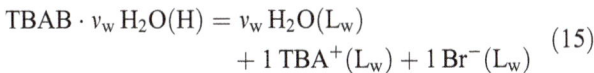

$$TBAB \cdot v_w H_2O(H) = v_w H_2O(L_w) \\ + 1\, TBA^+(L_w) + 1\, Br^-(L_w) \quad (15)$$

The complete modeling of the eNRTL model (Kwaterski and Herri, 2011) has been implemented in the GasHyDyn software (Herri *et al.*, 2011) with the reference parameters from Table 9.

From Equation (14), we can determine analytically the minimum occupancy $\theta_{min}$ which stabilizes the overall structure, independently of the chemical nature of the species in the cavities. It could be a single gas $N_{type\,g} = 1$ or a gas mixture ($N_{type\,g} > 1$), but the sum of their respective contributions $\sum_{j=1}^{N_{type\,g}} \theta_j$ needs to be at least equal to $\theta_{min}$. The value of $\theta_{min}$ is given in Figure 10 for type B hydrate, as a function of temperature, given as the difference between the operative temperature and $T_{cgr}^{HL_w}$.

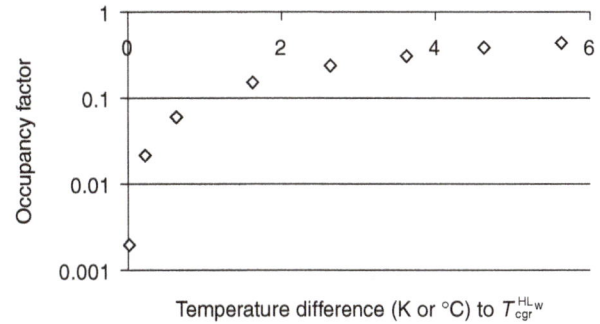

From Figure 10, we can determine a correlation to determine the occupancy factor $\theta_{min}$ of the cavities:

$$\theta_{min} = -3.719 \cdot 10^{-3} \left( T - T_{cgr}^{HL_w} \right)^2 \\ + 9.926 \cdot 10^{-1} \left( T - T_{cgr}^{HL_w} \right) \quad (16)$$

The consequence of Equation (16) is that the minimum gas content in the clathrate structure decreases as the temperature reaches the temperature $T_{cgr}^{HL_w}$ of formation of the pure TBAB semi-clathrate hydrate.

At thermodynamic equilibrium, the content of gas $j$ in the $5^{12}$ cavity of the semi-clathrate hydrate is expressed under the following form (van der Waals and Platteeuw, 1959)

$$\theta_j = \frac{C_{f,j} f_j(T,P)}{1 + \sum_j C_{f,j} f_j(T,P)} \quad (17)$$

$\theta_j$ is the occupancy factor ($\theta_j \in [0,1]$) of the cavities of type $5^{12}$ by the gas molecule $j = CO_2, N_2$. $C_{f,j}$ is the Langmuir constant of component $j$ in the cavity $5^{12}$. In this notation, the Langmuir constant is calculated by using a modified Kihara approach (*Eq. 18*) taking into account the fugacity $f$ as the driving force. It describes the interaction potential between the encaged guest molecule and the surrounding water molecules. $f_j$ is the fugacity of the component.

## 9 DISCUSSION

### 9.1 About the Operative Pressure

From correlation (16) and Equation (17), we can see that the operative pressure and gas content in the hydrate phase follow the same trend, the higher the gas content, the higher the operative pressure. If the operative pressure is lowered, the gas content is decreased also. In terms of process, it implies that we need to handle more hydrate volume to capture a same quantity of $CO_2$ gas. So the choice of an operative pressure becomes the economic optimum between the minimization of the cost of compression which implies to drop the pressure, and the minimization of the volume of the reactor which implies to increase the pressure.

### 9.2 About the Hydrate Composition

The results of this work show that TBAB enhances the selectivity of the $CO_2$ capture (*Fig. 9*). This enhancement can be explained theoretically from two points of view.

On the one hand, the surrounding of the $CO_2$ gas molecule encapsulated in the semi-clathrate hydrate structure (*Fig. 2*) is different from the surrounding in the clathrate hydrate structure because of the vincinity of the TBAB molecule. It affects necessary the Langmuir constant $C_{f,j}$. In fact, the Langmuir constant $C_{f,j}$ is evaluated by assuming a spherically symmetrical cage that can be described by a spherical symmetrical potential (van der Waals and Platteeuw, 1959):

$$C = \frac{4\pi}{kT} \int_0^\infty \exp\left(-\frac{w(r)}{kT}\right) r^2 dr \qquad (18)$$

where $w(r)$ is the interaction potential between the structure and the gas molecule according to the distance $r$ between the guest molecule and the water molecules over the structure. If the nature of the structure is modified, the Langmuir constant is modified also, especially if a water molecule at the vicinity of the gas molecules is replaced by a new molecule such as an anion of a halogenide type.

But this hypothesis is not sufficient at explaining the experimental measurements. In fact, from Equation (17), we can write that:

$$\frac{\theta_{CO_2}}{\theta_{N_2}} = \frac{z_{CO_2}}{z_{N_2}} = \frac{C_{f,CO_2} f_{CO_2}(T,P)}{C_{f,N_2} f_{N_2}(T,P)} \\ = \frac{C_{f,CO_2} Z(T,P,y_{CO_2}) y_{CO_2} P}{C_{f,N_2} Z(T,P,y_{CO_2}) y_{N_2} P} \qquad (19)$$

$y$ is the mole fraction of a chemical species in the gas phase, $z$ is the mole fraction of a chemical species in the cavities of the clathrate hydrate structure, and $Z$ is the compressibility factor. The equation becomes:

$$\frac{z_{CO_2} y_{N_2}}{z_{N_2} y_{CO_2}} = \frac{C_{f,CO_2}}{C_{f,N_2}} \qquad (20)$$

Because the Langmuir constants (*Eq. 18*) are only dependent on temperature, the left term of Equation (20) is theoretically constant at a given temperature, but it is not experimentally validated (*Tab. 7*). So, the semi-clathrate hydrate does not form under thermodynamic equilibirium.

So, if the thermodynamic consideration can not explain the enhancement of the selectivity of the $CO_2$ in the semi-clathrate hydrate, the other reason is necessary a kinetic one. Based on the fact that clathrates are non-defined component, Herri and Kwaterski (2012) have demonstrated that the composition of the solid is dependent on the relative rates of mass transfers, and on kinetic rates of the reaction of integration of the gaseous molecule in the solid structure under growing, following a modified expression of Equation (17) under the form:

$$\theta_j = \frac{C_{x,j} x_{j,\text{int}} \frac{k_j/G}{1+k_j/G}}{1 + \sum_{j' \in S_g} C_{x,j'} x_{j',\text{int}} \frac{k_j/G}{1+k_{j'}/G}} \qquad (21)$$

The Langmuir coefficient is calculated by taking into account the mole fraction $x$ as the driving force. $x_{j,\text{int}}$ is the mole fraction of component $j$ at the vicinity of the growing clathrate hydrate surface. $G$ is the growth rate. $k_j$ is to be regarded as an intrinsic kinetic constant of component $j$.

The composition of the semi-clathrate hydrate is so fixed by kinetic considerations also, based on the values of the $x_{j,\text{int}}$ which result on a mass balance between:
- the gas to liquid transfer rate that feeds the solution;
- the growth that consumes the gaseous component;
- the affinity of the cavities to adsorb the species, from a thermodynamic point of view ($C_{x,j}$) and a kinetic point of view ($k_j$).

The solving of the mass balance is given in Herri and Kwaterski (2012) and it implies to couple the mass balance with a population balance to describe the size and the number of crystals in the system.

## 10 PERSPECTIVES

So, the composition of the gas hydrate is thermodynamic and kinetic dependent. It can be orientated in a favorable direction (*i.e.* the capture of $CO_2$) by modifying the mass transfer rates to modify the growth rate $G$, with the use of specific reactor geometries. Also, the selectivity can be modified from a control on the growth constants $k_j$ with the use of kinetic additives. For example, surfactants such as Sodium Dodecyl Sulfate (SDS) present the characteristic to speed up $CH_4$ hydrate crystallization (Gayet *et al.*, 2005; Tajima *et al.*, 2010; Torre *et al.*, 2011), but not $CO_2$ hydrate crystallization, while the reverse effect is observed with fluorinated surfactants (Zhong and Rogers, 2000).

## CONCLUSION

The present work confirms the possibility to capture $CO_2$ from a technology based on the crystallization of semi-clathrate hydrate. A first preliminary costing had been previously evaluated by some authors of this paper in Duc et al. (2007) for an application to $CO_2$ capture in steel making plants. The specificity of the steel making industry is that the flue gases can be very $CO_2$ rich, from 23% to 25% .We observed that the main part of the cost is due to compressors. In this work, we add new considerations because we understand now that the pressure can be dropped to the atmospheric pressure. In this work, we have shown also experimentally that the selectivity can be orientated in a favorable direction from kinetic considerations. The hydrate composition is 100% $CO_2$ when the flue gas is above 20% $CO_2$. It opens the possibility to treat low $CO_2$ concentrated gases, for example from coal combustion.

## ACKNOWLEDGMENTS

This work has been carried out within the ANR $CO_2$ (French Research Agency) project called SECOHYA and the FP7 European project iCAP (GA n°241393).

## REFERENCES

Arjmandi M., Chapoy A., Tohidi B. (2007) Equilibrium Data of Hydrogen, Methane, Nitrogen, Carbon Dioxide, and Natural Gas in Semi-Clathrate Hydrates of Tetrabutyl Ammonium Bromide, *J. Chem. Eng. Data* **52**, 2153-2158.

Belandria V., Mohammadi A.H., Richon D. (2009) Volumetric properties of the (tetrahydrofuran + water) and (tetra-*n*-butyl ammonium bromide + water) systems: Experimental measurements and correlations, *J. Chem. Thermodyn.* **41**, 12, 1382-1386.

Bollas G.M., Chen C.C., Barton P.I. (2008) Refined electrolyte-NRTL model: Activity coefficient expressions for application to multi-electrolyte systems, *AIChE J.* **54**, 6, 1608-1624.

Brinchi L., Castellani B., Cotana F., Filipponi M., Rossi F. (2011) Investigation of a novel reactor for gas hydrate production, Proceedings of the *7th International Conference on Gas Hydrates (ICGH 2011)*, Edinburgh, Scotland, United Kingdom, 17-21 July.

Chen C.-C., Britt H.I., Boston J.F., Evans L.B. (1982) Local composition model for excess Gibbs energy of electrolyte systems. Part I: Single solvent, single completely dissociated electrolyte systems, *AIChE J.* **28**, 4, 588-596.

Chen C.-C., Evans L.B. (1986) A local composition model for the excess Gibbs energy of aqueous electrolyte systems, *AIChE J.* **32**, 3, 444-454.

Compingt A., Blanc P., Quidort A. (2009) Slurry for Refrigeration Industrial Kitchen Application, Proceedings of *8th IIR Conference on Phase Change Material and Slurries for Refrigeration and Air Conditioning 2009*, Karlsruhe, 3-5 June.

Danesh A. (1998) *PVT and Phase Behaviour of Petroleum Reservoir Fluids*, Elsevier.

Darbouret M. (2005) Étude rhéologique d'une suspension d'hydrates en tant que fluide frigoporteur diphasique. Résultats expérimentaux et modélisation, *PhD Thesis*, École Nationale Supérieure des Mines de Saint-Étienne, France.

Darbouret M., Cournil M., Herri J.M. (2005) Rheological study of TBAB hydrate slurries as secondary two-phase refrigerants, *Int. J. Refrig.* **28**, 5, 663-671.

Davidson D.W. (1973) Clathrate Hydrates, in *Water, a comprehensive treatise; Vol. 2; Water in crystalline hydrates; aqueous solutions of simple nonelectrolytes*, Franks F. (ed.), Plenum Press, New York, pp. 140-146.

Delahaye A., Fournaison L., Marinhas S., Chatti I., Petitet J.-P., Dalmazzone D., Fürst W. (2006) Effect of THF on equilibrium pressure and dissociation enthalpy of $CO_2$ hydrates applied to secondary refrigeration, *Industrial Engineering Chemistry Research* **45**, 1, 391-397.

Deppe G., Tam S.S., Currier R.P., Young J.S., Anderson G.K., Le L.A., Spencer D.F. (2002) A high pressure carbon dioxide separation process in an IGCC Plant, Proceedings of the *Future Energy System and Technology for CO2 abatement*, Antwerpen, Belgique, 18-19 Nov.

Deschamps J., Dalmazzone D. (2009) Dissociation enthalpies and phase equilibrium for TBAB semi-clathrates of $N_2$, $CO_2$, $N_2$ + $CO_2$ and $CH_4$ + $CO_2$, *J. Therm. Anal. Calorim.* **98**, 1, 113-118.

Douzet J. (2011) Conception, construction, expérimentation et modélisation d'un banc d'essai grandeur nature de climatisation utilisant un fluide frigoporteur diphasique à base d'hydrates de TBAB, *PhD Thesis*, École Nationale Supérieure des Mines de Saint-Étienne, France.

Douzet J., Brantuas P., Herri J.-M. (2011) Cristallisation and flowing of high concentrated slurries of quaternary ammonium semi-clathrates. Application to air conditioning and $CO_2$ capture, Proceedings of the *7th International Conference on Gas Hydrates (ICGH 2011)*, Edinburgh, Scotland, United Kingdom, 17-21 July.

Douzet J., Kwaterski M., Lallemand A., Chauvy F., Flick D., Herri J.M. (2013) Prototyping of a real size air-conditioning system using a tetra-*n*-butylammonium bromide semiclathrate hydrate slurry as secondary two-phase refrigerant - Experimental investigations and modelling, *International Journal of Refrigeration* **36**, 6, 1616-1631.

Du J., Liang D., Li D. (2010) Experimental Determination of the Equilibrium Conditions of Binary Gas Hydrates of Cyclopentane plus Oxygen, Cyclopentane plus Nitrogen, and Cyclopentane plus Hydrogen, *Industrial Engineering Chemistry Research* **49**, 22, 11797-11800.

Duc N.G., Chauvy F., Herri J.-M. (2007) $CO_2$ Capture by Hydrate Crystallization - A Potential Solution for Gas Emission of Steelmaking Industry, *Energy Conversion and Management* **48**, 1313-1322.

Dyadin Yu.A, Udachin K.A. (1984) Clathrate formation in water-peralkylonium systems, *Journal of Inclusion Phenomena* **2**, 61-72.

Dyadin Yu.A, Udachin K.A. (1987) Clathrate polyhydrates of peralkylonium salts and their analogs, *Translated from Zhurnal Strukturnoi Khimii* **28**, 3, 75-116, May-June. *J. Structural Chemistry* **28**, 3, 394-432.

Dyadin Yu.A, Bondaryuk I.V., Aladko L.S. (1995) Stoichiometry of clathrates, *Journal of Structural Chemistry* **36**, 6, 995-1045.

Fan S.S., Liang D.Q., Guo K.H. (2001) Hydrate Equilibrium Conditions for Cyclopentane and a Quaternary Cyclopentane-Rich Mixture, *J. Chem. Eng. Data* **46**, 4, 930-932.

Galindo A., Gil-Villegas A., Jackson G., Burgess A.N. (1999) SAFT-VRE: Phase behavior of electrolyte solutions with the statistical associating fluid theory for potentials of variable range, *J. Phys. Chem. B* **103**, 46, 10272-10281.

Gayet P., Dicharry C., Marion G., Graciaa A., Lachaise J., Nesterov A. (2005) Experimental determination of methane hydrate dissociation curve up to 55 MPa by using a small amount of surfactant as hydrate promotor, *Chemical Engineering Science* **60**, 21, 5751-5758.

Gibert V. (2006) Ice Slurry, Axima Refrigeration Experience, *7th Conference on Phase Change Materials and Slurries for Refrigeration and Air Conditioning*, Dinan, France, 13-15 Sept.

Herri J.-M., Bouchemoua A., Kwaterski M., Fezoua A., Ouabbas Y., Cameirao A. (2011) Gas Hydrate Equilibria from $CO_2$-$N_2$ and $CO_2$-$CH_4$ gas mixtures – Experimental studies and Thermodynamic Modelling, *Fluid Phase Equilibria* **301**, 2, 171-190.

Herri J.M., Kwaterski M. (2012) Derivation of a Langmuir type of model to describe the intrinsic growth rate of gas hydrates during crystallization from gas mixtures, *Chemical Engineering Science* **81**, 28-37.

Holder G.D., Zetts S.P., Pradhan N. (1988) Phase Behavior in Systems Containing Clathrate Hydrates: A Review, *Reviews Chemical Engineering* **5**, 1-4, 1-70.

Herslund, P.J., Thomsen, K., Abildskov, J., von Solms, N., Galfré, A., Brântuas, P., Kwaterski, M., Herri, J.M., Thermodynamic promotion of carbon dioxide–clathrate hydrateformation by tetrahydrofuran, cyclopentane and their mixtures, *International Journal of Greenhouse Gas Control* **17**, 397–410.

Jeffrey G.A. (1984) Hydrate inclusion compounds, in *Inclusion compounds*, Atwood J.L., Davies J.E.D., MacNicol D.D. (eds), I, 135-190, Academic Press, New York.

Kamata Y., Oyama H., Shimada W., Ebinuma T., Takeya S., Uchida T., Nagao J., Narita H. (2004) Gas separation method using tetra-*n*-butyl ammonium bromide semi-clathrate hydrate, *Jpn J. Appl. Phys.* **43**, 362-365.

Kang S.-P., Lee H. (2000) Recovery of $CO_2$ from Flue Gas Using Gas Hydrate: Thermodynamic Verification through Phase Equilibrium Measurements, *Environ. Sci. Technol.* **34**, 4397-4400.

Kang S.-P., Lee H., Lee C.-S., Sung W.-M. (2001) Hydrate phase equilibria of the guest mixtures containing $CO_2$, $N_2$ and tetrahydrofuran, *Fluid Phase Equilibria* **185**, 1-2, 101-109.

Kwaterski M., Herri J.M. (2011) Thermodynamic modelling of gas semi-clathrate hydrates using the electrolyte NRTL model, Proceedings of the *7th International Conference on Gas Hydrates (ICGH 2011)*, Edinburgh, Scotland, United Kingdom, 17-21 July.

Li S., Fan S., Wang J., Lang X., Liang D. (2009) $CO_2$ capture from binary mixture *via* forming hydrate with the help of tetra-*n*-butyl ammonium bromide, *Journal of Natural Gas Chemistry* **18**, 1, 15-20.

Li S., Fan S., Wang J., Lang X., Wang Y. (2010) Clathrate Hydrate Capture of $CO_2$ from Simulated Flue Gas with Cyclopentane/Water Emulsion, *Chinese Journal of Chemical Engineering* **18**, 2, 202-206.

Li X.-S., Xu C.-G., Chen Z.-Y., Wu H.-J. (2011) Hydrate-based pre-combustion carbon dioxide capture process in the system with tetra-*n*-butyl ammonium bromide solution in the presence of cyclopentane, *Energy* **36**, 3, 1394-1403.

Li X.-S., Xu C.-G., Chen Z.-Y., Jing C. (2012) Synergic effect of cyclopentane and tetra-*n*-butyl ammonium bromide on hydrate-based carbon dioxide separation from fuel gas mixture by measurements of gas uptake and X-ray diffraction patterns, *International Journal of Hydrogen Energy* **37**, 1, 720-727.

Lin W., Delahaye A., Fournaison L. (2008) Phase equilibrium and dissociation enthalpy for semi-clathrate hydrate of $CO_2$ + TBAB, *Fluid Phase Equilibria* **264**, 1-2, 220-227.

Linga P., Kumar R., Englezos P. (2007a) Gas hydrate formation from hydrogen/carbon dioxide and nitrogen/carbon dioxide gas mixtures, *Chemical Engineering Science* **62**, 16, 4268-4276.

Linga P., Kumar R., Englezos P. (2007b) The clathrate hydrate process for post and pre-combustion capture of carbon dioxide, *Journal of Hazardous Materials* **149**, 3, 625-629.

Linga P., Kumar R., Lee J.-D., Ripmeester J., Englezos P. (2010) A new apparatus to enhance the rate of gas hydrate formation: Application to capture of carbon dioxide, *International Journal of Greenhouse Gas Control* **4**, 4, 630-637.

Lipkowski J., Komorov V.Y., Rodionova T.V., Dyadin Y.A., Aladko L.S. (2002) The Structure of Tetrabutylammonium Bromide Hydrate, *Journal of Supramolecular Chemistry* **2**, 435-439.

McMullan R., Jeffrey G.A. (1959) Hydrates of the Tetra *n*-butyl and Tetra *i*-amyl Quaternary Ammonium Salts, *Journal of Chemical Physics* **31**, 5, 1231-1234.

Meunier F., Rivet P., Terrier M.F. (2007) *Froid Industriel*, Ed. Dunod, Paris.

Mizukami T. (2010) Thermal Energy Storage system with clathrate hydrate slurry, *Keio University "Global COE Program" International Symposium, Clathrate Hydrates and Technology Innovations, Challenges Toward a Symbiotic Energy Paradigm*, Yokohama, Japan 15, March.

Mohammadi A.H., Richon D. (2009) Phase equilibria of clathrate hydrates of methyl cyclopentane, methyl cyclohexane, cyclopentane or cyclohexane + carbon dioxide, *Chemical Engineering Science* **64**, 24, 5319-5322.

Mohammadi A.H., Richon D. (2010) Clathrate hydrate dissociation conditions for the methane + cycloheptane/cyclooctane + water and carbon dioxide + cycloheptane/cyclooctane + water systems, *Chemical Engineering Science* **65**, 10, 3356-3361.

Mohammadi A.H., Richon D. (2011) Phase equilibria of binary clathrate hydrates of nitrogen + cyclopentane/cyclohexane/methyl cyclohexane and ethane + cyclopentane/cyclohexane/methyl cyclohexane, *Chemical Engineering Science* **66**, 20, 4936-4940.

Nakajima M., Ohmura R., Mori Y.H. (2008) Clathrate Hydrate Formation from Cyclopentane-in-water emulsion, *Ind. Eng. Chem. Res.* **47**, 22, 8933-8939.

Ogoshi H., Matsuyama E., Miyamoto H., Mizukami T. (2010) Clathrate Hydrate Slurry, CHS Thermal Energy Storage System and Its Applications, Proceedings of *2010 International Symposium on Next-generation Air Conditioning and Refrigeration Technology*, Tokyo, Japan, 17-19 Feb.

Obata Y., Masuda N., Joo K., Katoh A. (2003) Advanced Technologies Towards the New Era of Energy Industries, *NKK Tech. Rev.* **88**, 103-115.

Oyama H., Shimada W., Ebinuma T., Kamata Y. (2005) Phase diagram, latent heat, and specific heat of TBAB semiclathrate hydrate crystals, *Fluid phase Equilib.* **234**, 1-2, 131-135.

Oyama H., Ebinuma T., Nagao J., Narita H. (2008) Phase Behavior of TBAB Semiclathrate Hydrate Crystal under several Vapor Components, Proceedings of the *6th International Conference on Gas Hydrates (ICGH 2008)*, Vancouver, British Columbia, Canada, 6-10 July.

Paricaud P. (2011) Modeling the Dissociation Conditions of Salt Hydrates and Gas Semiclathrate Hydrates: Application to Lithium Bromide, Hydrogen Iodide, and Tetra-*n*-butylammonium Bromide + Carbon Dioxide Systems, *J. Phys. Chem. B* **115**, 2, 288-299.

Riesco N., Trusler J.P.M. (2005) Novel optical flow cell for measurements of fluid phase behavior, *Fluid Phase Equilibria* **228-229**, 233-238.

Sabil K.M., Witkamp G.-J., Peters C.J. (2010) Phase equilibria in ternary (carbon dioxide + tetrahydrofuran + water) system in hydrate-forming region: Effects of carbon dioxide concentration and the occurrence of pseudo-retrograde hydrate phenomenon, *Journal of Chemical Thermodynamics* **42**, 1, 8-16.

Sato K., Tokutomi H., Ohmura R. (2013) Phase Equilibrium of Ionic Semiclathrate Hydrates formed with Tetrabutylammonium Bromide and Tetrabutylammonium Chloride, *Fluid Phase Equilibria* **337**, 115-118.

Seo Y.-T., Kang S.-P., Lee H. (2001) Experimental determination and thermodynamic modeling of methane and nitrogen hydrates in the presence of THF, propylene oxide, 1,4-dioxane and acetone, *Fluid Phase Equilibria* **189**, 1-2, 99-110.

Seo Y., Kang S.-P., Lee S., Lee H. (2008) Experimental Measurements of Hydrate Phase Equilibria for Carbon Dioxide in the Presence of THF, Propylene Oxide, and 1,4-Dioxane, *J. Chem. Eng. Data* **53**, 2833-2837.

Shimada W., Ebinuma T., Oyama H., Kamata Y., Takeya S., Uchida T., Nagao J., Narita H. (2003) Separation of Gas Molecule Using Tetra-*n*-butyl Ammonium Bromide Semi-Clathrate Hydrate Crystals, *Jpn J. Appl. Phys.* **42**, L129-L131.

Shimada W., Ebinuma T., Oyama H., Kamata S., Narita H. (2005a) Free-growth forms and growth kinetics of tetra-*n*-butyl ammonium bromide semi-clathrate hydrate crystals, *J. Cryst. Growth* **274**, 246-250.

Shimada W., Shiro M., Kondo H., Takeya S., Oyama H., Ebinuma T., Narita H. (2005b) Tetra-*n*-butylammonium bromide-water (1/38), *Acta Cryst.* **C61**, o65-o66.

Sloan E.D. (1998) *Clathrate Hydrates of Natural Gases*, 2nd ed., Marcel Dekker, New York.

Sloan E.D., Koh C.A. (2008) *Clathrate hydrates of natural gases*, 3rd ed., CRC Press, Boca Raton.

Spencer D.F. (1997) Methods of selectively separating $CO_2$ from a multicomponent gaseous stream, US Patent 5700311.

Spencer D.F. (2000) Methods of selectively separating $CO_2$ from a multicomponent gaseous stream, US Patent 6106595.

Suginaka T., Sakamoto H., Iino K., Sakakibara Y., Ohmura R. (2013) Phase Equilibrium for Ionic Semiclathrate Hydrate Formed with $CO_2$, $CH_4$, or $N_2$ plus Tetrabutylphosphonium Bromide, *Fluid Phase Equilibria* **344**, 108-111.

Tajima H., Yamasaki A., Kiyono F. (2004) Energy consumption estimation for greenhouse gas separation processes by clathrate hydrate formation, *Energy* **29**, 11, 1713-1729.

Tajima, H., Kiyono, F., Yamasaki, A. (2010) HYPERLINK "/full_record.do?product = UA&search_mode = GeneralSearch&qid = 2&SID = Z1iFau3bbGfvwrcdbBP&page = 1&doc = 5". Direct Observation of the Effect of Sodium Dodecyl Sulfate (SDS) on the Gas Hydrate Formation Process in a Static Mixer, *Energy & Fuels* **24**, 432-438.

Takao S., Ogoshi H., Matsumato S. (2001) Air conditioning and thermal storage systems using clathrate hydrate slurry, US Patent 6560971 B2.

Takao S., Ogoshi H., Matsumato S. (2002) Air conditioning and thermal storage systems using clathrate hydrate slurry, US Patent 083720 A1.

Takao S., Ogoshi H., Fukushima S., Matsumato H. (2004) Thermal storage medium using a hydrate and apparatus thereof, and method for producing the thermal storage medium, US Patent 20050016200.

Thiam A., Bouchemoua A., Chauvy F., Herri J.-M. (2008) Gas Hydrates Crystallization from $CO_2$-$CH_4$ gas mixtures: Experiments and modelling, Proceedings of the *6th International Conference on Gas Hydrates (ICGH 2008)*, Vancouver, British Columbia, Canada, 6-10 July.

Tohidi B., Danesh A., Todd A.C., Burgass R.W., Østergaard K.K. (1997) Equilibrium data and thermodynamic modelling of cyclopentane and neopentane hydrates, *Fluid Phase Equilibria* **138**, 1-2, 241-250.

Torre J.P., Dicharry C., Ricaurte M., Daniel-David D., Broseta D. (2011) $CO_2$ capture by hydrate formation in quiescent conditions: In search of efficient kinetic additives, *Energy Procedia* **4**, 621-628.

van der Waals J.H., Platteeuw J.C. (1959) Clathrate solutions, *Adv. Chem. Phys.* **2**, 1-57.

Xu C.G., Li X.S., Lv Q.N., Chen Z.Y., Cai J. (2012) Hydrate-based $CO_2$ (carbon dioxide) capture from IGCC (integrated gasification combined cycle) synthesis gas using bubble method with a set of visual equipment, *Energy* **44**, 358-366.

Yang H., Fan S., Lang X., Yang Y. (2011) Phase Equilibria of Mixed Gas Hydrates of Oxygen + Tetrahydrofuran, Nitrogen + Tetrahydrofuran, and Air + Tetrahydrofuran, *Journal of Chemical and Engineering Data* **56**, 11, 4152-4156.

Zhang J.S., Lee J.W. (2009a) Equilibrium of Hydrogen + Cyclopentane and Carbon Dioxide + Cyclopentane Binary Hydrates, *Journal of Chemical and Engineering Data* **54**, 2, 659-661.

Zhang J.S., Lee J.W. (2009b) Enhanced Kinetics of $CO_2$ Hydrate Formation under Static Conditions, *Industrial Engineering Chemistry Research* **48**, 13, 5934-5942.

Zhang J., Yedlapalli P., Lee J.W. (2009) Thermodynamic analysis of hydrate-based pre-combustion capture of $CO_2$, *Chemical Engineering Science* **64**, 22, 4732-4736.

Zhong Y., Rogers R.E. (2000) Surfactant effects on gas hydrate formation, *Chemical Engineering Science* **55**, 19, 4175-4187.

# 10

# Formation and Destruction of NDELA in 30 wt% MEA (Monoethanolamine) and 50 wt% DEA (Diethanolamine) Solutions

Hanna Knuutila*,  Naveed Asif,  Solrun Johanne Vevelstad and  Hallvard F. Svendsen

*Norwegian University of Science and Technology (NTNU), Department of Chemical Engineering, Sem Saelands vei 4,*
*Trondheim NO 7491 - Norway*
*e-mail: hanna.knuutila@ntnu.no*

\* Corresponding author

**Résumé — Formation et destruction de NDELA dans des solutions de 30%m de MEA (monoéthanolamine) et de 50%m de DEA (diéthanolamine)** — La formation de nitrosodiéthanolamine (NDELA) dans une installation pilote de laboratoire a été étudiée en injectant des quantités contrôlées d'oxyde d'azote et de dioxyde d'azote dans le flux de gaz entrant dans l'absorbeur. La destruction par irradiation UV de la NDELA présente dans le solvant a aussi été étudiée sur la même installation pilote. Deux campagnes de mesure ont été menées, la première utilisant une solution de 30 %m de monoéthanolamine (MEA) et la seconde utilisant une solution de 50 %m de diéthanolamine (DEA). Durant la campagne de mesure sur la DEA, la destruction de la NDELA dans les eaux de nettoyage a aussi été testée. De plus, la dégradation thermique d'échantillons de solution dégradée prélevés dans l'installation pilote a été testée. Les résultats indiquent que de la NDELA se forme en présence d'oxyde d'azote et de dioxyde d'azote. Il a été constaté que la destruction de la NDELA par la lumière UV dans la boucle de solvant était lente. La destruction de la NDELA par la lumière UV dans le compartiment des eaux de nettoyage a été démontrée. La dégradation de la NDELA durant des études de dégradation thermique à 135 °C a été établie.

*Abstract — Formation and Destruction of NDELA in 30 wt% MEA (Monoethanolamine) and 50 wt% DEA (Diethanolamine) Solutions — The formation of nitrosodiethanolamine (NDELA) in a lab scale pilot was studied by feeding known amounts of nitrogen oxide and nitrogen dioxide into the gas entering the absorber. In the same pilot, the destruction by UV-irradiation of NDELA present in the solvent was studied. Two campaigns were performed, one with 30 wt% monoethanolamine (MEA) and one with 50 wt% diethanolamine (DEA). During the DEA campaign the destruction of NDELA in the water wash section was also tested. Additionally, degraded solution samples withdrawn from the pilot were tested for thermal degradation. The results show that NDELA was formed when nitrogen oxide and nitrogen dioxide were present. Destruction of NDELA with UV-light in the solvent loop was found to be slow. In the water wash section, the UV-light destroyed the NDELA effectively. NDELA was found to degrade during the thermal degradation studies at 135°C.*

## INTRODUCTION

Global warming caused by anthropogenic $CO_2$ emissions is one of the most severe problems presently. Carbon capture and storage may offer a route to significantly reducing these emissions. Of the capture technologies, reactive absorption seems to be the most viable option. However, in order to operate absorption processes on a global scale, one has to make certain that the processes are benign and do not create additional environmental problems. One of the issues that could be detrimental to the application of this technology is the formation and potential emissions of nitrosamines when using amines or amino acids as absorption reagents.

Amine processes have been in use on modest scale for many years. One of the most used amines, MEA, has been a popular reagent for capture of $CO_2$ from power plant exhaust gases, *e.g.* the Warrior Run plant with ABB-Lummus technology (Kohl and Nielsen, 1997). Formation or emissions of nitrosamines from these plants have not been reported in the open literature.

The formation of nitrosamines from absorption plants can stem from two sources. One is atmospheric formation from emissions of solvent amines from the plant, as discussed by Bråten *et al.* (2008), Wisthaler (2010) and Nielsen *et al.* (2010). However, a proper wash process can limit the emitted exhaust gas amine content below 0.1 vppm (Graff *et al.*, 2013). Thus this problem seems to be under control, at least for MEA, and is not the focus of the present work.

However, nitrosamines may also be formed in the process itself. Fostås *et al.* (2010) found that gas containing $NO_x$ and oxygen led to the formation of levels of 20-50 µg/g of DEA in their AMINOX plant after 25-75 hours. The formation of nitrosamines was also studied and NDELA (Nitrosodiethanolamine) was the main product, forming at a rate of 200-700 ng/g after 25-100 hours. The end concentration of NDELA was 0.5 µg/g in the solvent solution. Traces of two volatile nitrosamines, nitrosodimethylamine (NDMA) and nitrosomorpholine (NMOR) were also detected.

Most of the nitrosamines formed in the plant will most likely stay in the solvent loop, but nitrosamines have been detected in the gas leaving the water wash section located above the absorber (Kolderup *et al.*, 2012). Even though these measurements were performed in a research pilot that was not designed to minimize nitrosamine or amine slip, it is reasonable to consider that nitrosamines would be found in the water wash solutions also in other plants if they are formed in the solvent liquid. The volatile nitrosamines will penetrate to the water wash section in gaseous form, whereas droplets and aerosols might transfer non-volatile nitrosamines from the absorber into the water wash.

The formation rate of nitrosamines under normal operation is not known, but will depend on temperature, liquid phase composition and the oxygen and $NO_x$ content of the gas phase. $NO_2$ is believed to be the critical component in $NO_x$ and will dissolve in the liquid phase and may disproportionate into nitrate and nitrite. The nitrite formed will be reactive toward secondary amine groups and can form nitrosamines, typically NDELA (Fostås *et al.*, 2011). This is one possible route to nitrosamine formation, but there may be others. Direct UV radiation could be an option to destroy the nitrosamines in both the solvent and water wash liquids.

Direct UV photolysis is currently used to remove NDMA from drinking water and treated wastewater and most of the literature available on destruction of nitrosamines with UV-light is related to water treatment applications (Sharma, 2012; Nawrocki and Andrzejewski, 2011). Some literature is available for amine applications. Jackson and Attala (2012) have a patent on treating an amine solvent with UV-radiation.

In this paper, the formation and destruction of NDELA in pilot conditions are studied. The formation of NDELA is studied by feeding known amounts of NO and $NO_2$ in the pilot, and the destruction of NDELA is studied by irradiating the solvent with UV-light. Two campaigns were made: one with 30 wt% MEA and another with 50 wt% DEA. During the DEA campaign the destruction of NDELA in the water wash section was also tested. Additionally degraded solution samples withdrawn from the pilot are tested for thermal degradation to determine if NDELA degrades at high temperatures.

## 1 EXPERIMENTAL APPARATUSES

### 1.1 UV-Light Reactor

The setup shown in Figure 1 was used to study the destruction of nitrosamines in the pilot plant. The reactor setup contained a centrifugal circulation pump, a valve to control the circulation rate, a commercial UV-light reactor (*Sterilight* silver S8Q-PA) and sampling points at the UV-reactor inlet and outlet. Main technical data of the commercial UV-light reactor with lamp effect of 37 W are presented in Table 1.

### 1.2 Lab Scale Pilot Plant

The lab scale pilot plant located at the Gløshaugen campus in Trondheim was used in the experiments and a

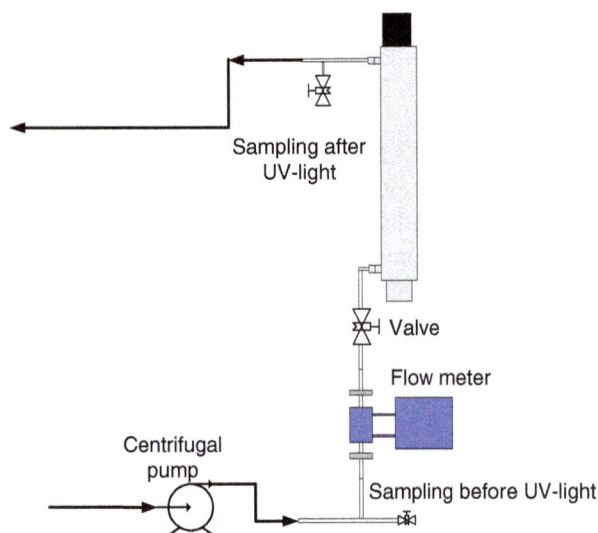

Figure 1

UV-reactor, connected to a centrifugal pump and flow meter.

TABLE 1

Technical information about Sterilight silver S8Q-PA UV-light reactor

|  | Value |
|---|---|
| Power consumption | 46 W |
| Lamp power | 37 W |
| Max. flow rate | 37.9 L/min |
| Chamber material | 304 stainless steel |
| Chamber length | 90.0 cm |
| Chamber diameter | 6.4 cm |
| Lamp | Sterilume-EX model S810RL |
| Sleeve | Quartz Model QS-810 |

in the figure). The condensate returns to the reboiler, and the $CO_2$ is mixed with the outlet gas from the water wash column and sent back as feed gas to the absorber. It should be noted that the complete plant is run as a closed system, thus all $CO_2$ that is stripped is transferred back to the absorber.

From the outlet of the absorber the gas enters either the water wash section, or in case this section is bypassed, goes directly to the fan and into the absorber. The fan is dimensioned to overcome the pressure drop in the absorber column and water wash sections. In the water wash section, wash water is circulated from the water tank to the top of the water wash column, and trickles counter currently to the gas downwards to the water tank.

Samples, for liquid analysis to obtain liquid species concentrations, were withdrawn at the inlets and outlets of the absorber and desorber and at the outlet of the reboiler as shown in Figure 2. The concentration of $CO_2$ was determined in the inlet and outlet gas of the absorber by 2-channel IR analyzers. The $CO_2$ analysers were calibrated with known mixtures of $N_2$ and $CO_2$ on regular basis. The oxygen level was measured at the absorber inlet with a Servoflex 5200 multipurpose oxygen analyser from *Servomex* (accuracy 0.2%). During this campaign, a factory calibrated portable DX4000 FTIR by *Gasmet* was connected to the inlet of the absorber and was mainly used to measure the NO and $NO_2$ levels.

While running the pilot, the liquid and gas flows, the temperature profiles in the packed columns, the $CO_2$ concentrations in and out the absorber, the reboiler heat duty and temperatures and pressures in the pipes were all continuously logged. A more detailed description of the plant is found in Tobiesen *et al.* (2007).

During this campaign, the UV-reactor setup shown in Figure 1 was connected either to the lean solvent outlet from the mixing tank or to the outlet from the mixing tank in the water wash system towards the end of the campaigns. In both cases the liquid was returned to the mixing tank after exposure to UV-light. Since the same UV-reactor setup was used in both the solvent loop and the water wash loop, no experiments were made where UV-light was simultaneously applied to the solvent circulation and the water wash circulation loop. The flow rate through the UV-light reactor was 3 L/min in all the experiments.

Known amounts of NO, $NO_2$, air and $CO_2$ were added to the gas leaving the absorber before the fan/water wash column using Bronkhorst HI-TEC mass flow controllers. The mass flow controllers were calibrated before the campaign (estimated accuracy of calibration $\pm 2\%$). A total of 4 mass flow controllers were used,

simplified flow sheet of the fully automated pilot plant is given in Figure 2. In Table 2 some additional information about the pilot is given. Lean amine exits the reboiler, flows through the cross flow heat exchanger HEX1, into a mixing tank, and is pumped through the heat exchanger (cooler) HEX2 to cool the solvent if the temperature into the mixing tank is too high. After passing through the absorption column, the rich amine exits the absorber, and is pumped through the heat exchanger HEX1 entering the top of the stripper.

The vapor from the top of the stripper enters two water-cooled condensers in series (one condenser shown

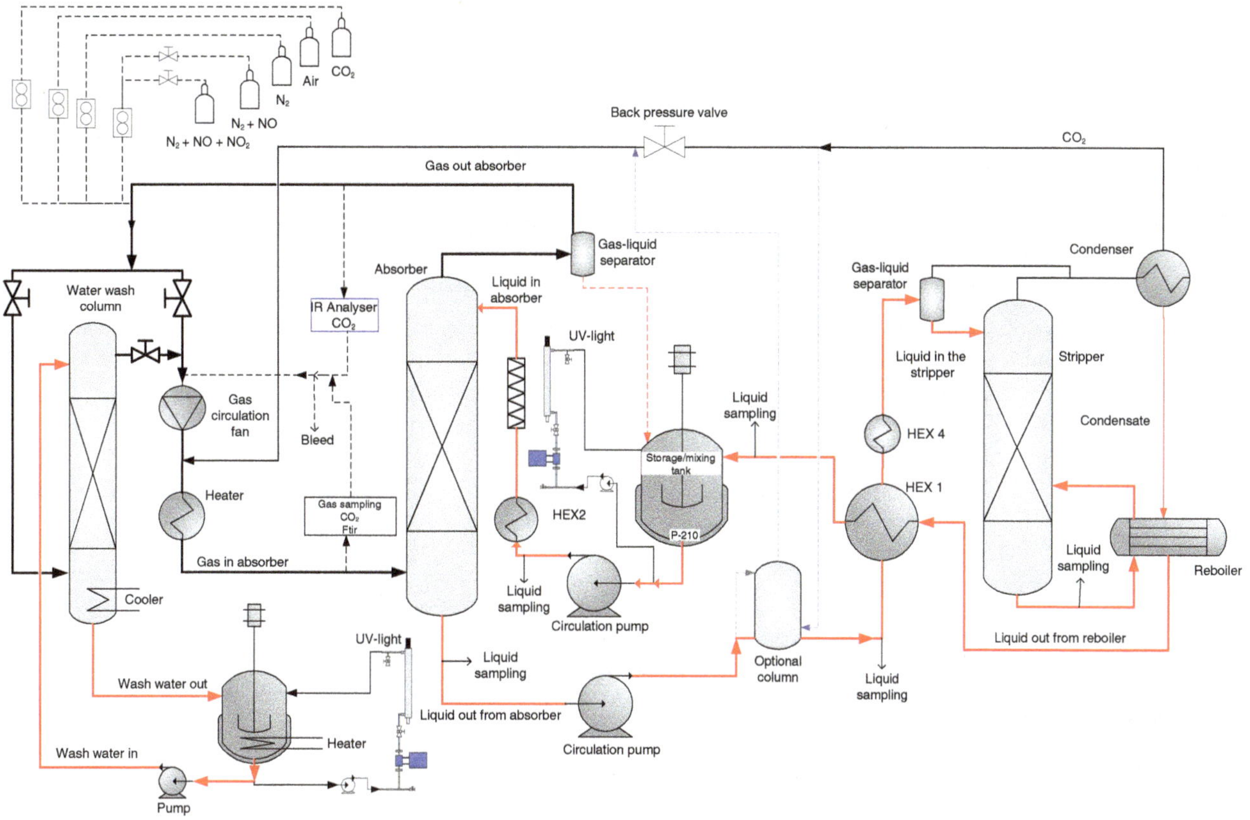

Figure 2

Flow diagram of the modified pilot plant. The UV-light reactor system with a pump, flow meter and UV-light reactor was used. The UV-light reactor system was connected either to the lean amine mixing tank or the water wash loop.

TABLE 2

Brief description of typical process parameters in the used pilot plant

| Packing type | |
|---|---|
| Wash water section | Mellapak250Y |
| Absorber section | Mellapak250Y |
| Stripper section | Mellapak250Y |
| Wash water diameter/height (m) | 0.15/2.10 |
| Absorber diameter/height (m) | 0.15/4.39 |
| Stripper diameter/height (m) | 0.1/3.89 |
| Temperature difference between rich amine inlet and lean amine outlet in LRX | Typically (-6)-(-8) |
| Reboiler heat duty (kW) | Up to 18 kW |
| Solvent flow (L/min) | 3-9 |
| Gas flow (m$^3$/h) | 100-150 |

one for air, one for $CO_2$ (99.999% pure) and one for $N_2$ (99.999% pure). The fourth flow controller was used to control the flow of premixed $N_2/NO$ or $N_2/NO/NO_2$ gas mixtures. The $N_2/NO$ gas mixture contained 1 800 ppm NO (relative uncertainty 2%) mixed with $N_2$ and the $N_2/NO/NO_2$ mixture contained 1 800 ppm NO (relative uncertainty 2%) and 200 ppm $NO_2$ (relative uncertainty 5%). With these flow controllers it was possible to control the amounts of NO, $O_2$ and $NO_2$ fed to the system. Small amounts of $CO_2$ were added to the gas phase to compensate for the $CO_2$ leaving through the bleed. The amount of $O_2$ was manually recorded whereas NO and $CO_2$ levels in the gas phase at the absorber inlet were recorded electronically.

The FTIR was able to measure the amount of NO in the gas phase but did not detect $NO_2$ in the gas phase during the campaign. This was probably due to the water condensate in the absorber inlet pipe absorbing the $NO_2$ from the gas phase before the FTIR. When the premixed $NO/NO_2/N_2$ was feed to FTIR, the FTIR was able to detect the $NO_2$. However since a ready mixture of

$N_2/NO/NO_2$ was used together with calibrated flow controllers, it was possible to calculate the amount of NO and $NO_2$ feed to the system.

## 1.3 Analytical Methods

The total alkalinity of the solutions was determined by acid titration (0.1 M $H_2SO_4$) and $CO_2$ concentrations were measured using the $BaCl_2$ method (Ma'mun et al., 2007). IC was used to measure nitrite, nitrate and formate with method described in Vevelstad et al., (2013) and LC-MS was used to analyse for DEA, MEA, nitrosamines, HEI, HEF, OZD, HEA, HEPO, HeGly and BHEOX. Methylamine, dimethylamine, ethylamine, diethylamine and ammonia were analysed using GC-MS. More thorough descriptions of the analytical methods for LC-MS and GC-MS can be found in da Silva et al. (2012) and Lepaumier et al. (2011).

## 2 PILOT PLANT EXPERIMENTS

Two pilot campaigns were run, one with 30 wt% MEA and one with 50 wt% DEA. The main objective of the campaigns was to study the formation and destruction of NDELA with NO and $NO_2$ present in the gas phase. For that reason the goal was to operate the pilot under stable operating conditions. Both campaigns were divided into three phases:

- the campaigns started with feeding small amounts of $NO/N_2$ mixture and $O_2$ to have a NO concentration around 100 ppm and an $O_2$ level between 5-7 vol% in the gas entering the absorber;
- after some time, the $NO/N_2$ mixture was switched to a $NO/NO_2/N_2$ mixture. The goal was to keep the NO level around 100 ppm, which would give $NO_2$ level of about 11 ppm if the same amount of NO and $NO_2$ were absorbed. The $O_2$ feeding was continued;
- in the third phase, the feed of all gases was stopped and the effect of UV-irradiation was measured in the solvent loop during the campaigns. In the 50 wt% DEA campaign, UV-irradiation was also tested in the water wash loop.

During the campaigns the solvent degradation and formation of NDELA was monitored by regular liquid sampling.

### 2.1 Pilot Operation During 30 wt% MEA and 50 wt% DEA Campaigns

During both of the campaigns the lean solvent flow rate was 3 L/min and the gas flow rate was around 120 $m^3$/h.

The pressure in the stripper and reboiler was held around 2 bara. Typical operating conditions during the MEA and DEA campaigns are presented in Table 3. As seen from the table for MEA, the rate of stripped and absorbed $CO_2$ was 5.0 kg/h giving a heat demand of 4.49 kJ/kg. This is in very good agreement with previous MEA campaigns performed in the same pilot (Tobiesen et al., 2008). In Figure 3, the temperature profiles in the stripper and absorber are presented for the conditions presented in Table 3. During the MEA campaign, some adjustments to both reboiler duty (varied from 3 kW to 7 kW) and lean loading were made due to operational problems. However the data presented in Table 3 gives a good picture of the operational conditions during the campaigns. During the DEA campaign, there were no operational problems.

The 30 wt% MEA solution used in this campaign was previously used for 700 hours in the same pilot. During those 700 hours no NO, $NO_2$ or $O_2$ was added to the system. In the present campaign, 30 wt% MEA was tested in the lab pilot for 990 hours as shown in Table 4. During the first 670 hours of the campaign, only NO and $O_2$ was added to the gas as shown in Figure 4. During this time, the NO feed was on for 528 hours. Starting from campaign hour 674, a pre-mixed gas containing both NO and $NO_2$ was fed into the gas phase until campaign hour 890 as can be seen in Figure 4. During this time, both NO and $NO_2$ were fed for 187 hours. The objective of the combined NO and $NO_2$ feeding was to monitor the formation of nitrosamines and degradation products. After 890 campaign hours, the NO, $NO_2$, $O_2$ feeds were stopped and the UV-light reactor was connected to the lean solvent mixing tank for UV-light tests.

The lab pilot campaign with 50 wt% DEA was similar to the MEA present campaign where lab pilot was operated for 990 hours. However, the DEA campaign lasted only 410 hours. As for the MEA campaign, the DEA campaign was started by feeding small amounts of the $NO + N_2$ mixture as well as $O_2$ and $CO_2$ to the gas phase. After 150 hours of $NO + N_2$ feed, the feed was switched to a mixture of $NO + NO_2 + N_2$ as shown in Table 4. The 50 wt% DEA solution was exposed to $NO_2$ for 100 hours. The NO and $NO_2$ feed as a function of campaign hours is shown in Figure 4b. After the NO and $NO_2$ feeds were stopped UV-radiation was tested for NDELA degradation.

The total amount of gas added to the system was between 1-1.5 L/min and assuming that the system was completely closed and stable, the added amount would only leave through the system bleed. Based on the mass flow controller and the composition of the added

TABLE 3
Typical operational parameters during MEA and DEA campaign

| | | 30 wt% MEA (27.4.2011) | 50 wt% DEA (24.5.2011) |
|---|---|---|---|
| Absorber | | | |
| *Flow conditioning column* | | | |
| Gas | $(m^3/h)$ | 120 | 110 |
| Liquid | (L/min) | 3 | 3 |
| Superf. velocity inlet ABS | (m/s) | 1.9 | 1.8 |
| Gas load | $(m^3/m^2, h)$ | 6 800 | 6 230 |
| Liquid load | $(m^3/m^2, h)$ | 10 | 10 |
| Gas/liquid ratio | | 666 | 610 |
| Absorbed $CO_2$ | (kg/h) | 5.0 | 4.2 |
| *Pressure* | | | |
| P_gas upstream Fl.meter (a) | (kPa a) | 107 | 103 |
| *Concentration liquid* | | | |
| Amine group | (mole MEA/L) | 5.6 | 5.0 |
| Lean loading | (mole/mole) | 0.35 | 0.17 |
| Rich loading | (mole/mole) | 0.45 | 0.26 |
| *Concentration gas* | | | |
| $CO_2$ inlet | (vol% dry) | 8.2 | 8.4 |
| $CO_2$ outlet | (vol% dry) | 5.4 | 6.0 |
| $CO_2$ inlet | (vol% wet) | 6.9 | 7.7 |
| $CO_2$ outlet | (vol% wet) | 4.6 | 5.5 |
| *Temperature* | | | |
| Gas inlet | (°C) | 55 | 49 |
| Gas outlet | (°C) | 57 | 45 |
| Liquid inlet | (°C) | 41 | 38 |
| Liquid outlet | (°C) | 54 | 48 |
| Reboiler | | | |
| Reboiler duty | (kW) | 6.8 | 5.6 |
| *Temperature* | | | |
| T-06 (R-solvent downstream EX01) | (°C) | 109 | 108 |
| T-07 (R-solvent inlet DES) | (°C) | 106 | 105 |
| T-09 (DES, swamp, L-solvent outlet) | (°C) | 110 | 106 |
| T-11 (steam out DES) | (°C) | 103 | 100 |

(continued)

TABLE 3 *(continued)*

|  |  | 30 wt% MEA (27.4.2011) | 50 wt% DEA (24.5.2011) |
|---|---|---|---|
| TK (overflow L-solvent reboiler) | (°C) | 116 | 113 |
| T-14 (steam reboiler) | (°C) | 116 | 113 |
| *Flow* |  |  |  |
| Flowrate condensate (Coriolis) | (kg/h) | 2.48 | 1.80 |
| Reflux ratio | (kg $H_2O$/kg $CO_2$) | 0.45 | 0.48 |

Legend:

- TA1--TAB_column (50 wt% DEA)
- Gas top ABS, T-15 (50 wt% DEA)
- Lean solvent ABS inlet, T-05 (50 wt% DEA)
- TA1--TAB_column (30 wt% MEA)
- Gas top ABS, T-15 (30 wt% MEA)
- Lean solvent ABS inlet, T-05 (30 wt% MEA)
- TT-01 (R-solvent inlet Col.2)
- Gas inlet pipe, TAB (50 wt% DEA)
- Gas pipe ABS outlet T-03 (50 wt% DEA)
- Rich solvent ABS sump, TA0 (50 wt% DEA)
- Gas inlet pipe, TAB (30 wt% MEA)
- Gas pipe ABS outlet T-03 (30 wt% MEA)
- Rich solvent ABS sump, TA0 (30 wt% MEA)
- TS1--TS5_column (50 wt% DEA)
- Steam reboiler, T-14 (50 wt% DEA)
- Lean solvent DES sump, T-09 (50 wt% DEA)
- TS1--TS5_column (30 wt% MEA)
- Steam reboiler, T-14 (30 wt% MEA)
- Lean solvent DES sump, T-09 (30 wt% MEA)
- Lean solvent reboiler, Tk (50 wt% DEA)
- Steam out DES, T-11 (50 wt% DEA)
- Rich solvent inlet DES, T-07 (50 wt% DEA)
- Lean solvent reboiler, Tk (30 wt% MEA)
- Steam out DES, T-11 (30 wt% MEA)
- Rich solvent inlet DES, T-07 (30 wt% MEA)

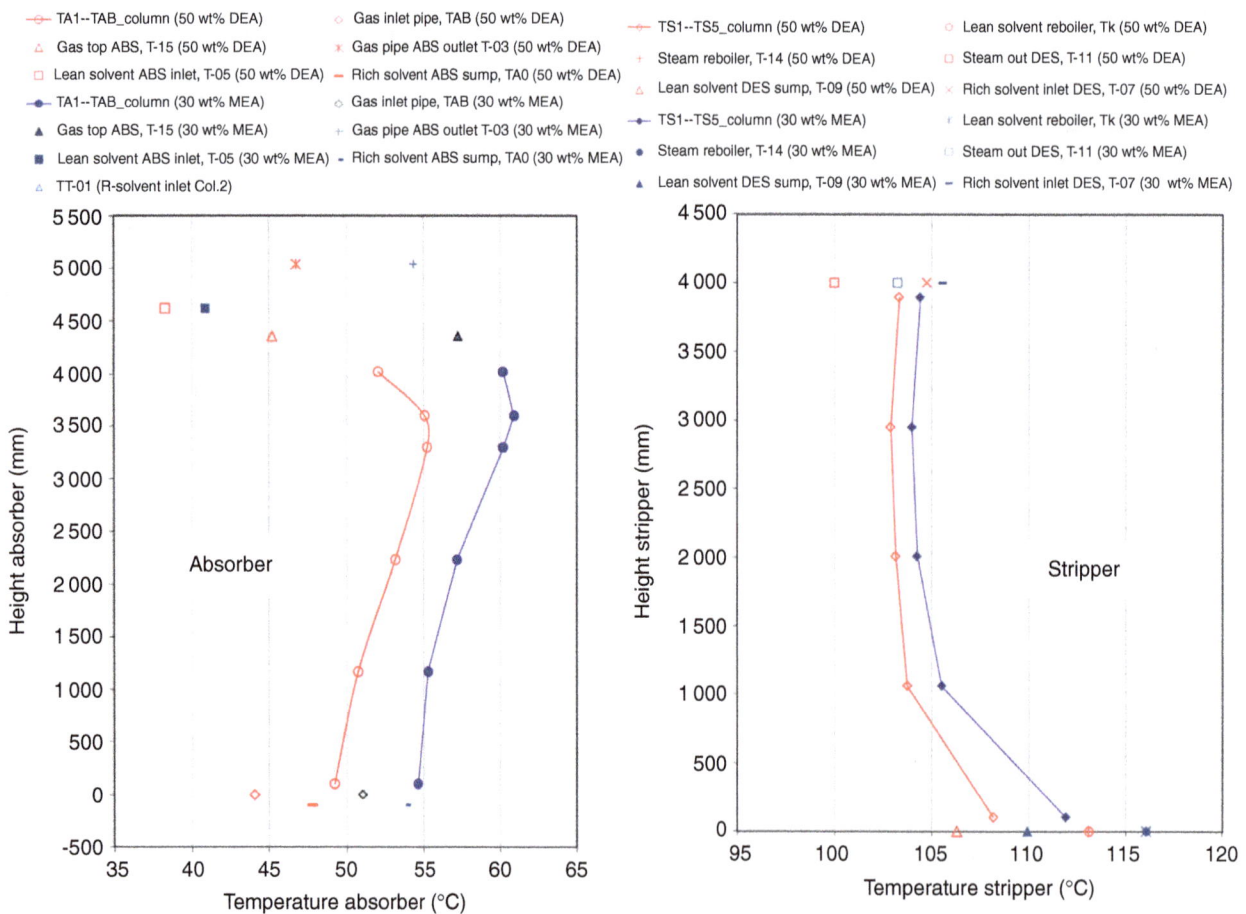

Figure 3

The temperature profiles in the absorber and stripper during 30 wt% MEA on (in blue) and 50 wt% DEA campaigns on (in red).

$N_2 + NO$ and $N_2 + NO + NO_2$ gas mixtures provided by the supplier, the amount of added NO and $NO_2$ could be calculated.

During the 30 wt% MEA campaign, the NO concentrations measured were between ~70-110 ppm (dry) in the absorber inlet. The oxygen level at the absorber inlet

TABLE 4

Basic information about the 30 wt% MEA and 50 wt% DEA pilot campaigns

|  | 30 wt% MEA | 50 wt% DEA |
| --- | --- | --- |
| Campaign duration (h) | 990 | 410 |
| NO feed (actual feeding hours) (h) | 715 | 250 |
| $NO_2$ feed (actual feeding hours) (h) | 187 | 100 |
| UV-light radiation in the main solvent circulation (h) | 37 | 48 |
| UV-light radiation in water wash section (h) | - | 5 |

varied from 5 to 10 vol% (dry). During DEA campaign the NO concentration at the absorber inlet was ~110 ppm (dry), whereas the oxygen level decreased during the first 70 campaign hours from above 10 to 6 vol% (dry). After that the oxygen level stayed at a very constant level around 5.5 vol% (dry).

## 3 RESULTS FOR THE 30 WT% MEA CAMPAIGN

### 3.1 Solvent Degradation During 30 wt% MEA Campaign

Degradation of the solution was monitored during the whole campaign by taken liquid samples regularly. The liquid samples were analysed with IC and LC-MS. The analysed concentrations of HEI, HEF, OZD, HEA, HEPO, HeGly and BHEOX during the campaign are shown in Figure 5. From the figure it can be seen that the solvent was partially degraded from the campaign start because of the earlier use for approximately 700 hours where no NO/NO$_2$ was added to the system. The results at 0 hours thus show the degradation products accumulated without NO/NO$_2$ in the gas phase after approximately 700 hours. It can be seen that the order of these degradation products with respect to relative amounts formed, does not seem to be affected by the NO and NO$_2$ addition. The DEA concentration was quite stable over the whole campaign as seen from Figure 6. Additionally, from Figure 6 it can be seen that the formate and nitrate concentrations increased throughout the campaign. Formate is a degradation product and also a reactant further to HEF. The formate concentration is clearly increasing during the UV-radiation of 34 hours (campaign hours from 958 to 992). The nitrate concentration seems to be slowly increasing. Nitrate concentration in lean amine sample withdrawn after the mixing tank and sample withdrawn at UV-reactor outlet seems to increase during the 34 hour UV-radiation (sample at 958 and 992 hours). Nitrite concentration was below detection limit throughout the campaign.

Selected samples were also analysed for methylamine, dimetylamine, ethylamine and diethylamine, from which only methylamine was detected in mg/L whereas ammonia, another volatile degradation compound, was detected at 200 times higher concentrations.

Based on analysis of LC-MS scan N-(2-hydroxyethyl)-ethylenediamine (HEEDA) and 1-(2-Hydroxyethyl)-2-imidazolidinone (HEIA) were identified based on mass and retention time. These compounds have been previously reported by da Silva et al. (2012) to be formed in thermal degradation experiments. Similarly, 2-(2-hydroxyethylamino)-2-oxoacetic acid (HEO) has been reported by da Silva et al. (2012) and a peak at the right mass was also seen in the scans.

The samples were also analysed for following nitrosamines: NDMA, NDEA, NPIP, NMEA, NPYR, NMOR, NDPA, NDBA and NDELA. During addition of NO gas, no formation of NDELA or any other nitrosamine was detected. After combined NO and NO$_2$ addition 230 ng/mL of NDELA was detected (see Sect. "Formation of NDELA"). The other nitrosamines (NDMA, NDEA, NPIP, NMEA, NPYR, NMOR, NDPA and NDBA) were not detected. However, total nitrosamine after 940 hours was analysed externally by Henkel AG & Co with a method described in Langenohl et al. (2011) and it showed much higher nitrosamine concentration than that of NDELA. This indicates that unknown nitrosamines were present. This agrees with Einbu et al. (2012) who reported that only few percent of nitrosamines present during their tests were found to be NDELA.

### Formation of NDELA

The formation of NDELA during the campaign is presented in Figure 7. Additionally in the figure, the accumulative amounts of NO and NO$_2$ are presented. During NO feeding the NDELA concentration was below the Limit Of Quantification (LOQ).

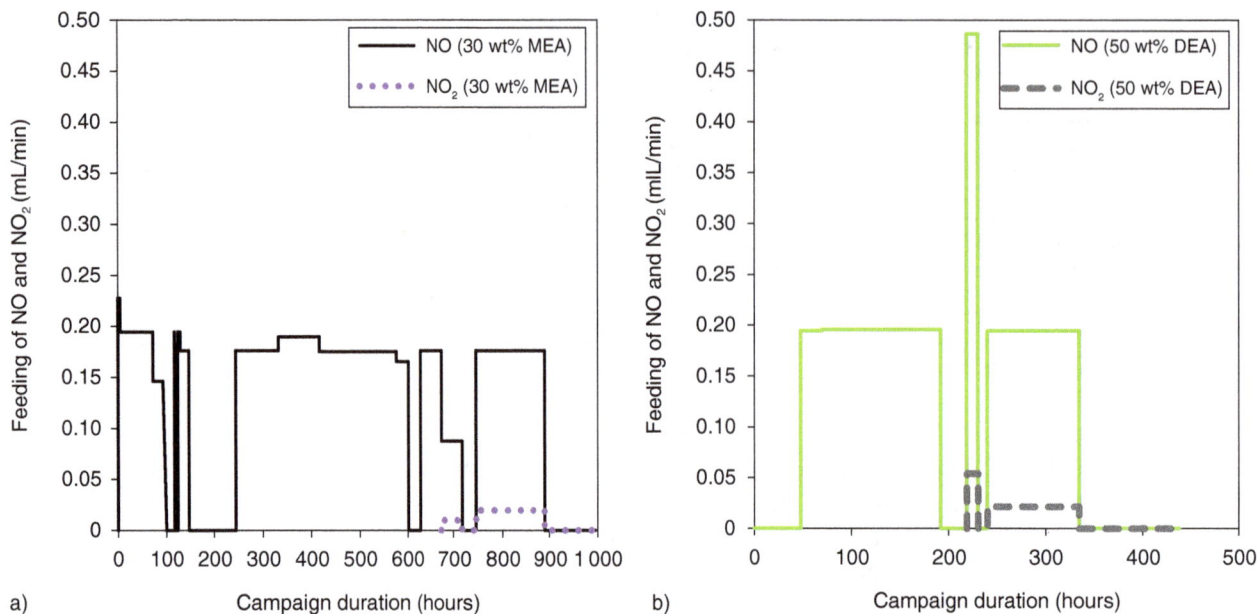

Figure 4

Added NO and NO$_2$ during a) 30 wt% MEA and b) 50 wt% DEA campaigns. Oxygen was added at the same times, when NO and/or NO$_2$ was added.

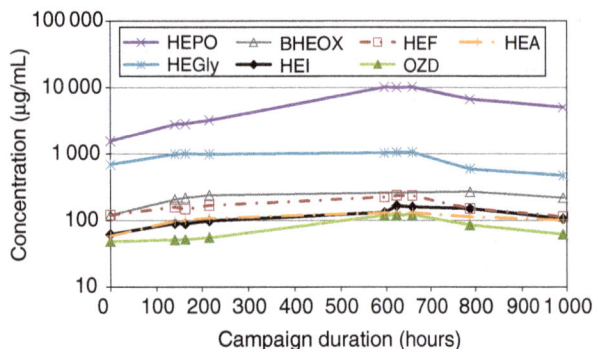

Figure 5

Concentration of degradation products during MEA campaign.

After 660 campaign hours the NO$_2$ feeding was started (NO and O$_2$ feed were continued) and after 740 campaign hours NDLEA was detected at a concentration of 1.7 nmol/mL (230 ng/mL). Pedersen *et al.* (2010) reported a NDELA concentration of 500 ng/g solvent after 95 hours with NO$_x$ levels of 25-50 ppmv. They did not report what was the NO/NO$_2$ ratio during the tests. Even though the results are not exactly the same,

they are of the same order of magnitude and considering that experimental conditions were different, the results agree reasonable well with each other. After 890 hours the solution was spiked with NDELA to ensure high enough NDELA concentrations during the UV-light tests (shown with a black line) and the feed of NO, NO$_2$, O$_2$ and CO$_2$ were stopped.

## 3.2 Destruction of NDELA with UV-Light

As mentioned above, after 890 hours the solvent was spiked with NDELA to increase the concentration with ~300 ng/mL$_{solution}$ of NDELA (2.24 nmol/mL) to ensure high enough NDELA concentration for UV-light tests. At campaigning hours between 932 and 935, the circulating solution was exposed to UV-light for 3 hours. The start NDELA concentration was 550 ng/mL. Three sampling points were used: one at the UV-reactor inlet, one at the UV-reactor outlet and the third sampling was the return of lean solvent from the reboiler. The sampling points before and after the UV-light reactor tell about the effect of UV light whereas the out of reboiler sample tells more how the UV-light effects the NDELA concentration in the amine solution in the whole pilot. However 3 hours of UV-radiation had no significant effect on the NDELA concentration in the solution. Therefore a longer UV-light campaign was started two

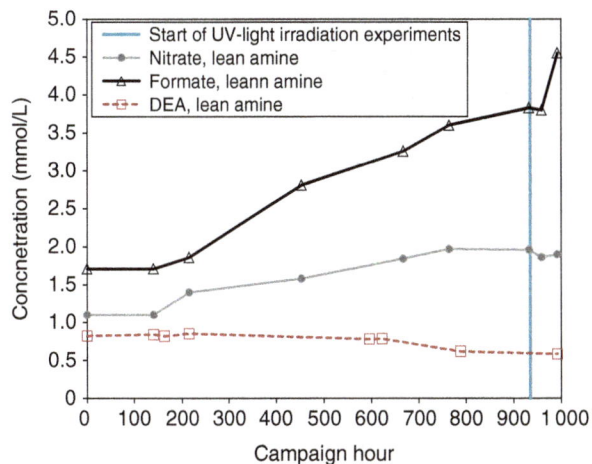

Figure 6

Concentration of DEA, formate and nitrate during 30 wt% MEA campaign. The measurement at campaign hour 932, is a lean solvent after first 3 hours of UV-irradiation; the sample at campaign hour 958 is taken just before starting the second UV-irradiation and the sample at campaign hour 992 is taken after 34 hours of UV-radiation.

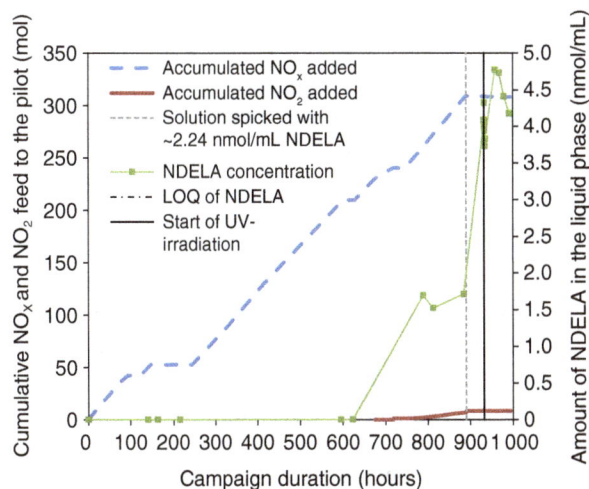

Figure 7

Cumulative NO and $NO_2$ feed together with NDELA concentration during the 30 wt% MEA campaign. For the first 660 hours, the concentration of $NO_x$ is only NO since no $NO_2$ is added.

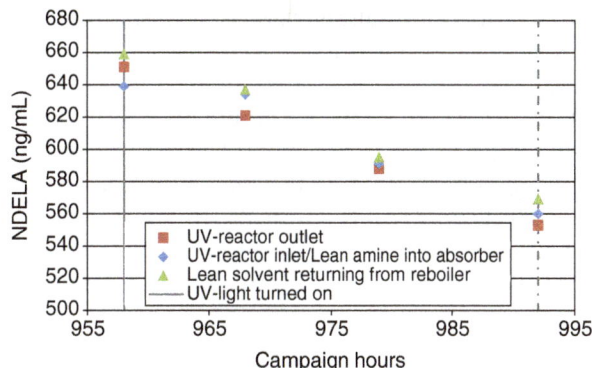

Figure 8

NDELA concentration during the 34 h UV-radiation.

days later. This time the UV-light was on for 34 hours and from Figure 8, it can be seen that the concentration of NDELA decreased approximately 10%. Additionally it can be seen from the results that the NDELA concentration after the UV-light reactor is very close to the NDELA concentration entering the UV-light. This indicates that the UV radiation was not very effective. Interesting point here is also that at the start of the first UV

campaign the NDELA concentration was 550 ng/mL where as two days later the concentration was 640-660 ng/mL. However NO and $NO_2$ feed to the system was turned off two day before both of these tests. This indicated that the reactions towards NDELA continued after feeding of NO and $NO_2$ is stopped. One final remark is that the first samples (956 hours) for the long term UV-radiation tests should be equal. The analytical results vary from 639 to 659 ng/mL which are within 3%. This value indicates a good analytical accuracy.

During the long term radiation the destruction of NDELA seems to be 0-order reaction. The solvent in the pilot campaign was coloured and had a low penetration depth throughout the campaign (below 0.5 cm). The low penetration depth can also explain why short time irradiation of UV-light did not degrade NDELA significantly. The short penetration depth can explain why we see 0-order reaction. NDELA had a limited access to UV-light and the light was absorbed by other components in the solution. This is in line with results from Knuutila et al. (2012) with fresh and used MEA solutions, where 1st order reaction behaviour was seen in fresh solutions but where coloured solutions would exhibit 0th-order behaviour. In Knuutila et al. (2012), the total liquid amount was ~30 L, where in the pilot plant there is around 180 L of solvent. In this paper and in Knuutila et al. (2012), the same UV-reactor and same liquid flow was used. So during the pilot campaign the solvent passed the UV-reactor once an hour (since also the solvent circulation rate was 3 L/min) whereas in the batch scale experiments presented in Knuutila et al. (2012) the 30 wt% MEA passed the UV-reactor every 10 minutes. This difference can also explain some of the difference in the degradation rates between Knuutila et al. (2012) and this work.

In the literature, nitrite and nitrate has been reported to be formed during decomposition of NDMA (Plumlee and Reinhard, 2007; Lee *et al.*, 2005). In this work, the amount of degraded NDELA during the 34 hour UV-light campaign was 0.59 nmol/mL (from 639 ng/mL to 560 ng/mL). Even if all this amount would give nitrate, the 0.59 nmol/mL is too small to be seen in the nitrate concentration shown in Figure 6, when compering to the increase in nitrate seen throughout the campaign.

## 4 PILOT TESTS WITH 50 WT% DEA

### 4.1 Solvent Degradation During 50 wt% DEA Campaign

During the DEA campaign, some liquid samples were analysed for degradation products using LC-MS. The results are shown in Figure 9. The concentration of NDELA is also shown in Figure 9. The decrease in the NDELA concentration from 70 to 22 µg/mL from 337 to 385 hours is due to the UV-radiation, which will be discussed more detailed in the next chapter. From the figure, it can be seen that NDELA is the most common degradation compound from those that were analysed followed by HEPO, HEGly and BHEOX. NDELA was below LOQ (50 ng/mL) before the campaign, but already after 150 hours, the concentration was 24 µg/mL. The samples were also analyzed for other nitrosamines (NDMA, NDEA, NPIP, NMEA, NYPR, NMOR, NDPA and NDBA) but all concentrations were below the detection limit. Formate was below LOQ in all samples analysed. Nitrate was detected only in the sample taken after UV-light campaign (after 385 campaign

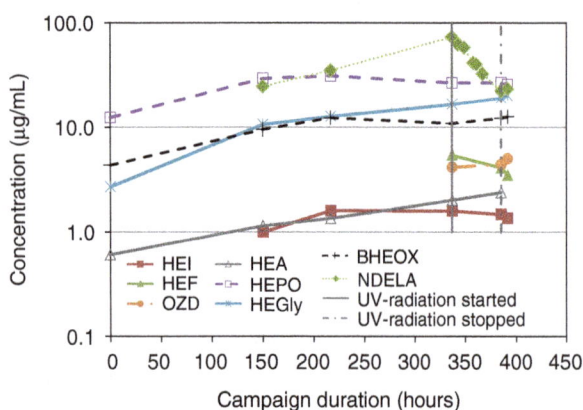

Figure 10

Cumulative $NO_x$ and $NO_2$ feed (on the left hand axis) together with NDELA concentration (on the right hand axis) during 50 wt% DEA campaign. For the first 220 hours the concentration of $NO_x$ is only NO since no $NO_2$ is added.

hours). HEI was below limit of quantification (LOQ = 1 µg/mL) until 150 hours. OZD and HEF were below LOQ (LOQ = 1 µg/mL) until 337 hours.

### Formation of NDELA

The accumulative amounts of NO and $NO_2$ feed into the system together with the formation of NDELA during the campaign are presented in Figure 10. The figure clearly shows that the NDELA started to form already during the NO feeding and after 220 hours the NDELA concentration was 260 nmol/mL. During the 30 wt% MEA campaign, no NDELA was detected during 660 hours of NO feeding. The reason is likely to be the very different DEA concentrations (5 000 mmol/L in DEA campaign, and 0.6 mmol/L in MEA campaign, since NDELA is formed from DEA. After 220 hours feeding of $NO_2$ and NO was started. The formation of NDELA continues to increase. It is however difficult to conclude based on the data, if the formation rate of NDELA increased during feeding of $NO/NO_2$ mixture.

### 4.2 UV-Light Tests with 50 wt% DEA

The UV-light reactor was connected to the solvent mixing tank in the same manner as during the MEA campaign. The liquid flow rate through the UV-light reactor was 3 L/min and the total experimental time was 48 hours. Since the NDELA concentration was

Figure 9

Formation of degradation products as a function of campaign hours. After campaign hour 385 the UV-irradiation was connected to the water wash system and no samples from the main loop were analysed for degradation after that.

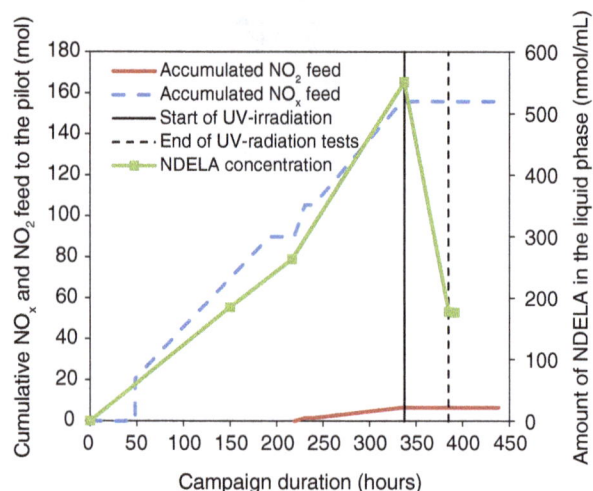

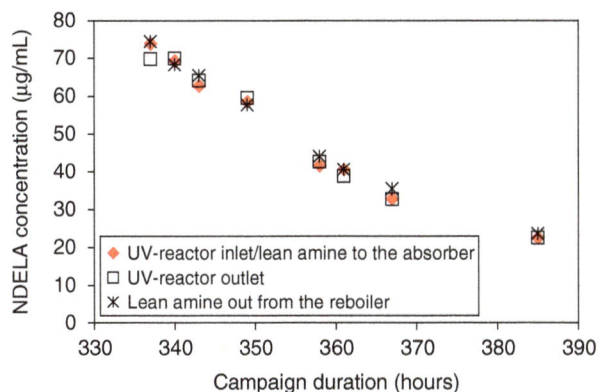

Figure 11

NDELA concentration during UV-radiation.

not known before starting the UV-light tests, the solution was spiked with NDELA. The concentration that would have been reached with the spiked NDELA was estimated to ~0.3 μg/mL. After the first UV-light campaign, the NDELA concentrations for the whole campaign were analysed and it was found that the starting concentration was 70 μg/mL from which 0.3 μg/mL was spiked NDELA. The decomposition of NDELA during the UV-radiation campaign is shown in Figure 11.

During 48 hours approximately 70% of the NDELA was decomposed. This is much faster than with MEA. The penetration depth of the 50 wt% DEA solution was below 0.5 cm and very close to that of 30 wt% MEA during the MEA campaign. However visual comparison of the MEA and DEA solution showed a colour difference, with 30 wt% MEA being much darker colour. Maybe, the fact that the NDELA concentration during the DEA campaign was more than 10 times higher, could be part of the explanation, why degradation of NDELA was faster in DEA solution.

Before UV-light was started no nitrite or nitrate were detected with IC analyses. After the UV-light campaign in the lean amine sample 0.43 mmol/L of nitrate was detected. This value is very close to the limit of quantification and samples taken from other parts of the pilot were below LOQ. During the UV-light campaign and concentration of NDELA decreased from 74 to 22.5 μg/mL which corresponds to 0.4 mmol/L, which is close to the value of nitrate found in the solution. It could be speculated if this detected nitrate is due to the decomposition of NDELA into nitrate. Nitrate and nitrite have been reported in literature to be formed under NDMA photodegradation in weakly acidic water solution (Plumlee and Reinhard, 2007). However literature findings also suggest

that in alkaline solutions, other inorganic nitrogen products than nitrate and nitrite like $N_2$ and $N_2O$ are formed (Xu et al., 2009a; Stefan and Bolton, 2002).

During the pilot campaigns, the penetration depth for both of the amine solutions was low, and even during 50 wt% DEA campaign which was started with fresh 50 wt% DEA the penetration depth was around 0.5 cm after 180 hours of running. The low penetration depth can explain the slow degradation of NDELA in the DEA solution. The results are similar to those reported in Knuutila et al. (2012) where the 50 wt% DEA was tested in a batch reactor was tested and the decomposition of NDELA was found to be slow.

## UV-Light in the Water Wash Circulation

The effect of UV-light in the water wash section was tested at the end of DEA campaign. One day before the experiment the water wash section was spiked with a small amount of DEA and NDELA solution to ensure an NDELA concentration that could be detected by LC-MS and also to have DEA concentration representative of what can be expected in a water wash, about 0.5 wt%. The UV-light was connected to a mixing tank as shown in Figure 2 and the liquid flow through the UV-light reactor was approximately 3 L/min. The total amount of liquid in the water wash section is approximately 35 L.

The decomposition of NDELA is shown in Figure 12. The destruction of NDELA in the water wash solution is much faster compared to destruction rates found in the solvent loops. Here, the NDELA concentration decreases from 360 ng/mL to 6 ng/mL in 90 minutes. This can be explained by the colourless solution and low amine concentration which both decreases the

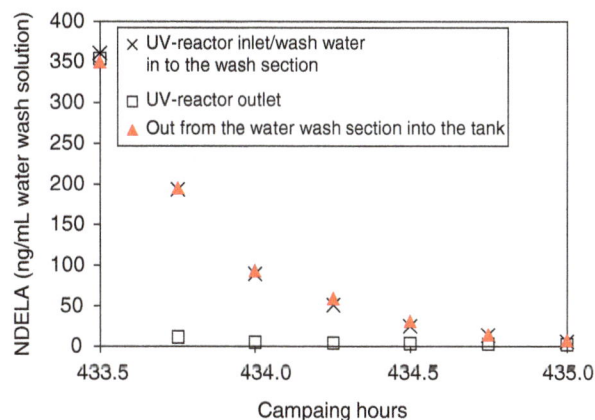

Figure 12

NDELA concentration in the water wash circulation loop during the UV-light reactor experiment.

penetration depth and thereby also the effect of the UV-light (Knuutila et al., 2012). From the results with UV-irradiation, one can see that when the solution goes through the UV-light reactor, almost all the NDELA is decomposed during one pass through. Additionally the destruction seems to be 1st order in respect to NDELA. These results clearly indicate that the use of UV-light is much more efficient in the water wash section compared to the solvent loop.

## 5 THERMAL DEGRADATION OF NDELA

Pilot solutions after the 30 wt% MEA and 50% DEA campaigns were tested for thermal degradation. Thermal degradation metal cylinders (316 stainless steel tubes) were used. The cylinders were filled with the end solutions from 30 wt% MEA and 50 wt% DEA and closed in a glove box under a nitrogen atmosphere. For each experiment 10 cylinders were prepared. To check for leakages, the cylinders including solution were weighed before and after the experiment.

The cylinders were stored in a thermostatic chamber at an upright position at 135°C and were not moved or agitated until sampled. Every week, 2 cylinders with 30 wt% MEA and 50 wt% DEA were sampled. The solution was analysed to determine the degradation of NDELA, MEA and DEA and the concentration of nitrate, nitrite, formate, HEA, HEI, HEPO, OZD, HEF, BHEOX and HeGly. The total experimental time was 5 weeks. Cylinders opened for sampling were not returned to the chamber. Parallel cylinders removed from the thermostatic chamber day 21 and 35 were both analysed. From other samplings days only one cylinder was send to analyses.

### 5.1 Experiment with MEA Taken from the Pilot

The MEA solution was analysed for amine and $CO_2$ before starting the experiments. The amine concentration was 28 wt% and loading was 0.35 $mol_{CO_2}/mol_{amine}$. The concentration of MEA during the thermal degradation tests are shown in Figure 13. It can be seen that the MEA concentrations decreases from 5 to 2.88 mol/L, which equals a 43% loss of MEA. This is quite similar to the results presented by da Silva et al. (2012). Additionally it can be seen that DEA concentration increases throughout the experiment, which was expected since DEA is a known product of MEA degradation (Fostås et al., 2011). In Figure 14, the NDELA concentration is shown to decrease under quantification limit (100 ng/mL) during the first week and it stays under limit of quantification until the end of the experiment.

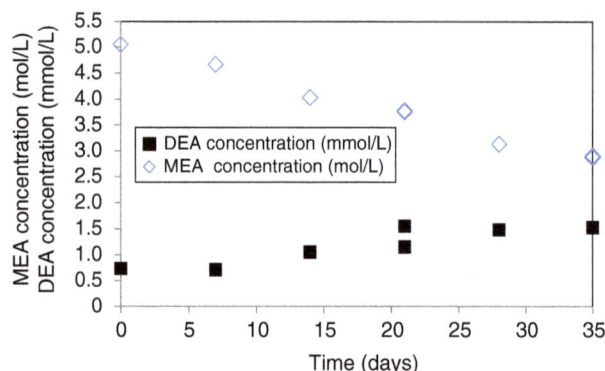

Figure 13

Concentration of MEA and DEA during thermal degradation test with 30 wt% MEA previously used in a pilot plant.

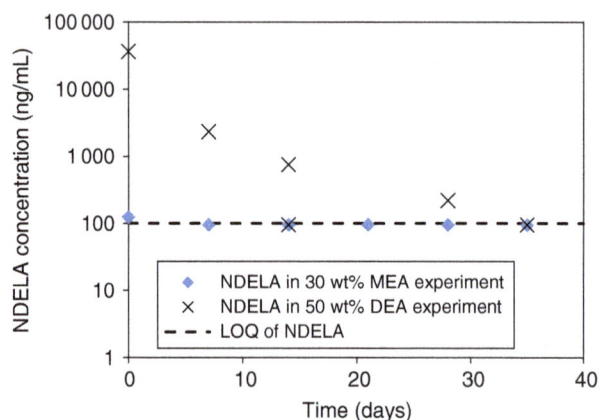

Figure 14

NDELA concentrations during MEA and DEA experiments. NDELA concentrations below 100 ng/mL are under quantification limit.

HEPO, HEA, HEF and HEI, concentrations stayed constant during the experiment whereas the HeGly decreased throughout the experiment. HEA, HEF, HEI and OZD were detected at levels of hundreds of µg/mL, while HEPO concentration was 10 times higher compared to HEA, HEF and HEI. The concentration of formate, doubled during the experiment ending up to 0.6 $g/kg_{solution}$. Nitrate, nitrite and oxalate were only detected in the original sample (day 0), but the concentrations were too low to be quantified.

### Thermal degradation of DEAThermal degradation of DEA

The DEA previously used in the pilot was analysed for amine and $CO_2$ loading prior the experiment. The

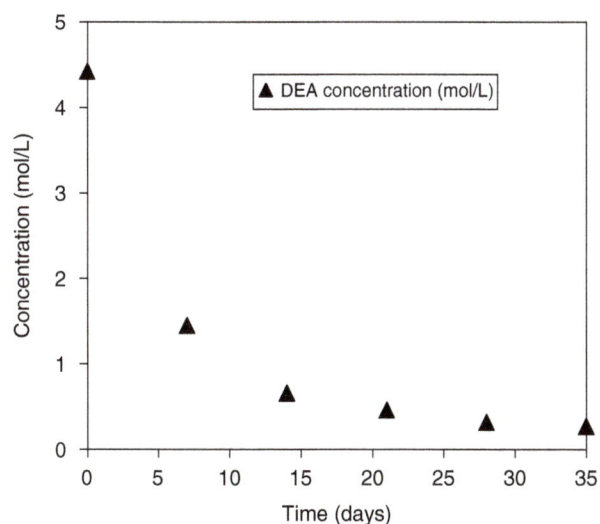

Figure 15

Concentration of DEA during thermal degradation test with 50 wt% DEA previously used in a pilot plant.

DEA concentration was 48.5 wt% and loading was 0.16 $mol_{CO_2}$/$mol_{amine}$. DEA degraded almost fully (93%) during the thermal degradation experiment as shown in Figure 15. This is in a good agreement with results from Eide-Haugmo (2011) who found that 96% of DEA degraded during thermal degradation tests with fresh DEA at a $CO_2$ loading of 0.5 $mol_{CO_2}$/mol amine. The concentration of NDELA is shown in Figure 14. From the figure, it is clear, that NDELA decomposes at 135°C. After one week, the concentration of NDELA comes down from 36 000 ng/mL to 2 400 ng/mL and after 4 weeks, the concentration is down to 280 ng/mL and after 5 weeks, it is below the limit of quantification (100 ng/mL).

HEPO was detected at levels of few hundred micrograms/mL and HeGly concentration was around 4 times smaller compared to HEPO. The concentrations of HEA and HEF were below the limit of quantification (10 μg/mL), except for the first sample for HEA and last sample of HEF which were just above it. OZD, HEI and BHEOX (10, 10 and 100 μg/mL, respectively) were below the limit of quantification throughout the experiment. Similarly as with MEA, the analysed amount of formate increased throughout the experiment ending up to 500 μg/$g_{solution}$. Oxalate and nitrate were detected in the start sample, but the concentrations were below limit of quantification. Nitrite was not detected at all.

## CONCLUSIONS

During the pilot campaigns UV-destruction was tested for both 30 wt% MEA solution and 50 wt% DEA

solution. In both cases, the UV-light was working, even though the rate of decay was low. For the DEA solution the decrease in NDELA concentration was higher compared to 30 wt% MEA campaign. This is believed to be related to the high NDELA concentration during the DEA campaign. The small scale experiments presented in Knuutila et al. (2012) support and agree well with the pilot results.

During 50 wt% DEA campaign destruction of NDELA in water wash loop was found to be fast. 98% of NDELA decomposed during 90 minutes of UV-irradiation. This is much faster compared to the UV-irradiation tests done in the solvent loop indicating the optimal location for UV-radiation would be in the water wash section. However, it should be remembered that the destruction rates are dependent on the nitrosoamine (Knuutila et al., 2012; Plumlee and Reinhard, 2007; Xu et al., 2009b) and in this work, only NDELA was tested.

During the pilot campaign, the formation of nitrosamines was also studied. From the nitrosamines analysed, only NDELA was detected. During the MEA campaign 660 hours of NO feeding did not give a detectable NDELA concentration. After, this NO and $NO_2$ were added for 100 hours, and the NDELA concentration rose to around 220 ng/mL. During the DEA campaign, the NDELA started to form already during NO feeding and no change in the formation rate was seen when the solvent was exposed to both NO and $NO_2$. In total, after 250 hours of $NO_x$ feed, the NDELA concentration was 70 μg/mL. Other degradation products were also monitored. The relative order of the analyzed degradation compounds analyzed during the 30 wt% MEA and 50 wt% DEA campaigns did not depend on the presence of NO and $NO_2$ in the gas phase.

The thermal degradation experiments done with degraded MEA and DEA solutions at 135°C, showed that NDELA degrades at high temperatures.

## REFERENCES

Bråten H.B., Bunkan A.J., Bache-Andreassen L., Solimannejad M., Nielsen C. (2008) Final report on a theoretical study on the atmospheric degradation of selected amines, Oslo/kjeller (NILU OR 77/2008).

da Silva E.F., Lepaumier H., Grimstvedt A., Vevelstad S.J., Einbu A., Vernstad K., Svendsen H.F., Zahlsen K. (2012) Understanding 2-Ethanolamine Degradation in Postcombustion $CO_2$ Capture, Industrial Engineering Chemistry Research 51, 41, 13329-13338.

Eide-Haugmo I. (2011) Environmental impacts and aspects of absorbents used for $CO_2$ capture, Doctoral theses, NTNU Trondheim.

Einbu A., DaSilva E., Haugen G., Grimstvedt A., Lauritsen K. G., Zahlsen K., Vassbotn T. (2012) A new test rig for studies of degradation of $CO_2$ absorption solvents at process conditions; comparison of test rig results and pilot plant data for degradation of MEA, *11th International Conference on Greenhouse Gas Technologies, GHGT11*, Kyoto, Japan.

Fostås B.F., Sjøvoll M., Pedersen S. (2010) Flue gas degradation of amines, *Climitdagene Soria Moria*, Oslo, October.

Fostås B., Gangstad G., Nenseter B., Pedersen S., Sjøvoll M., Sørensen A.L. (2011) Effects of $NO_x$ in the flue gas degradation of MEA, *Energy Procedia* 4, 1566-1573.

Graff O.F., Bade O.M., Gorset O., Woodhouse S. (2013) Amin utslippskontroll, *Norwegian patent 332812*.

Jackson P., Attalla M.I. (2012) Solvent treatment process, *European patent EP 2536483*.

Knuutila H., Svendsen H., Asif N. (2012) Destruction of nitrosoamines with UV-light, *11th International Conference on Greenhouse Gas Technologies, GHGT11*, Kyoto, Japan.

Kohl A.L., Nielsen R.B. (1997) *Gas Purification*, Fifth ed., Gulf Publishing Company, Houston.

Kolderup H., Hjarbo K.W., Mejdell T., Huizinga A., Tuinman I., Zahlsen K., Vernstad K., Hyldbakk A., Holten H., Kvamsdal H.M., van Os P., da Silva E.F., Goetheer E., Khakharia P. (2012) Emission studies at the Maasvlakte $CO_2$ capture pilot plant, *University of Texas conference on CCS*, 25 Jan.

Langenohl N., Frings M., Herrmann R., Frischmann M. (2011) Control of Nitrosamine Formation in $CO_2$ Capture Plants, *The IEA Nitrosamine workshop*, 1-2 Feb.

Lee C., Choi W., Yoon J. (2005) UV Photolytic Mechanism of N-Nitrosodimethylamine in Water: Roles of Dissolved Oxygen and Solution pH, *Environmental Science Technology* 39, 24, 9702-9709.

Lepaumier H., Grimstvedt A., Vernstad K., Zahlsen K., Svendsen H.F. (2011) Degradation of MMEA at absorber and stripper conditions, *Chemical Engineering Science* 66, 3491-3498.

Ma'mun S., Svendsen H.F., Hoff K.A., Juliussen O. (2007) Selection of new absorbents for carbon dioxide capture, *Energy Conversion Management* 48, 251-258.

Nawrocki J., Andrzejewski P. (2011) Nitrosamines and water, *Journal Hazardous Materials* 189, 1-2, 1-18.

Nielsen C.J., D'Anna B., Dye C., George C., Graus M., Hansel A., Karl M., King S., Musabila M., Müller M., Schmidbauer N., Stenstrøm Y., Wisthaler A. (2010) Atmospheric degradation of amines; Summary report, NILU OR 8/2010.

Plumlee M.H., Reinhard M. (2007) Photochemical Attenuation of N-Nitrosodimethylamine (NDMA) and other Nitrosamines in Surface Water, *Environmental Science Technology* 41, 17, 6170-6176.

Sharma V.K. (2012) Kinetics and mechanism of formation and destruction of N-nitrosodimethylamine in water – A review, *Separation Purification Technology* 88, 1-10.

Stefan M.I., Bolton J.R. (2002) UV direct photolysis of N-nitrosodimethylamine (NDMA): Kinetic and product study, *Helvetica Chimica Acta* 85, 5, 1416-1426.

Tobiesen F.A., Juliussen O., Svendsen H.F. (2007) Experimental validation of a rigorous model for $CO_2$ post combustion capture using monoethanolamine (MEA), *AIChE J.* 53, 4, 846-865.

Tobiesen F.A., Juliussen O., Svendsen H.F. (2008) Experimental validation of a rigorous desorber model for $CO_2$ post-combustion capture, *Chem. Eng. Sci.* 63, 2541-2656.

Vevelstad S.J., Grimstvedt A., Elnan J., da Silva E.F., Svendsen H.F. (2013) Oxidative degradation of 2-ethanolamine; the effect of oxygen concentration and temperature on product formation, Submitted to the *Journal of Greenhouse Gas Control*.

Wisthaler A. (2010) Atmospheric Degradation of amines (ADA-2010), *CLIMIT dagene* (2010), Soria Moria, Oslo, Norway, 12-13 Oct.

Xu B., Chen Z., Qi F., Shen J., Wu F. (2009a) Factors influencing the photodegradation of N-nitrosodimethylamine in drinking water, *Front. Environ. Sci. Eng. China* 3, 1, 91-97.

Xu B., Chen Z., Qi F., Shen J., Wu F. (2009b) Rapid degradation of new disinfection by-products in drinking water by UV irradiation: *N*-nitrosopyrrolidine and *N*-nitrosopiperidine, *Sep. Purif. Technol.* 69, 126-133.

## APPENDIX

## ABBREVIATIONS AND CAS-NUMBERS

| | | |
|---|---|---|
| BHEOX | *N'*-bis(2-hydroxyethyl)oxalamide | 1871-89-2 |
| DEA | Diethanolamine | 111-42-2 |
| DMA | Dimethylamine | 124-40-3 |
| HEA | *N*-(2-hydroxyethyl)acetamide | 142-26-7 |
| HEEDA | *N*-(2-hydroxyethyl)ethylenediamine | 111-41-1 |
| HEF | *N*-(2-hydroxyethyl)formamide | 693-06-1 |
| HeGly | N-(2-hydroxyethyl) glycine | 5835-28-9 |
| HEI | *N*-(2-hydroxyethyl)imidazole | 1615-14-1 |
| HEIA | *N*-(2-hydroxyethyl)imidazolidinone | 3699-54-5 |
| HEO | 2-(2-hydroxyethylamino)-2-oxoacetic acid | 5270-73-5 |
| HEPO | 4-(2-hydroxyethyl)piperazin-2-one | 23936-04-1 |
| MEA | Monoethanolamine | 141-43-5 |
| MNPZ | N-Nitrosopiperazine | 5632-47-3 |
| NDBA | Nitrosodibutylamine | 924-16-3 |
| NDEA | N-Nitrosodiethylamine | 55-18-5 |
| NDELA | N-Nitrosodiethanolamine | 1116-54-7 |
| NDMA | N-nitrosodimethylamine | 62-75-9 |
| NMEA | Nitrosomethylethylamine | 10595-95-6 |
| NMOR | Nitrosomorpholine | 59-89-2 |
| NPIP | N-Nitrosopiperidine | 100-75-4 |
| NPYR | N-Nitrosopyrrolidine | 930-55-2 |
| OZD | 2-oxazolidinone | 497-25-6 |
| PZ | Piperazine | 110-85-0 |

# 11

# Post-Combustion CO$_2$ Capture by Vacuum Swing Adsorption Using Zeolites – a Feasibility Study

G. D. Pirngruber*, V. Carlier and D. Leinekugel-le-Cocq

*IFP Energies nouvelles, Rond-point de l'échangeur de Solaize, BP 3, 69360 Solaize - France*
*e-mail: gerhard.pirngruber@ifpen.fr - vincent.carlier27@gmail.com - damien.leinekugel@ifpen.fr*

* Corresponding author

**Résumé** — **Captage du CO$_2$ en postcombustion par adsorption modulée en pression avec désorption sous vide sur zéolithes** — **étude de faisabilité** — Les résultats de simulation issus de la littérature suggèrent que les procédés d'adsorption modulée en pression avec désorption sous vide (*Vacuum Swing Absorption*, VSA) employant un physisorbant sont nettement moins efficaces d'un point de vue énergétique que le procédé de référence de captage de CO$_2$ par absorption aux amines. La plupart des études considèrent la zéolithe NaX comme adsorbant. La NaX présente une très forte affinité pour le CO$_2$, mais elle est difficile à régénérer et est très sensible à la présence d'eau dans les fumées. Jouer sur la polarité de l'adsorbant doit permettre de trouver un meilleur compromis entre capacité d'adsorption, caractère régénérable et sensibilité à l'eau. C'est pourquoi, dans cette contribution, nous avons testé une série de zéolithes afin d'évaluer leur performances comme physisorbants dans un procédé de captage de CO$_2$ par VSA. Les adsorbants sont testés par adsorption de fumées modèles séches et humides sur un lit d'adsorbant, en perçage et en conditions cycliques. Le matériau le plus intéressant, la zéolithe EMC-1, est choisi pour la simulation d'un cycle VSA complet, en comparaison avec la NaX. Les deux adsorbants atteignent les objectifs en termes de rendement (> 90 %) et de pureté du CO$_2$ (> 95 %), mais le très faible niveau de pression nécessaire à la régénération de l'adsorbant sera un handicap de poids pour l'utilisation de cette technologie à grande échelle.

*Abstract* — *Post-Combustion CO$_2$ Capture by Vacuum Swing Adsorption using Zeolites — a Feasibility Study* — *Simulation results in the literature suggest that Vacuum Swing Adsorption (VSA) processes using physisorbents might largely outperform the current state-of-the-art post-combustion CO$_2$ capture technologies based on amine solvents in terms of energy consumption. Most studies consider the zeolite NaX as adsorbent. NaX has a very strong affinity for CO$_2$ but is difficult to regenerate and very sensitive to the presence of water in the flue gas. By tuning the polarity of the adsorbent, it might be possible to find a better compromise between adsorption capacity, regenerability and sensitivity to H$_2$O. In the present contribution, we therefore screen the performance of a series of zeolites as physisorbents in a VSA process for CO$_2$ capture. The adsorbents are tested by breakthrough experiments of a dry and wet model flue gas, in once-through and cyclic operation. The most interesting material, zeolite EMC-1, is selected for numerical simulations of a full VSA cycle, in comparison with zeolite NaX. Both solids satisfy the performance targets in terms of recovery (> 90%) and purity of CO$_2$ (> 95%) but the very low pressure required for regeneration of the adsorbents will be a serious handicap for the deployment of this technology on a large scale.*

## INTRODUCTION

There is now general consensus that the concentration of $CO_2$ in the atmosphere needs to be stabilized at a level of 450 ppm in order to limit the extent of global warming. In order to achieve this target, a series of measures will be necessary: increasing energy efficiency, shifting to energy sources that produce little or no $CO_2$ emissions (wind, solar, nuclear, biomass, etc.) and, last but not least, employing CCS technology (Carbon Capture and Storage) to large $CO_2$ emitters, in particular to coal firing power plants. Scenarios show that a ~ 20% contribution of CCS is needed in order to stabilize the $CO_2$ concentration at the desired level. There are different ways to deploy CCS: one distinguishes oxy-combustion, pre-combustion $CO_2$ capture and post-combustion $CO_2$ capture. We are here concerned with post-combustion $CO_2$ capture, which is the technology that can be most easily applied in existing coal/petroleum firing power plants, refineries, etc., by "simply" adding a $CO_2$ capture unit to the plant. Post-combustion $CO_2$ capture acts on the flue gases produced by combustion. $CO_2$ in these flue gases is diluted (10-15%) and at low pressure (total pressure is ~ 1 bar), which makes $CO_2$ removal difficult.

The current state-of-the-art post-combustion capture technology is absorption of $CO_2$ by an aqueous Mono-EthanolAmine (MEA) solution. MEA has a very strong affinity to $CO_2$ and captures $CO_2$ efficiently even at low partial pressure but the regeneration of the solvent by heating requires a large amount of energy [1, 2]. Moreover, degradation of the amine by oxygen in the flue gas causes corrosion problems and leads to a high net consumption of solvent (> 1.4 kg MEA/t $CO_2$ captured, i.e. > 4 700 t per year for a 600 MW power plant). Therefore an intensive search for alternative solutions is going on. One of the options is to use Pressure (or Vacuum) Swing Adsorption (PSA or VSA) technology in combination with solid sorbents. PSA/VSA technology is proven and robust, it avoids the difficulties associated with the handling of liquids. The challenge is, however, to find appropriate sorbents. Post-combustion flue gases typically contain 10-15% $CO_2$, 8-15% $H_2O$, 3-4% $O_2$, traces of $NO_x$ and $SO_x$ (depending on the DeNO$x$ and DeSO$x$ treatment), the balance being $N_2$. The objective is to capture a high fraction of the $CO_2$ (90%) and to recover the $CO_2$ at high purity. The fraction of non-condensable impurities (i.e. $N_2$, $O_2$) must be below 4% in order to allow the compression of $CO_2$ for its transport to the storage site. We therefore need highly selective $CO_2$ sorbents. That is why many amine-based solid sorbents have been studied for post-combustion $CO_2$ capture [3-9]. Amines can be immobilized on a solid support either by impregnation or by grafting [10]. The acid-base interaction between amino groups and $CO_2$ assures a very high selectivity towards sorption of $CO_2$. The presence of water in the flue gas does not interfere with this acid-base interaction. In some cases, the $CO_2$ capacity is even

enhanced in the presence of water because the formation of bicarbonates becomes possible [11, 12]. Immobilized amines mimic the chemistry of amine based solvents, hence, we find their strengths but also their weaknesses. The strong chemisorption leads to a high $CO_2$ adsorption capacity at low pressure, the $CO_2/N_2$ selectivity is quasi infinite but regeneration is rather difficult and degradation may occur. There are also health and safety concerns about the contamination of the decarbonized flue gas by volatilization of the amine or its degradation products.

Another option is the use of physisorbents, like zeolites or Metal Organic Frameworks (MOF). It is well known that polar zeolites like NaX or zeolite 5A are very good adsorbents for $CO_2$, even at low partial pressures. The adsorption capacity at 0.15 bar is about 3 mol/kg [13], which is comparable to the best amine-based solid sorbents. The $CO_2/N_2$ selectivity of NaX is also extremely high, above 100 [14]. Similar results are obtained for zeolite 5A [15]. The high capacity and selectivity are a result of the high polarity of the zeolites NaX and 5A. The quadrupole moment of $CO_2$ strongly interacts with the electric field generated by the extra-framework cations of the zeolite. Metal organic frameworks belonging to the CPO-27 structure type have even higher adsorption capacities than zeolites NaX or 5A under the conditions relevant for $CO_2$ capture, i.e. at a partial pressure of $CO_2$ of about 0.1 bar [16]. The CPO-27 structure is characterized by the presence of coordinatively unsaturated metal centers, which have a strong affinity for $CO_2$, similar to the role of the extra-framework cations in zeolites.

A fundamental problem in the design of physisorbents for $CO_2$ capture is that these materials systematically prefer the adsorption of $H_2O$ over the adsorption of $CO_2$. Flue gases fatally contain humidity (at low temperatures they are even saturated with humidity). The $CO_2$ adsorption capacity of NaX, for example, decreases by 99% in humid streams [17], because water is preferentially adsorbed over $CO_2$. The MOF structures CPO-27-Ni and HKUST-1 are somewhat less sensitive to water than zeolites NaX and 5A but the inhibition of $CO_2$ adsorption remains very important [16]. Only purely siliceous, defect-free zeolites have been reported to be entirely hydrophobic [18] but these solids are entirely devoid of polarity and therefore have a very low adsorption capacity for $CO_2$ [19], as well. The dilemma is that a certain polarity of the adsorbent is needed to assure a high adsorption capacity of $CO_2$ and a high $CO_2/N_2$ selectivity. Since polarity and hydrophilicity are intimately linked with each other, a high polarity necessarily implies that $H_2O$ will be adsorbed even more strongly and hence strongly inhibit the adsorption of $CO_2$.

The challenge, therefore, is to tune the polarity of the adsorbent to an intermediate level, so as to find the best compromise between $CO_2$ adsorption capacity, $CO_2/N_2$ selectivity, easy of regeneration under vacuum and inhibition by $H_2O$. Palomino et al. [20] nicely demonstrated that the $CO_2$

adsorption and separation properties can be finely tuned by varying the Si/Al ratio, *i.e.* the polarity, of the framework. In the present study, we have therefore studied the performance of a series of zeolites with varying Si/Al ratios. The adsorption capacity and regenerability of the materials was evaluated by breakthrough experiments with a $CO_2/N_2$ feed mixture. Also, the water tolerance of the materials was tested. The most promising material was then selected for numerical simulations of a full VSA cycle, in comparison with zeolite NaX.

# 1 EXPERIMENTAL SECTION

## 1.1 Adsorbents

The adsorption measurements were carried out on commercial zeolites purchased from Zeolyst *(Tab. 1)*: two mordenites with different Si/Al ratios and two USY zeolites with different Si/Al ratios. Of these four samples, only the CVB10a was in the $Na^+$ form, the other zeolites have $NH_4^+$ as extra-framework cation. The $NH_4^+$ zeolites were transformed into $Na^+$ zeolites by a three-fold ion exchange with a 0.5 molar solution of NaCl at 363 K. A reference sample of zeolite NaX was provided by CECA.

In addition to the commercial zeolites, a synthetic EMC-1 (FAU-type zeolite) with a Si/Al ratio of 3.9 was tested. This zeolite was prepared according to the recipe described by Chatelain *et al.* [21]. A starting gel having the following molar composition: $10\ SiO_2 : 1\ Al_2O_3 : 2.1\ Na_2O : 0.5$ (15-Crown-5) : $100\ H_2O$ was prepared by mixing 55.5 g of 40 wt% silica sol (Ludox HS40, *Dupont*), 6.9 g of sodium aluminate (56% $Al_2O_3$, 37% $Na_2O$, *Carlo Erba*), 4.1 g of crown ether (15-crown-5, 98%, *Aldrich*), 2.56 g of sodium hydroxide (98%, *Fluka*) and 30.9 g of water (desionized water, 18 MΩ). This zeolite was synthesised hydrothermally

TABLE 1

Global and framework Si/Al ratio, Na-content, surface area and micropore volume of the samples

| Sample | Si/Al | Si/Al$_f$ | Wt% Na | $S_{BET}$ (m²/g) | $V_{micro}$ (cm³/g) |
|---|---|---|---|---|---|
| Na-MOR-6 (CBV10a) | 6.2 | 8 | 4.8% | 415 | |
| $NH_4$-MOR-10 (CBV21a)[a] | 10 | | 0.06% | 424 | 0.17 |
| Na-MOR-10 | | | 2.78% | | |
| $NH_4$-USY-6 (CBV712)[a] | 6 | 11 | 0.04% | 791 | 0.27 |
| Na-USY-6 | | | 1.3% | | |
| $NH_4$-USY-15 (CBV720)[a] | 15.2 | 25 | 0.02% | 805 | 0.25 |
| Na-USY-15 | | | 1.07% | | |
| EMC-1 | 3.9 | | 6.02% | 780 | 0.33 |
| NaX | 1.3 | 1.3 | 9.68% | 816 | 0.34 |

[a] $NH_4$-zeolites are transformed into H-zeolites by thermal activation before testing.

in a 100 mL stainless steel autoclave at 110°C for 8 hours. After cooling to room temperature, the crystals were filtered on a Buchner funnel, and washed thoroughly with deionized water. The products were dried overnight at 100°C and calcined at 550°C under air during 8 hours.

The Si/Al ratio of the zeolites was determined by X-ray fluorescence. The USY zeolites contain quite a high fraction of extra-framework Al. The actual framework Si/Al ratio was therefore estimated from $^{29}$Si-NMR and IR spectroscopy. The Na-content of the materials was determined from atomic absorption spectroscopy.

The porosity of the adsorbents was characterized by their $N_2$ adsorption isotherm at 77 K (measured on a Micromeretics ASAP 2010). BET surface areas and micropore volumes (obtained from a t-plot) are compiled in Table 1.

## 1.2 Breakthrough Experiments

The adsorbents were filled into a column that was placed into an oven. A gas distribution system allowed us to feed the column either with pure $N_2$ or with a mixture of $CO_2$ in $N_2$ and to switch rapidly between the two. Moreover, we had the possibility to introduce water vapour into the feed stream by passing *via* a saturator. The pressure was always atmospheric plus the (small) pressure drop over the column (between 0.1 and 1 bar, depending on the adsorbent). The sorbents were initially treated in $N_2$ at 623 K for 2 h and then brought to the adsorption temperature (323 K) and a breakthrough curve $N_2 \rightarrow N_2 + CO_2$ was recorded by means of a TCD detector at the column exit. Adsorption isotherms of $CO_2$ were measured by repeating the breakthrough experiments at different $CO_2$ concentrations (from 5 to 30%), each time followed by a regeneration at 623 K for 2 h. Note that these isotherms are in fact $CO_2/N_2$ co-adsorption isotherms but the amount of adsorbed $N_2$ was not quantified. In addition to the adsorption isotherms, cyclic adsorption-desorption experiments were performed. In that case, the $CO_2$ concentration was fixed at 15% and the feed was alternated between $CO_2 + N_2$ and $N_2$, while keeping the temperature constant. These cyclic adsorption experiments give us information on the regenerability of the samples.

For evaluating the influence of humidity on adsorption, the following procedure was used. The $N_2$ stream was passed through a saturator that was filled with distilled water and thermostated at 306 K (which corresponds to a vapour pressure of 5 kPa) for 1 h before starting the breakthrough experiment. During the whole breakthrough experiment the partial pressure of $H_2O$ was maintained at 5 kPa by passing through the saturator. At the column exit water was condensed in a cool bath (at 273 K) so as to enter with a gas of constant humidity into the TCD detector. After breakthrough the feed was switched back to humid $N_2$ and the breakthrough experiment with humid feed was repeated at intervals of several hours.

Breakthrough experiments for measuring the $CO_2/N_2$ selectivity were carried out on a different experimental setup. Activation was carried out in a non-adsorbable carrier gas (He) and the breakthrough curve was triggered by switching from He to a 50/50 mixture of $CO_2/N_2$. The $CO_2$ and $N_2$ concentrations at the column exit were measured by a mass spectrometer, following the procedure described in [22]. The experiments were carried out at 303 K. At least four break-through/desorption measurements were carried out with each adsorbent:

- a step from He to $CO_2/N_2$, followed by a step back to He;
- a step from pure $CO_2$ to $CO_2/N_2$, followed by a step back to He.

The adsorbed amount of $CO_2$ was determined from the step He → $CO_2/N_2$, the amount of $N_2$ was obtained by averaging the results of the four breakthrough/desorption curves. For an equimolar feed mixture of $CO_2/N_2$ the adsorption selectivity α simply corresponds to the ratio of the adsorbed amounts of $CO_2$ and $N_2$.

## 2 RESULTS

### 2.1 Adsorption Isotherms

Figure 1 compares the adsorption isotherms of $CO_2$ over three different mordenite samples. It clearly shows the influence of Si/Al ratio and of the extra-framework cation on the adsorption properties. The sample with the lower Si/Al ratio has the higher adsorption capacity. Moreover,

$Na^+$-exchanged samples adsorb more $CO_2$ than H-mordenite (activation transforms $NH_4$-mordenite into H-mordenite). It has already been shown before that Na-ZSM-5 has a higher heat of adsorption than H-ZSM-5 [23]. The electrostatic interaction between the $Na^+$ cations and the quadrupole moment of $CO_2$ is stronger than that of a largely covalent Bronsted OH-group.

The trends observed for mordenite are confirmed with the faujasite adsorbents (Fig. 2). The $CO_2$ adsorption capacity decreases with increasing Si/Al ratio and $NH_4$-exchanged samples adsorb much less $CO_2$ than Na-faujasites. Zeolite NaX has by far the highest adsorption capacity because its Si/Al ratio is much lower than that of the other faujasites. At similar Si/Al ratio, the $CO_2$ isotherms of the Na-faujasite samples are not as steep as the isotherms of Na-mordenites. The strong initial adsorption of $CO_2$ in mordenite may be related to the strong confinement in the side pockets of the mordenite structure.

In order to rationalize the adsorption data on the series of faujasite and mordenite samples, the isotherms were fitted by a Langmuir model:

$$q = q_{sat} \cdot \frac{b \cdot p_{CO_2}}{1 + b \cdot p_{CO_2}}$$

The Henry constants of the isotherms were then calculated from $K_{Henry} = b \times q_{sat}$.

Figure 3 shows that the Henry constants are correlated with the Na content of the zeolites, which indicates that the $Na^+$ cations are the preferred adsorption sites of $CO_2$. The Na content can, therefore, be regarded as a good semi-quantitative indicator of the polarity of the zeolite. The slope of the curve

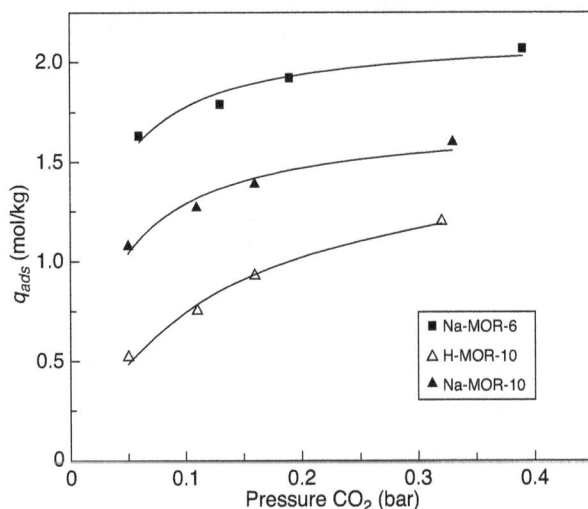

Figure 1

Adsorption isotherms of $CO_2$ (in mixture with $N_2$) at 323 K over three different mordenite samples.

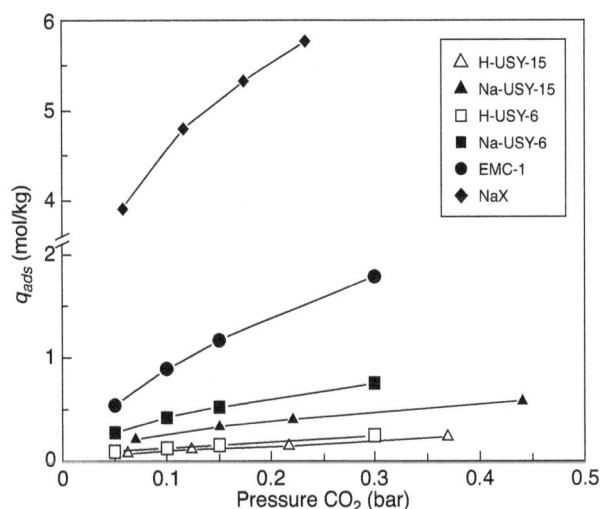

Figure 2

Adsorption isotherms of $CO_2$ (in mixture with $N_2$) at 323 K over different faujasite samples.

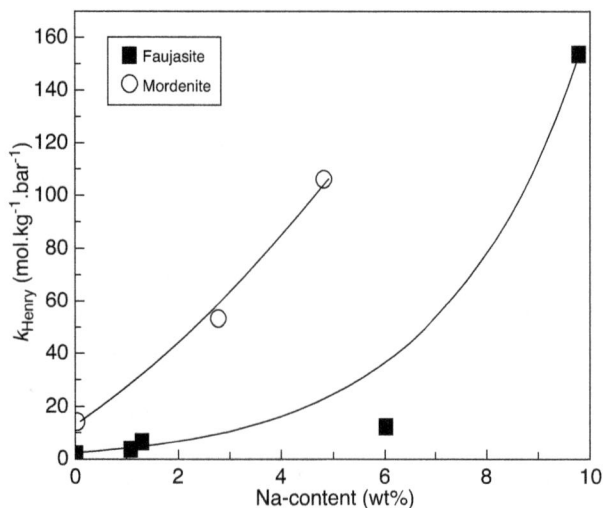

Figure 3

Henry constants of the $CO_2$ isotherms at 323 K as a function of the Na content of the zeolites.

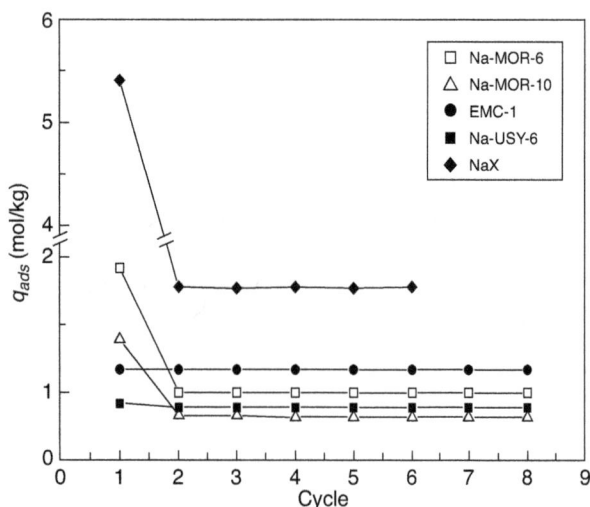

Figure 4

Adsorption capacity over several adsorption-desorption cycles at 323 K. $CO_2$ concentration 15% in $N_2$, desorption by $N_2$ purge for 30 min.

$K_{Henry}$ *vs* Na-content is much higher on mordenite than on faujasite. In mordenite, the electrostatic effect of the $Na^+$ cations is exacerbated by a stronger confinement, *i.e.* a smaller pore size.

## 2.2 Adsorption — Desorption Cycles

A high adsorption capacity at the working pressure of the process is not a sufficient criterion for the selection of an adsorbent. The adsorbent must also be easily regenerable

upon depressurisation and desorption under vacuum. The quantity that can be adsorbed and desorbed over many adsorption/desorption cycles is called the working capacity of the adsorbent. Since our experimental setup did not allow us to regenerate the adsorbents under vacuum, we established a qualitative ranking of their regenerability, by desorbing $CO_2$ *via* a $N_2$ purge gas for 30 min at 323 K (*i.e.* under isothermal conditions). The results are shown in Figure 4. With the exception of NaX, the faujasite adsorbents were perfectly regenerable by the $N_2$ purge and the adsorption capacity remained constant during 8 adsorption/desorption cycles. The two Na-mordenites lost a significant fraction of their initial adsorption capacity when going from the 1st to the 2nd cycle, *i.e.* they were not fully regenerable by a $N_2$ purge. Also in the case of NaX, 30 min were largely insufficient to desorb $CO_2$ from the column. As expected, the regenerability was related to the shape of the isotherm at low pressure. Very steep isotherms (*i.e.* NaX, Na-MOR-6, Na-MOR-10) mean difficult regeneration. In spite of its bad regenerability, the best cyclic capacity was obtained with NaX (1.8 mol/kg), followed by EMC-1 (1.2 mol/kg).

## 2.3 $CO_2/N_2$ Selectivity

The $CO_2/N_2$ selectivity was measured by breakthrough experiments with 50/50 mixtures of $CO_2/N_2$.

Figure 5 compares the breakthrough curves of EMC-1, Na-USY-15 and Na-MOR-10. One can immediately see that the quality of the $CO_2/N_2$ separation was less good on Na-USY-15. The selectivity values are reported in Table 2[1]. Both Na-mordenites as well as NaX and EMC-1 had selectivities above 40 (all selectivities above 40 were considered to be equal because of the large experimental error). In the case Na-USY-15, which is the least polar sample that was tested, the selectivity dropped significantly to 11. This suggests that there is a minimum polarity above which there is no measureable gain in the $CO_2/N_2$ selectivity any more. For faujasite samples, this threshold is somewhere between Na-USY-15 and EMC-1.

TABLE 2

Results of breakthrough experiments of a 50/50 $CO_2/N_2$ mixture at 303 K

|  | $Q_{ads} CO_2$ (mol/kg) | $Q_{ads} N_2$ (mol/kg) | Selectivity |
|---|---|---|---|
| NaX | 5.46 | $0.10 \pm 0.04$ | $54 \pm 22$ |
| EMC-1 | 2.84 | $0.06 \pm 0.03$ | $44 \pm 22$ |
| Na-USY-15 | 1.16 | $0.11 \pm 0.03$ | $11 \pm 3$ |
| Na-MOR-6 | 3.23 | $0.08 \pm 0.04$ | $42 \pm 25$ |
| Na-MOR-10 | 2.42 | $0.03 \pm 0.01$ | $80 \pm 30$ |

---

[1] The adsorbed amount of $N_2$ is very small. We, therefore, averaged results from several tests to obtain fairly reliable results.

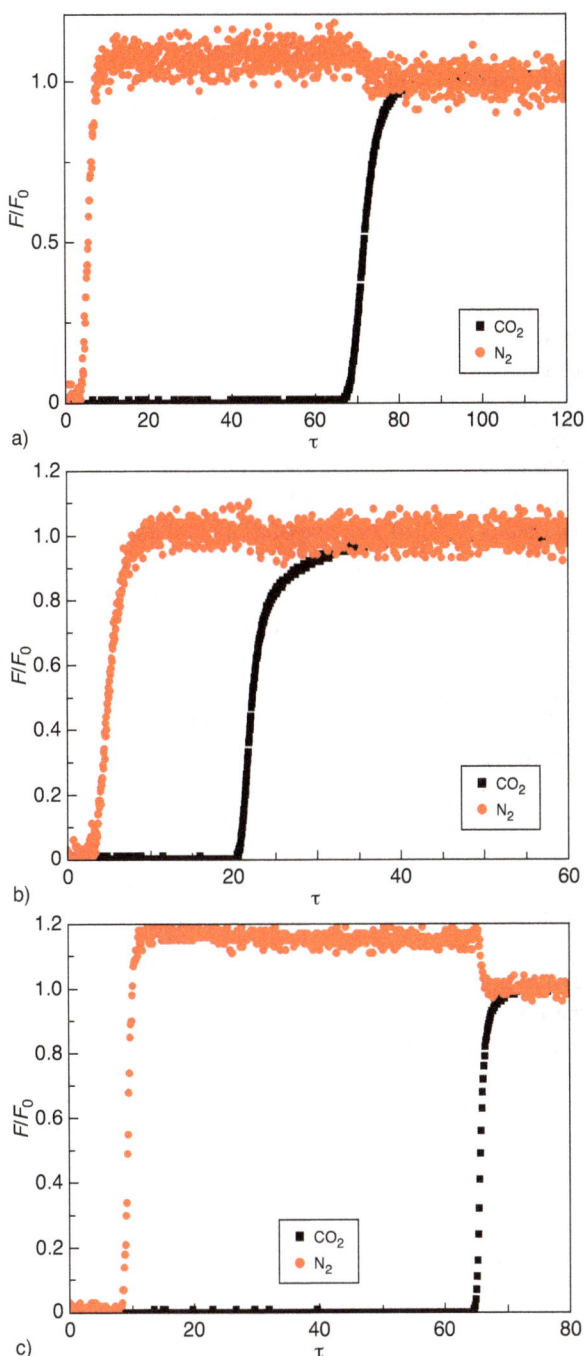

a)

b)

c)

Figure 5

Breakthrough curves of a 50/50 mixture of $CO_2/N_2$ over a) EMC-1, b) Na-USY-15, c) Na-MOR-10. $F/F_0$ = normalized molar flow rate at column exit ($F_0$ = feed flow rate). $\tau$ = time/contact time.

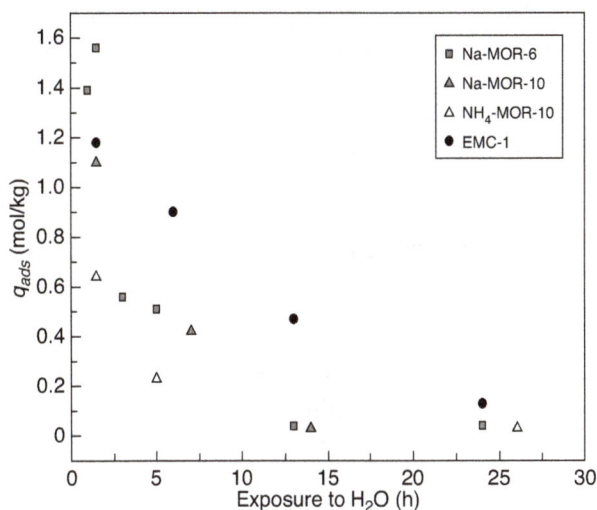

Figure 6

$CO_2$ adsorption capacity in a humid feed (15% $CO_2$, 5% $H_2O$) as a function of the total exposure time of the adsorbent to a humid gas.

## 2.4 Inhibition by H2O

As a next step, we evaluated the effect of humidity on $CO_2$ adsorption. Breakthrough experiments were carried out with humid gases, *i.e.* by switching from a mixture $N_2 + H_2O$ to a mixture of $N_2 + CO_2 + H_2O$. The $H_2O$ content was regulated by a saturator, which was thermostated at 306 K. This corresponds to a vapor pressure of 0.05 bar and a relative humidity of 43% at 323 K. The adsorbent was exposed to the $N_2 + H_2O$ mixture for 1 h before adding $CO_2$ to the feed. For the three mordenite samples, the $CO_2$ adsorption capacity dropped by 0.3 mol/kg compared to a dry feed. For EMC-1, no significant change was observed. We noticed, however, that the $CO_2$ capacity dropped more strongly when the initial exposure to the $N_2 + H_2O$ mixture was prolonged *(Fig. 6)*. That made us suspect that the solids were not yet saturated with $H_2O$. We therefore carried out several successive breakthrough experiments, at increasing total exposure times to $H_2O$. Figure 6 shows that the $CO_2$ adsorption capacity rapidly decreases with increasing exposure time to humid feed. Already after 15 h of exposure to 5% $H_2O$, the adsorption capacity of $CO_2$ has dropped almost to zero. 10 h corresponds roughly to the time needed to saturate the pore volume of mordenite with water under our conditions (concentration and flow rate). We can deduce that the pore volume of mordenite is indeed gradually filled with $H_2O$. As soon as the whole column is saturated with $H_2O$ hardly any $CO_2$ can be adsorbed any more. In the case of EMC-1, the $CO_2$ adsorption capacity decreases more gradually, which is probably due to the higher of pore volume of EMC-1, *i.e.* it takes more time to fill the pores with $H_2O$ but the principle remains the same.

These results confirm that even zeolites with moderate polarity like H-MOR-10 largely prefer the adsorption of $H_2O$ over the adsorption of $CO_2$. This does not mean that it is impossible to run a $CO_2$ adsorption process with zeolites in the presence of $H_2O$ but a large part of the bed will be monopolized by the adsorption of $H_2O$ and not be effective at all for $CO_2$ capture. Since water has a very strong affinity for zeolites its regeneration will be very difficult. It makes more sense to remove the bulk of the water content by "weaker" adsorbents, like silica gel, alumina or activated carbons [24].

## 3 MODELING OF THE VSA $CO_2$ CAPTURE PROCESS

After having presented the adsorption properties of a series of zeolites, we now turn to the chemical engineering aspects of a $CO_2$ capture process by VSA. We first discuss the technical boundary conditions, then describe the numerical modeling of the VSA process and finally discuss the industrial feasibility of $CO_2$ by VSA.

### 3.1 Boundary Conditions

Any process for $CO_2$ capture from flue gases must fulfill two technical targets:

– achieve a high $CO_2$ recovery;
– produce a $CO_2$ of high purity that is suited for sequestration.

The optimum recovery level is defined by the trade-off between recovery and cost, *i.e.* by minimizing the price of avoided $CO_2$. The energy consumption (per captured $CO_2$) of a VSA process increases with increasing recovery (as will be discussed later). On the other hand, the capital expenses for constructing a $CO_2$ capture plant do not depend much on the recovery and will, therefore, contribute more heavily to the overall cost, when the recovery is low. Many studies consider a recovery level of 90%, so we started out by fixing our target recovery to this value.

The technical constraint in terms of purity of the $CO_2$ is stricter. The level of non-condensable impurities, *i.e.* $N_2$ and $O_2$, must be below 4% in order to facilitate the compression of $CO_2$ to a supercritical state, for the purpose of transport and sequestration.

Purity strongly depends on the selectivity of the adsorbent. At the end of the adsorption step, some $N_2$ (and $O_2$) is co-adsorbed on the zeolite and some is present in the interstitial volume of the bed. In order to achieve 96% purity, a large fraction of the $N_2$ must be removed from the bed before the $CO_2$ recovery step. This can be achieved by modifications of the VSA cycle [25-28]. A co-current depressurisation of the bed, at the end of the adsorption step, releases a large fraction of the $N_2$ from the interstitial volume and allows advancing the concentration front of $CO_2$ towards to end of the bed, without sacrificing recovery, since the effluent of the depressurisation step is recycled for repressurisation of another bed. Additionally, a $CO_2$ rinse step, in which part of the recovered $CO_2$ is recycled to the bed, may be added in order to displace co-adsorbed $N_2$ from the adsorbent. Depressurisation and $CO_2$ rinse may be combined. The lower the pressure of the $CO_2$ rinse step $P_R$ is chosen, the smaller is the amount of $CO_2$ required for purging the column and the smaller is the energy consumption for compressing the $CO_2$ recycle stream to $P_R$. In order to maintain a high recovery of $CO_2$, the effluent of the depressurization step is usually used to repressurize another column to an intermediate pressure $P_I$ *(Fig. 7)*. In that case, the condition $P_R > P_I$ puts an upper limit on the amount of gas that can be used for repressurization, which indirectly imposes a lower limit on $P_R$.

After having removed a large fraction of $N_2$ and $O_2$ from the column by depressurization and purge, $CO_2$ is recovered by lowering the pressure, *i.e.* by pulling vacuum. In conventional PSA processes, the column is further regenerated

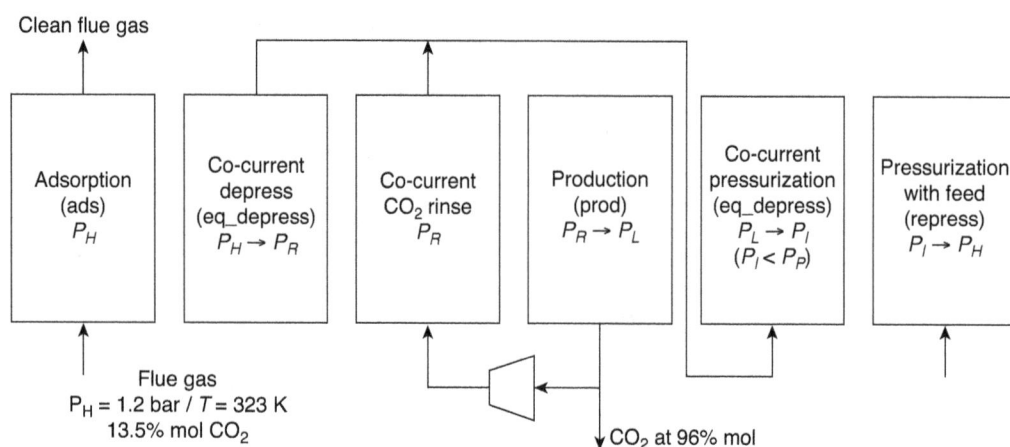

Clean flue gas

Adsorption (ads) $P_H$

Co-current depress (eq_depress) $P_H \rightarrow P_R$

Co-current $CO_2$ rinse $P_R$

Production (prod) $P_R \rightarrow P_L$

Co-current pressurization (eq_depress) $P_L \rightarrow P_I$ $(P_I < P_P)$

Pressurization with feed (repress) $P_I \rightarrow P_H$

Flue gas $P_H$ = 1.2 bar / $T$ = 323 K 13.5% mol $CO_2$

$CO_2$ at 96% mol

Figure 7

VSA cycle for $CO_2$ capture from flue gas.

by a purge with the light product, so as to perfectly clean the light product end of the column but, in the present case, it is not possible to employ a purge with a non-condensable gas like $N_2$ (*vide infra*), because of the stringent purity requirements in $CO_2$ capture. Therefore, at the end of the $CO_2$ recovery step, the light product end of the column is in equilibrium with almost pure $CO_2$ at the partial pressure of regeneration (plus the pressure drop in the column). In the following adsorption step, $CO_2$ cannot be adsorbed beyond that equilibrium partial pressure, leading to a permanent slip of $CO_2$ that limits the recovery. For example, if the adsorption is carried out at a pressure of 1.2 bar and the $CO_2$ concentration is 15%, the pressure of regeneration must be below 0.018 bar in order to be able to achieve a recovery of 90%. An examination of earlier simulation and experimental work on VSA for $CO_2$ capture confirms that recovery is limited by the ratio between the partial pressures of $CO_2$ in the feed step and in the regeneration step [29, 30], if the column is not purged with light product.

In order to be able to circumvent the use of such low pressures for regeneration, one would have to find another way to clean the light product end of the bed before starting a new adsorption step. Many VSA simulation studies employ a low pressure purge with light product ($N_2$) but, as already mentioned above, a $N_2$ purge necessarily pollutes the $CO_2$ product and makes it impossible to achieve the 96% purity [27, 28, 31]. An alternative is to repressurize the column from the light product end with $N_2$ (instead of repressurizing with feed). This method can be effective for adsorbents with a low $CO_2/N_2$ selectivity, like activated carbons. On an activated carbon $N_2$ competes with $CO_2$ for adsorption sites and a repressurization by $N_2$ can push the concentration front of $CO_2$ back into the column [32]. In polar zeolites, the co-adsorption of $N_2$ is very weak, as a consequence repressurizing with $N_2$ hardly affects the $CO_2$ concentration front. It is therefore impossible to avoid the very low regeneration pressures in VSA cycles with highly selective adsorbents. Since recovery is intimately related to the regeneration pressure and since the power consumption of the VSA process increases with decreasing regeneration pressure, we can now understand why a higher recovery leads to higher operating costs.

## 3.2 VSA Simulations

From our experimental screening of zeolites, two materials emerged as the most attractive candidates for a $CO_2$ capture application: zeolite NaX has a very high $CO_2$ affinity and a high adsorption capacity but is difficult to regenerate; zeolite EMC-1 has a moderate adsorption capacity but exhibits excellent regenerability. Both have a high $CO_2/N_2$ selectivity. We have, therefore, decided to confront the behavior of EMC-1 and NaX in a VSA cycle by numerical simulations. The flue gas was represented by a mixture of 13.5% $CO_2$ in $N_2$. For the sake of simplicity, a dry feed was used, *i.e.* it was

assumed that water has been removed from the flue gas upstream of the VSA. The single component adsorption isotherms of $CO_2$ and $N_2$ on NaX were taken from the literature [23], the isotherms of EMC-1 were in-house data. The experimental isotherms were modeled by a dual or single site Langmuir model:

$$q_j = \sum_i q_{sat,j} \cdot \frac{b_{j,i} \cdot p_j}{1 + b_{j,i} \cdot p_j}$$

$$b_{j,i} = b_{j,i,0} \cdot \exp\left(\frac{-h_{ads,j,i}}{RT}\right)$$

The parameters are given in Table 3.

TABLE 3

Parameters of the single or dual site Langmuir model used to fit the experimental isotherms of $CO_2$ and $N_2$

|  |  | $q_{sat}$ (mol/kg) | $b_0$ (Pa$^{-1}$) | $h_{ads}$ (kJ/mol) |
|---|---|---|---|---|
| NaX | $CO_2$ | 2.09 | $3.84 \times 10^{-10}$ | $-40$ |
|  |  | 4.11 | $8.75 \times 10^{-12}$ | $-40$ |
|  | $N_2$ | 6.25 | $3.81 \times 10^{-10}$ | $-19$ |
| EMC-1 | $CO_2$ | 3.34 | $3.24 \times 10^{-9}$ | $-25$ |
|  |  | 4.15 | $4.70 \times 10^{-10}$ | $-25$ |
|  | $N_2$ | 3.81 | $4.35 \times 10^{-10}$ | $-18$ |

A simple extended Langmuir equation was used for modeling the co-adsorption of $CO_2$ and $N_2$:

$$q_j = q_{sat,j,1} \cdot \frac{b_{j,1} \cdot p_j}{1 + \sum_m b_{m,1} \cdot p_m} + q_{sat,j,2} \cdot \frac{b_{j,2} \cdot p_j}{1 + \sum_m b_{m,2} \cdot p_m}$$

This is not the most precise co-adsorption model but is sufficiently accurate for a qualitative comparison of NaX and EMC-1. Since we were mainly interested in identifying the effect of the adsorption isotherms on the performance, we chose conditions under which mass transfer effects are totally negligible, *i.e.* a fairly long contact time of 0.4 s. The maximum gas velocity was fixed *via* the constraint of avoiding fluidization of the adsorbent particles and the column height was limited by the pressure drop (which should be below 0.2 bar). The adsorber column was considered to be adiabatic and the pressure drop in the column was neglected in the simulation. Further details of the model and its underlying assumptions are given in the Appendix. The VSA cycle was chosen based on the considerations above: an adsorption step, at $P_H = 1.2$ bar (to overcome the pressure drop in the column) and 323 K was followed by a co-current depressurization to $P_R$. The objective of the depressurization step was to remove $N_2$ from the interstitial volume of the column. The depressurization was followed by a rinse with part of the recovered $CO_2$ stream. In the next step, $CO_2$ was produced by lowering the pressure to $P_L$, *i.e.* by pulling vacuum on the column. Part of the $CO_2$ that is recovered during this step is recycled to another column, for performing the $CO_2$ rinse. For this purpose, $CO_2$ must be compressed from $P_L$ to $P_R$. Finally the column is repressurized, first with effluent of a column in depressurization and finally with the feed.

The most influential operating parameters of the VSA cycle were:

– the value of $P_L$: as explained above, it had a huge influence on recovery;

– the value of $P_R$: it had a large influence on the purity of $CO_2$;

– the duration of the adsorption and rinse steps.

These four parameters were varied in order to achieve our target values of 90% recovery and 96% purity. We found that, thanks to the high selectivity of NaX and EMC-1, a $CO_2$ rinse step was actually not necessary. The 96% purity of $CO_2$ could be achieved simply by optimizing the final pressure of the depressurization step.

Table 4 shows the conditions under which the target recovery and purity were achieved for NaX and EMC-1. The values of the key parameters $t_{ads}$, $P_P$ and $P_L$, were surprisingly close for the two materials. As a consequence, productivity and power consumption for the evacuation step were also quite similar. The working capacities, *i.e.* the amounts of $CO_2$ adsorbed in each cycle, were 0.70 and 0.75 mol/kg for NaX and EMC-1, respectively. Remember that our screening experiments had yielded cyclic capacities of 1.8 and 1.2 mol/kg for NaX and EMC-1, respectively. Two factors may degrade the working capacity in the VSA simulations compared to the cyclic capacity obtained in our screening. First of all, temperature rise in adsorption and the temperature drop in desorption diminishes the capacity of the bed compared to isothermal operation. Secondly, the equilibrium capacity of the bed is not fully exploited because the mass transfer zone of $CO_2$ must be contained in the bed in order to achieve a high recovery of $CO_2$.

In order to evaluate the importance of these two effects, we have calculated the theoretical working capacity at equilibrium, first under isothermal and then under adiabatic conditions. For calculating the theoretical isothermal working capacity, we assumed that the whole adsorbent is in equilibrium with $P_H \times 0.135 = 0.16$ bar $CO_2$ and with $P_L = 0.014$ bar $CO_2$ at the end of the adsorption step and of the desorption step, respectively. The difference between these two equilibrium adsorbed amounts yielded the theoretical working capacity. The theoretical capacities were 1.7 and 1.3 mol/kg for NaX and EMC-1, respectively, *i.e.* close to the experimental cyclic capacities *(Tab. 5)*. Co-adsorption of $N_2$ had a negligible impact on the theoretical working capacity. The good agreement between theoretical and experimental working capacities implies that our screening method was pertinent. As a next step, the adsorption isotherms were transposed into an adiabates [33], by accounting for the temperature rise caused by the exothermicity of adsorption, and the associated decrease of the adsorption capacity. For calculating the theoretical adiabatic working capacity, we assumed that the whole adsorbent is in equilibrium with $P_H \times 0.135 = 0.16$ bar $CO_2$ at 323 K + $\Delta T/2$ at the end of the adsorption step, and that the whole adsorbent is in equilibrium with $P_L = 0.014$ bar at 323 K – $\Delta T/2$. $\Delta T$ is the temperature swing between adsorption and desorption. The adiabatic capacities were 0.7 and 0.9 mol/kg for NaX and EMC-1, respectively. The working capacity of NaX dropped by 60% when the thermal effects were taken into account. The impact of adiabatic operation on the working capacity of EMC-1 was much smaller because its heat of adsorption is lower. The VSA simulations confirmed that the temperature swing between adsorption and desorption was indeed higher in the case of NaX *(Fig. 8)*.

TABLE 5

Working capacities of NaX and EMC-1: experimental estimation by adsorption – desorption experiments, theoretical calculation under isothermal and adiabatic conditions and values obtained in the VSA simulations

| Working capacity (mol/kg) | NaX | EMC-1 |
|---|---|---|
| Experimental screening | 1.8 | 1.2 |
| Theoretical isothermal | 1.75 | 1.35 |
| Theoretical adiabatic | 0.7 | 0.9 |
| VSA simulations | 0.7 | 0.75 |

TABLE 4

Results of the simulations of a VSA cycle for $CO_2$ capture, using either NaX or EMC-1 as adsorbent

| | NaX | EMC-1 |
|---|---|---|
| $t_{ads}$[(a)] (s) | 450 | 480 |
| $P_P$ (bar) | 0.55 | 0.4 |
| $P_L$ (bar) | 0.014 | 0.015 |
| Purity $CO_2$ | 96.7% | 95.9% |
| Recovery $CO_2$ | 89.9% | 89.8% |
| Power consumption[(b)] (kJ/mol $CO_2$) | 11.7 | 11.0 |

[a] Feed rate = 60 000 Nm³/h, column volume = 25.13 m³.

[b] Power consumption during the $CO_2$ production step.

The theoretical adiabatic working capacities were fairly close to the simulated working capacities, which means that they are a good qualitative indicator of the performance of an adsorbent. Still, the theoretical values predict that EMC-1 should perform slightly better than NaX in adiabatic operation, but in the VSA simulations both solids were equivalent. The explanation can be found in the concentration profiles *(Fig. 9)*.

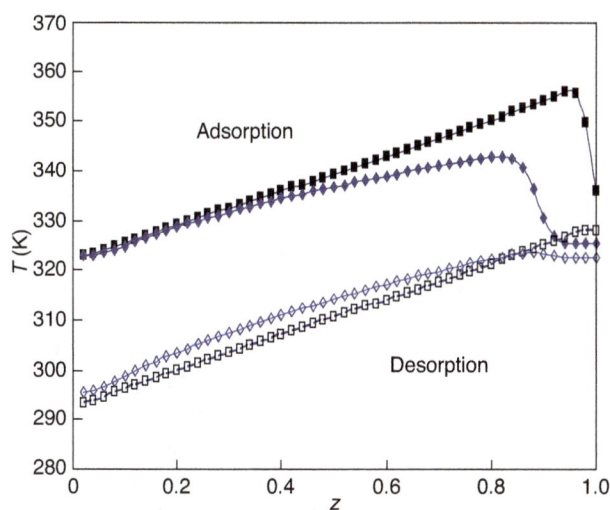

Figure 8

Temperature profile along the column axis $z$ (adimensionalized) at the end of the adsorption step (full symbols) and at the end of the desorption step (empty symbols). NaX (squares) $vs.$ EMC-1 (diamonds).

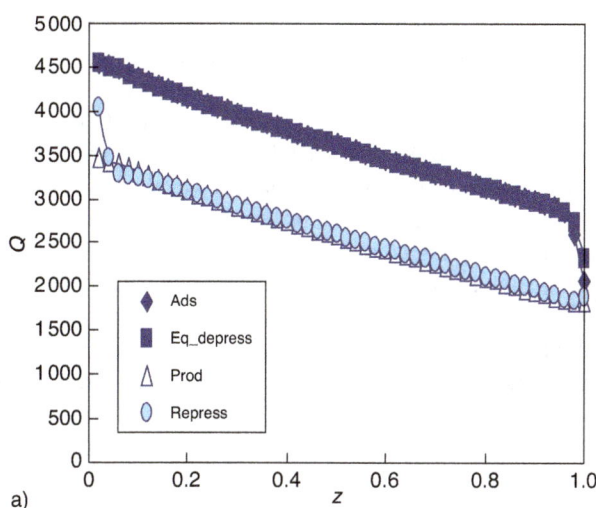

a)

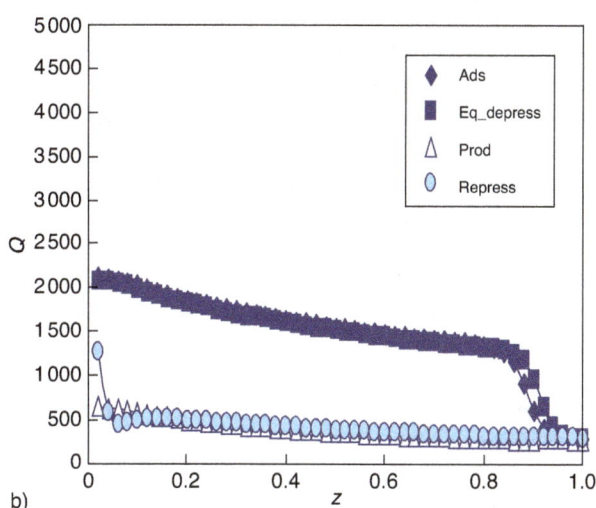

b)

Figure 9

Axial concentration profiles of $CO_2$ in the adsorbed phase $Q$ (in mol/m$^3$crystal) at the end of the different steps of the cycle: ads = adsorption, eq_depress = co-current depressurization, prod = production of $CO_2$, repress = repressurization with fedd. a) NaX (squares), b) EMC-1 (diamonds).

The mass transfer zone in the EMC-1 column was significantly larger than in the NaX column, $i.e.$ the capacity of the bed was less well exploited. The $CO_2/N_2$ co-adsorption isotherm of EMC-1 is less favorable than that of NaX and this leads to a stronger dispersion of the $CO_2$ concentration front at low concentrations.

In summary, the moderate $CO_2$ affinity of EMC-1 compared to NaX was an advantage, because it diminished the temperature excursions in the column and therefore led to a higher theoretical working capacity under adiabatic conditions. In practice part of this advantage was lost, because the less favorable $CO_2$ isotherm of EMC-1 led to a broader mass transfer zone, hence the bed was less well exploited. Overall, the effects cancelled each other and the performance of both solid was quasi equivalent. In practice, one would probably stick with NaX as the preferred adsorbent since its preparation is easier and cheaper.

## 3.3 Industrial Feasibility

The major asset of VSA technology for $CO_2$ capture is the fact that the energy consumption of process is, at least theoretically, very low. According to our simulations under idealized conditions, $i.e.$ under total absence of mass transfer limitations, the power consumption of the vacuum pump and the blower used for compressing the feed to 1.2 bar amounts to only 0.35 GJ/t $CO_2$. This theoretical value is lower than recently published estimates (also based on simulation data) [27], because in our case a high recovery and purity of $CO_2$ could be achieved without using a $CO_2$ rinse step (the recycling of $CO_2$ consumed additional energy). Yet, the simulated theoretical values are overly optimistic because they assume a 100% efficiency of the (vacuum) blowers. Experimentally measured energy consumptions are a factor of 4-5 higher than the theoretical values [34, 35]. Moreover, one has to be careful when comparing the electrical energy used by blowers with thermal energy used in temperature swing processes, as absorption by MEA. Electrical energy is produced from thermal energy with an efficiency of $\sim$40%, it is therefore more "expensive". When both of these factors are taken into account, the energy balance of the VSA process is not that favorable any more.

In addition, application of the VSA technology on a large scale (*i.e.* the scale required for a 600 MW power plant) faces serious technological problems, which are rarely taken into account in the literature. The flue gas flow rate in a 600 MW power plant is in the order of $1.8 \times 10^6$ Nm³/h. In order to be able to treat this enormous flow rate, while keeping the pressure drop over the column low, huge cross sections are needed. If we extrapolate our simulations to the above-mentioned feed flow rate, the total cross section of the adsorbers should be 1 500 m². The actual value should be lower because we carried out our simulations at fairly low gas velocities so as to avoid any mass transfer limitation. For comparison, Lively *et al.* [36] estimated the cross section of fixed bed adsorbers to be 850 m², if a pressure drop of 0.69 bar is authorized. These huge cross sections ask for new concepts in adsorber design, for example, structured adsorbent beds, where the adsorbents are packed into hollow fibers [34, 36, 37]. Such solutions are not yet industrial standard.

Another concern that must be raised about $CO_2$ capture by VSA is the very low pressure that must be used for desorption of $CO_2$ in order to assure a high recovery (see above). At 0.014 bar, the volume flow rate of the recovered $CO_2$ stream is 10 times higher than the volume flow rate of the feed. A large number of pumps must be installed in parallel in order to be able to cope with this huge volume flow and several compression stages are probably needed to raise the pressure of the recovered $CO_2$ stream back to 1 bar (pressure at which $CO_2$ is released in the reference MEA process). The high capital expenses associated with the installation of a large number of blowers, as well as the technological difficulty of dealing with the huge volume flow rates encountered during $CO_2$ recovery may render $CO_2$ capture by VSA processes economically unviable. A way to circumvent to low pressure desorption is to use dual-reflux systems [38-41]. In dual-reflux pressure swing separations, the feed is injected at an intermediate point of the column. Light product ($N_2$) is recycled to the top of the column and heavy product ($CO_2$) is recycled to the bottom of the column. In this way, both light and heavy product can be obtained at high recovery and purity, at desorption pressures as high as 0.3 bar [38, 41]. We are, however, not aware of any existing industrial unit that has been constructed according to a dual-reflux design.

## CONCLUSIONS

One of the objectives of the present work was to tune the adsorption properties of zeolites so as to optimize their performance in post-combustion $CO_2$ capture. Our results show that it is possible to modulate the $CO_2$ adsorption isotherms of zeolites by changing the content of extra-framework $Na^+$ cations and/or the zeolite structure (*i.e.* the pore size and pore geometry). Within a given zeolite structure, reducing the extra-framework cation content reduces the adsorption

capacity at low pressure but improves the regenerability. The $CO_2/N_2$ selectivity is not measurably affected as long as the $Na^+$ content is above a certain threshold (which is around 2% wt Na for faujasite zeolites). Among the series of faujasite and mordenite zeolites that we have tested, the faujasites NaX (Si/Al = 1.3) and Na-EMC-1 (Si/Al = 3.9) present the best working capacity in cyclic adsorption – desorption experiments. Zeolite NaX has a very high $CO_2$ affinity, a high heat of adsorption and a high adsorption capacity but is difficult to regenerate; zeolite EMC-1 has a moderate adsorption capacity (and a moderate heat of adsorption) but exhibits excellent regenerability. Both solids exhibit a high $CO_2/N_2$ selectivity. Under isothermal conditions, the working capacity of zeolite NaX is about 50% higher than that of zeolite EMC-1. In adiabatic operation, the ranking is inversed. The higher heat of adsorption of $CO_2$ on NaX (compared to EMC-1) leads to higher temperature excursions in the bed during adsorption and desorption, and thereby strongly degrades the adsorption capacity of NaX [25, 26], even below the value of EMC-1. Simulations of a VSA cycle for $CO_2$ capture show, however, that the small theoretical advantage of EMC-1 is lost because the less favorable $CO_2$ isotherm of EMC-1 leads to a broader mass transfer zone, hence the bed is less well exploited.

From an application point of view, the main asset of VSA technology for $CO_2$ capture is the fact that the energy consumption of the process could, at least theoretically, be very low. There are, however, many technological hurdles to be overcome before VSA for $CO_2$ capture can become an industrial reality. Vacuum blowers are needed that can cope with huge volumetric flow rates and adsorber design has to be reinvented in order to minimize the pressure drop in the system. An additional obstacle is that the flue gas must be dried within or upstream of the VSA system, because all of the most promising $CO_2$ adsorbents largely prefer the adsorption of $H_2O$ over the adsorption of $CO_2$. It is, therefore, highly uncertain whether $CO_2$ capture by VSA technology for large scale power plants can be economically and technically viable in the near future.

## ACKNOWLEDGMENTS

Charles Leroux and Nicolas Bats are gratefully acknowledged for the synthesis of zeolite EMC-1.

## REFERENCES

1 Raynal L., Bouillon P.A., Gomez A., Broutin P. (2011) From MEA to demixing solvents and future steps, a roadmap for lowering the cost of post-combustion carbon capture, *Chem. Eng. J.* **171**, 3, 742-752.

2 Knudsen J.N., Jensen P.J., Vilhelmsen P.J., Biede O. (2009) Experience with $CO_2$ capture from coal flue gas in pilot-scale: Testing of different amine solvents, *Energy Procedia* **1**, 783-790.

3 Serna-Guerrero R., Belmabkhout Y., Sayari A. (2010) Further investigations of $CO_2$ capture using triamine-grafted pore-expanded mesoporous silica, *Chem. Eng. J.* **158**, 3, 513-519.

4 Serna-Guerrero R., Belmabkhout Y., Sayari A. (2010) Triamine-grafted pore-expanded mesoporous silica for $CO_2$ capture: Effect of moisture and adsorbent regeneration strategies, *Adsorption* **16**, 6, 567-575.

5 Belmabkhout Y., Sayari A. (2010) Isothermal *versus* Non-isothermal Adsorption-Desorption Cycling of Triamine-Grafted Pore-Expanded MCM-41 Mesoporous Silica for $CO_2$ Capture from Flue Gas, *Energy Fuels* **24**, 5273-5280.

6 Drese J.H., Choi S., Lively R.P., Koros W.J., Fauth D.J., Gray M.L., Jones C.W. (2009) Synthesis-Structure-Property Relationships for Hyperbranched Aminosilica $CO_2$ Adsorbents, *Adv. Funct. Mater.* **19**, 23, 3821-3832.

7 Ebner A.D., Gray M.L., Chisholm N.G., Black Q.T., Mumford D.D., Nicholson M.A., Ritter J.A. (2011) Suitability of a Solid Amine Sorbent for $CO_2$ Capture by Pressure Swing Adsorption, *Ind. Eng. Chem. Res.* **50**, 9, 5634-5641.

8 Li B., Jiang B., Fauth D.J., Gray M.L., Pennline H.W., Richards G.A. (2011) Innovative nano-layered solid sorbents for $CO_2$ capture, *Chem. Commun.* **47**, 6, 1719-1721.

9 Gray M.L., Hoffman J.S., Hreha D.C., Fauth D.J., Hedges S.W., Champagne K.J., Pennline H.W. (2009) Parametric Study of Solid Amine Sorbents for the Capture of Carbon Dioxide, *Energy Fuels* **23**, 4840-4844.

10 Jones C.W., Maginn E.J. (2010) Materials and Processes for Carbon Capture and Sequestration, *Chemsuschem.* **3**, 8, 863-864.

11 Hiyoshi N., Yogo K., Yashima T. (2005) Adsorption characteristics of carbon dioxide on organically functionalized SBA-15, *Micropor. Mesopor. Mater.* **84**, 1-3, 357-365.

12 Gray M.L., Champagne K.J., Fauth D., Baltrus J.P., Pennline H. (2008) Performance of immobilized tertiary amine solid sorbents for the capture of carbon dioxide, *Int. J. Greenhouse Gas Control* **2**, 1, 3-8.

13 Sircar S., Golden T.C. (1995) Isothermal and Isobaric Desorption of Carbon-Dioxide by Purge, *Ind. Eng. Chem. Res.* **34**, 8, 2881-2888.

14 Kim J.-N., Chue K.-T., Kim K.I., Cho S.-H., Kim J.-D. (1994) Non-isothermal Adsorption of Nitrogen-Carbon Dioxide Mixture in a Fixed Bed of Zeolite-X, *J. Chem. Eng. Japan* **27**, 45-51.

15 Merel J., Clausse M., Meunier F. (2008) Experimental investigation on $CO_2$ post-combustion capture by indirect thermal swing adsorption using 13X and 5A zeolites, *Ind. Eng. Chem. Res.* **47**, 1, 209-215.

16 Liu J., Wang Y., Benin A.I., Jakubczak P., Willis R.R., Levan M.D. (2010) $CO_2/H_2O$ Adsorption Equilibrium and Rates on Metal-Organic Frameworks: HKUST-1 and Ni/DOBDC, *Langmuir* **26**, 17, 14301-14307.

17 Li G., Xiao P., Webley P., Zhang J., Singh R., Marshall M. (2008) Capture of $CO_2$ from high humidity flue gas by vacuum swing adsorption with zeolite 13X, *Adsorption* **14**, 2-3, 415-422.

18 Trzpit M., Soulard M., Patarin J., Desbiens N., Cailliez F., Boutin A., Demachy I., Fuchs A.H. (2007) The effect of local defects on water adsorption in silicalite-1 zeolite: A joint experimental and molecular simulation study, *Langmuir* **23**, 20, 10131-10139.

19 Dunne J.A., Mariwala R., Rao M., Sircar S., Gorte R.J., Myers A.L. (1996) Calorimetric heats of adsorption and adsorption isotherms .1. $O_2$, $N_2$, Ar, $CO_2$, $CH_4$, $C_2H_6$ and $SF_6$ on silicalite, *Langmuir* **12**, 24, 5888-5895.

20 Palomino M., Corma A., Rey F., Valencia S. (2010) New Insights on $CO_2$ – Methane Separation Using LTA Zeolites with Different Si/Al Ratios and a First Comparison with MOFs, *Langmuir* **26**, 3, 1910-1917.

21 Chatelain T., Patarin J., Soulard M., Guth J.L., Schultz P. (1995) Synthesis and characterization of high-silica EMT and FAU zeolites prepared in the presence of crown-ethers with either ethylene-glycol or 1,3,5-trioxane, *Zeolites* **15**, 90

22 Belmabkhout Y., Pirngruber G., Jolimaitre E., Methivier A. (2007) A complete experimental approach for synthesis gas separation studies using static gravimetric and column breakthrough experiments, *Adsorption* **13**, 3-4, 341-349.

23 Dunne J.A., Rao M., Sircar S., Gorte R.J., Myers A.L. (1996) Calorimetric heats of adsorption and adsorption isotherms. 2. $O_2$, $N_2$, Ar, $CO_2$, $CH_4$, $C_2H_6$, and $SF_6$ on NaX, H-ZSM-5, and Na-ZSM-5 zeolites, *Langmuir* **12**, 24, 5896-5904.

24 Li G., Xiao P., Webley P.A., Zhang J., Singh R. (2009) Competition of $CO_2/H_2O$ in adsorption based $CO_2$ capture, *Energy Procedia* **1**, 1, 1123-1130.

25 Zhang J., Webley P.A. (2008) Cycle development and design for $CO_2$ capture from flue gas by vacuum swing adsorption, *Environ. Sci. Technol.* **42**, 2, 563-569.

26 Zhang J., Webley P.A., Xiao P. (2008) Effect of process parameters on power requirements of vacuum swing adsorption technology for $CO_2$ capture from flue gas, *Energy Convers. Manage.* **49**, 2, 346-356.

27 Liu Z., Grande C.A., Li P., Yu J.G., Rodrigues A.E. (2011) Multi-bed vacuum pressure swing adsorption for carbon dioxide capture from flue gas, *Sep. Purifi. Technol.* **81**, 3, 307-317.

28 Park J.H., Beum H.T., Kim J.N., Cho S.H. (2002) Numerical analysis on the power consumption of the PSA process for recovering $CO_2$ from flue gas, *Ind. Eng. Chem. Res.* **41**, 16, 4122-4131.

29 Chue K.T., Kim J.N., Yoo Y.J., Cho S.H., Yang R.T. (1995) Comparison of Activated Carbon and Zeolite 13X for $CO_2$ Recovery from Flue-Gas by Pressure Swing Adsorption, *Ind. Eng. Chem. Res.* **34**, 2, 591-598.

30 Na B.K., Lee H., Koo K.K., Song H.K. (2002) Effect of rinse and recycle methods on the pressure swing adsorption process to recover $CO_2$ from power plant flue gas using activated carbon, *Ind. Eng. Chem. Res.* **41**, 22, 5498-5503.

31 Dantas T.L.P., Luna F.M.T. Silva I.J., Torres A.E.B., de Azevedo D.C.S., Rodrigues A.E., Moreira R.F.P.M. (2011) Carbon dioxide-nitrogen separation through pressure swing adsorption, *Chem. Eng. J.* **172**, 2-3, 698-704.

32 Kikkinides E.S., Yang R.T., Cho S.H. (1993) Concentration and recovery of $CO_2$ from flue-gas by pressure swing adsorption, *Ind. Eng. Chem. Res.* **32**, 11, 2714-2720.

33 Pirngruber G.D., Hamon L., Bourrelly S., Llewellyn P.L., Lenoir E., Guillerm V., Serre C., Devic T. (2012) A Method for Screening the Potential of MOFs as $CO_2$ Adsorbents in Pressure Swing Adsorption Processes, *Chemsuschem.* **5**, 4, 762-776.

34 Liu Z., Wang L., Kong X., Li P., Yu J., Rodrigues A.E. (2012) Onsite $CO_2$ Capture from Flue Gas by an Adsorption Process in a Coal-Fired Power Plant, *Ind. Eng. Chem. Res.* **51**, 21, 7355-7363.

35 Cho S.-H., Park J.H., Beum H.T., Han S.-S., Kim J.-N. (2004) A 2-stage PSA Process for the Recovery of $CO_2$ from Flue Gas and its Power Consumption, *Stud. Surf. Sci. Catal.* **153**, 405-410.

36 Lively R.P., Chance R.R., Koros W.J. (2010) Enabling Low-Cost $CO_2$ Capture *via* Heat Integration, *Ind. Eng. Chem. Res.* **49**, 16, 7550-7562.

37 Lively R.P., Chance R.R., Kelley B.T., Deckman H.W., Drese J.H., Jones C.W., Koros W.J. (2009) Hollow Fiber Adsorbents for $CO_2$ Removal from Flue Gas, *Ind. Eng. Chem. Res.* **48**, 15, 7314-7324.

38 Sivakumar S.V., Rao D.P. (2011) Modified Duplex PSA. 1. Sharp Separation and Process Intensification for $CO_2$-$N_2$-13X Zeolite System, *Ind. Eng. Chem. Res.* **50**, 6, 3426-3436.

39 Kearns D.T., Webley P.A. (2006) Modelling and evaluation of dual-reflux pressure swing adsorption cycles: Part I. Mathematical models, *Chem. Eng. Sci.* **61**, 22, 7223-7233.

40 Diagne D., Goto M., Hirose T. (1994) New PSA Process with Intermediate Feed Inlet Position Operated with Dual Refluxes – Application to Carbon-Dioxide Removal and Enrichment, *J. Chem. Eng. Japan* **27**, 1, 85-89.

41 Diagne D., Goto M., Hirose T. (1995) Parametric Studies on $CO_2$ Separation and Recovery by a Dual Reflux PSA Process Consisting of Both Rectifying and Stripping Sections, *Ind. Eng. Chem. Res.* **34**, 9, 3083-3089.

42 Leinekugel-le-Cocq D., Tayakout-Fayolle M., Le Gorrec Y., Jallut C. (2007) A double linear driving force approximation for non-isothermal mass transfer modeling through bi-disperse adsorbents, *Chem. Eng. Sci.* **62**, 15, 4040-4053.

43 Farooq S., Ruthven D.M., Boniface H.A. (1989) Numerical simulation of a pressure swing adsorption oxygen unit, *Chem. Eng. Sci.* **44**, 12, 2809-2816.

## APPENDIX

The cyclic PSA model is based on the non-isothermal adsorption column model described by Leinekugel-le-Cocq et al. [42]. It is summarized in Table A1 and Table A2. The main assumptions of the bed model are:

- the ideal gas law is obeyed;
- the pressure drop in the bed is negligible;
- the flow in the packed bed is represented by a cascade of CSTR (Continuous Stirred Tank Reactors) in series; 101 reactors have been used in the following simulations corresponding to a Péclet number of approximately 200;
- the adsorbent pellets have a bi-modal pore size distribution (micro- and macropores), and the intra-particle mass transfer is described by the modified double linear driving force model;
- chemical kinetic time constant is small compare to mass transfer time constant in the particle;
- adsorption equilibrium are modeled considering extended multi-component Langmuir isotherms.

During pressurization and blow-down steps, we consider that the total pressure in the bed changes linearly with time [43]:

$$P(t) = P_I^{step} + (P_F^{step} - P_I^{step}) \frac{t - t_I^{step}}{\Delta t_{step}} \tag{1}$$

The computing code based on this model has been developed in Fortran. The ordinary differential equations system obtained by spatial discretization by a cascade of CSTRs is solved using the DDASPK subroutine based on the Petzold-Gear BDF method.

TABLE A1

Cyclic PSA model

| Models | Equations |
|---|---|
| $CO_2$ bulk phase mass balance | $\dfrac{\partial c_{CO_2}}{\partial t} = -\dfrac{1}{\varepsilon} \cdot \dfrac{\partial (u \cdot c_{CO_2})}{\partial z} - \dfrac{1-\varepsilon}{\varepsilon} \varepsilon_p \cdot a_s^p \cdot N_{CO_2}^{mac}$ |
| Bulk phase energy balance | $\dfrac{\partial T_g}{\partial t} = -\dfrac{u}{\varepsilon} \cdot \dfrac{\partial T_g}{\partial z} + \dfrac{1}{C_{P,g}} \cdot \dfrac{\partial P}{\partial t} - \dfrac{1-\varepsilon}{\varepsilon} \cdot \dfrac{a_s^p \cdot k_{th}}{C_{P,g}} \cdot (T_g - T_s) - \dfrac{1}{C_{P,g}} \cdot \dfrac{2 \cdot k_w}{\varepsilon \cdot R_{int}} \cdot (T_g - T_w)$ <br><br> with $C_{P,g} = \sum\limits_{i=1}^{n_c} c_i \cdot C_{P,i}$ |
| Superficial $u$ velocity expression: overall mass balance | $\dfrac{1}{\varepsilon} \dfrac{\partial u}{\partial z} = -\dfrac{1-\varepsilon}{\varepsilon} \cdot \dfrac{a_s^p \cdot k_{th}}{C_{P,g}} \cdot \left(1 - \dfrac{T_s}{T_g}\right) - \dfrac{1}{C_{P,g}} \cdot \dfrac{2 \cdot k_w}{\varepsilon \cdot R_{int}} \cdot \left(1 - \dfrac{T_w}{T_g}\right)$ <br><br> $\qquad + \dfrac{\partial P}{\partial t} \cdot \left(\dfrac{1}{T_g} \dfrac{1}{C_{P,g}} - \dfrac{1}{P}\right) - \dfrac{1-\varepsilon}{\varepsilon} \cdot \varepsilon_p \cdot \left(\dfrac{R \cdot T_g}{P}\right) \cdot a_s^p \cdot N_T^{mac}$ |
| $CO_2$ macroporous phase mass balance | $\dfrac{\partial c^{mac}}{\partial t} = a_s^p \cdot N_{CO_2}^{mac} - \dfrac{1-\varepsilon_p}{\varepsilon_p} \cdot a_s^c \cdot N_{CO_2}^{ads}$ |
| $N_T^{mac}$ expression: overall mass balance in macroporous phase | $N_T^{mac} = \dfrac{1}{a_s^p}\left(-\dfrac{P}{R \cdot T_s^2} \cdot \dfrac{\partial T_s}{\partial t} + \dfrac{1}{R \cdot T_s} \cdot \dfrac{\partial P}{\partial t} + \dfrac{1-\varepsilon_p}{\varepsilon_p} \cdot a_s^c \cdot N_{CO_2}^{ads}\right)$ |
| $CO_2$ adsorbed phase mass balance | $\underbrace{\left(\dfrac{\partial q}{\partial c^*}\right)_{T_s} \cdot \dfrac{\partial c^*}{\partial t} + \left(\dfrac{\partial q}{\partial T_s}\right)_{c^*} \cdot \dfrac{\partial T_s}{\partial t}}_{\frac{\partial q}{\partial t}} = a_s^c \cdot N_{CO_2}^{ads}$ |
| Particle energy balance | $\dfrac{\partial T_s}{\partial t} = \dfrac{a_s^c}{(1-\varepsilon_p) \cdot \rho_s \cdot C_s} \cdot N_{CO_2}^{ads} \cdot (-\Delta H_{CO_2}^{ads}) + \dfrac{a_s^p \cdot k_{th}}{(1-\varepsilon_p) \cdot \rho_s \cdot C_s} \cdot (T_g - T_s)$ |

TABLE A2

Intra-granular mass transfer molar flux and equilibrium

| Models | Equations |
| --- | --- |
| $CO_2$ molar flow rate between bulk and macroporous phases | $$N_{CO_2}^{mac} = N_T^{mac} \cdot \frac{x_{CO_2}^{mac} \cdot e^{-N_T^{mac} \cdot \left( \frac{\delta_f}{D_{CO_2}^f \cdot c_T} + \frac{\delta_{mac}}{D_{CO_2}^{mac} \cdot c_T^{mac}} \right)} - x_{CO_2}}{e^{-N_T^{mac} \cdot \left( \frac{\delta_f}{D_{CO_2}^f \cdot c_T} + \frac{\delta_{mac}}{D_{CO_2}^{mac} \cdot c_T^{mac}} \right)} - 1}$$ |
| $CO_2$ molar flow rate between macroporous and adsorbed phases | $$N_{CO_2}^{ads} = \frac{5}{R_c} \cdot D_{CO_2}^c \cdot \left( \frac{\partial q}{\partial c^*} \right)_{T_g} \cdot \left( c^{mac} - c^* \right)$$ |
| Adsorption isotherm | $$\frac{q_{CO_2}}{q_{sat}} = \frac{b \cdot c^*}{1 + b \cdot c^*} \quad \text{with} \quad b = b_0 \cdot T_s \cdot e^{\frac{\Delta H_{CO_2}^{ads}}{R \cdot T_s}}$$ |

# 12

# ACACIA Project – Development of a Post-Combustion $CO_2$ Capture Process. Case of the DMX$^{TM}$ Process

A. Gomez[1]*, P. Briot[1], L. Raynal[1], P. Broutin[1]*, M. Gimenez[2], M. Soazic[3], P. Cessat[3] and S. Saysset[4]

[1] IFP Energies nouvelles, Rond-point de l'échangeur de Solaize, BP 3, 69360 Solaize - France
[2] Lafarge, IPC, 95 rue du Montmurier, BP 70, 38291 Saint-Quentin-Fallavier Cedex - France
[3] VEOLIA Environnement Recherche et Innovation, 291 avenue Dreyfous Ducas, Zone Portuaire de Limay, 78520 Limay - France
[4] GDF SUEZ, CRIGEN, 361 avenue du Président Wilson, BP 33, 93211 Saint-Denis-La-Plaine Cedex - France
e-mail: adrien.gomez@ifpen.fr - patrick.briot@ifpen.fr - ludovic.raynal@ifpen.fr - paul.broutin@ifpen.fr
michel.gimenez@pole-technologique.lafarge.com - soazic.mary@veolia.com - pascal.cessat@veolia.com
samuel.saysset@gdfsuez.com

* Corresponding authors

*Abstract* — *The objective of the ACACIA project was to develop processes for post-combustion $CO_2$ capture at a lower cost and with a higher energetic efficiency than first generation processes using amines such as MonoEthanolAmine (MEA) which are now considered for the first Carbon Capture and Storage (CCS) demonstrators. The partners involved in this project were:* Rhodia *(Solvay since then), Arkema, Lafarge, GDF SUEZ, Veolia Environnement, IFP Energies nouvelles, IRCE Lyon, LMOPS, LTIM, LSA Armines. To validate the relevance of the breakthrough processes studied in this project, techno-economic evaluations were carried out with comparison to the reference process using a 30 wt% MEA solvent. These evaluation studies involved all the industrial partners of the project, each partner bringing specific cases of $CO_2$ capture on their industrial facilities. From these studies, only the process using demixing solvent, $DMX^{TM}$, developed by IFPEN appears as an alternative solution to the MEA process.*

**Résumé** — **Projet ACACIA – Développement d'un procédé de captage du $CO_2$ post-combustion – Cas du procédé DMX$^{TM}$** — L'objectif du projet ACACIA était de développer des procédés de captage du $CO_2$ en postcombustion à moindre coût et plus efficaces que les procédés de première génération utilisant des amines de types monoéthanolamine (MEA) qui sont actuellement envisagées pour les premiers démonstrateurs industriels de captage du $CO_2$ en vue de son stockage géologique. Les partenaires impliqués dans ce projet étaient : *Rhodia (depuis Solvay), Arkema, Lafarge, GDF SUEZ, Veolia Environnement, IFP Energies nouvelles, IRCE-Lyon, LMOPS, LTIM, LSA, Armines.* Afin de valider la pertinence des procédés de rupture étudiés dans ce projet, des travaux d'évaluation technico-économiques ont été réalisés avec comparaison au procédé de référence mettant en œuvre comme solvant une solution aqueuse de MEA à 30 % poids. Ces études ont été réalisées avec les partenaires industriels du projet, chacun apportant des cas concrets de captage du $CO_2$ sur leurs installations industrielles. De ces travaux, seul le procédé liquide de lavage des gaz par solvant demixant, $DMX^{TM}$, qui est en cours de développement à IFPEN apparait comme une solution de rupture alternative au procédé de lavage à la MEA.

## ABBREVIATIONS

| | |
|---|---|
| MEA | MonoEthanolAmine |
| $DMX^{TM}$ | 2nd generation of chemical solvent using a novel demixing solvent |
| MDEA | MethylDiEthanolAmine |
| TEG | Tri Ethylene Glycol used for $CO_2$ stream dehydration |
| OPEX | OPerating EXpenses, or operating cost |
| CAPEX | CAPital EXpenditure, investments |
| PC plant | Pulverised Coal plant |
| NGCC plant | Natural Gas Combined Cycle Plant |

## INTRODUCTION

The ACACIA project, which was launched by the AXELERA cluster (AXELERA is one of the 71 French "competitiveness cluster" initiated by the French government in September 2004), is dedicated to the development of new processes for $CO_2$ capture on industrial facilities before geological storage. As considered by IEA [1], Carbon Capture and Storage (CCS) is one of the possible pathway in order to mitigate greenhouse gases emissions; it thus requires the development of high efficiency $CO_2$ capture technologies.

The ACACIA project partners have chosen to consider only the post-combustion capture pathway for which the $CO_2$ is extracted directly from the industrial flue gases [2]. The main objective of the project was the research and development of new technologies to reduce the cost of capture per ton of $CO_2$ and the impact of $CO_2$ capture on the cost of electricity or industrial products (cement, chemicals). It is known from process studies [3] or from pilot demonstration [4], that the energy penalty reduction, especially due to the energy required at reboiler, estimated about 3.7 $GJ/ton_{CO_2}$ for the MEA (MonoEthanolAmine) 30 wt% process, is the key issue for making $CO_2$ capture an attractive solution for carbon mitigation.

Conventional processes for $CO_2$ capture are based mostly on absorption by a chemical solvent. Chemical solvents used are primary amines, and in particular MEA. If the MEA can recover up to 98% $CO_2$ and obtain a purity of 99.9%, its use leads to high operating costs. In the medium term, to make CCS deployment possible, it is necessary to develop new capture processes with lower energy costs further reducing the cost of carbon capture. The purpose of the ACACIA project was to develop new processes with a cost of capture 50% lower than the cost of existing processes while allowing to capture at least 90% $CO_2$ in the treated gas and obtain a $CO_2$ purity near 95%. Such a purity level is necessary for transport and storage.

Five types of processes were studied in the ACACIA project:
- demixing solvents: use of amine solvents which either for high $CO_2$ loadings or for high temperature form two non-miscible phases. With this type of solvents, only the heavy $CO_2$-rich phase is regenerated which reduces the energy cost of carbon capture [2];
- hydrates: research on thermodynamic additives to improve the operating conditions of $CO_2$ capture by hydrates; the objective being to capture $CO_2$ at low temperature and moderate pressure and deliver $CO_2$ at high pressure with low energy inputs which would reduce the cost of regeneration and $CO_2$ compression;
- enzymes: use of enzymes, which are immobilized in porous materials, to enhance $CO_2$ absorption with in particular an increase in $CO_2$ absorption kinetics and an associated investment reduction;
- ionic liquids: optimizing the absorption of $CO_2$ by the use of some ionic liquids offering high solubilities;
- innovative chemistry: development of new solvents with innovative chemical routes for $CO_2$ capture with low enthalpy of formation requiring less energy at regeneration step.

Through these lines of research, the ACACIA project aimed to develop solutions applicable in priority to the industry (power plants, cement plants, incinerators, and chemical industry). To validate the relevance of the new processes studied in this project, a benchmark based on a techno-economical study between MEA technology and new processes was undertaken by the industrial partners of the project, each bringing specific cases of $CO_2$ capture corresponding to an industrial case. The pooling of these cases and appropriate technological solutions was a very important part of this project because it allowed the identification of viable pre industrial study technological solutions: a validation process

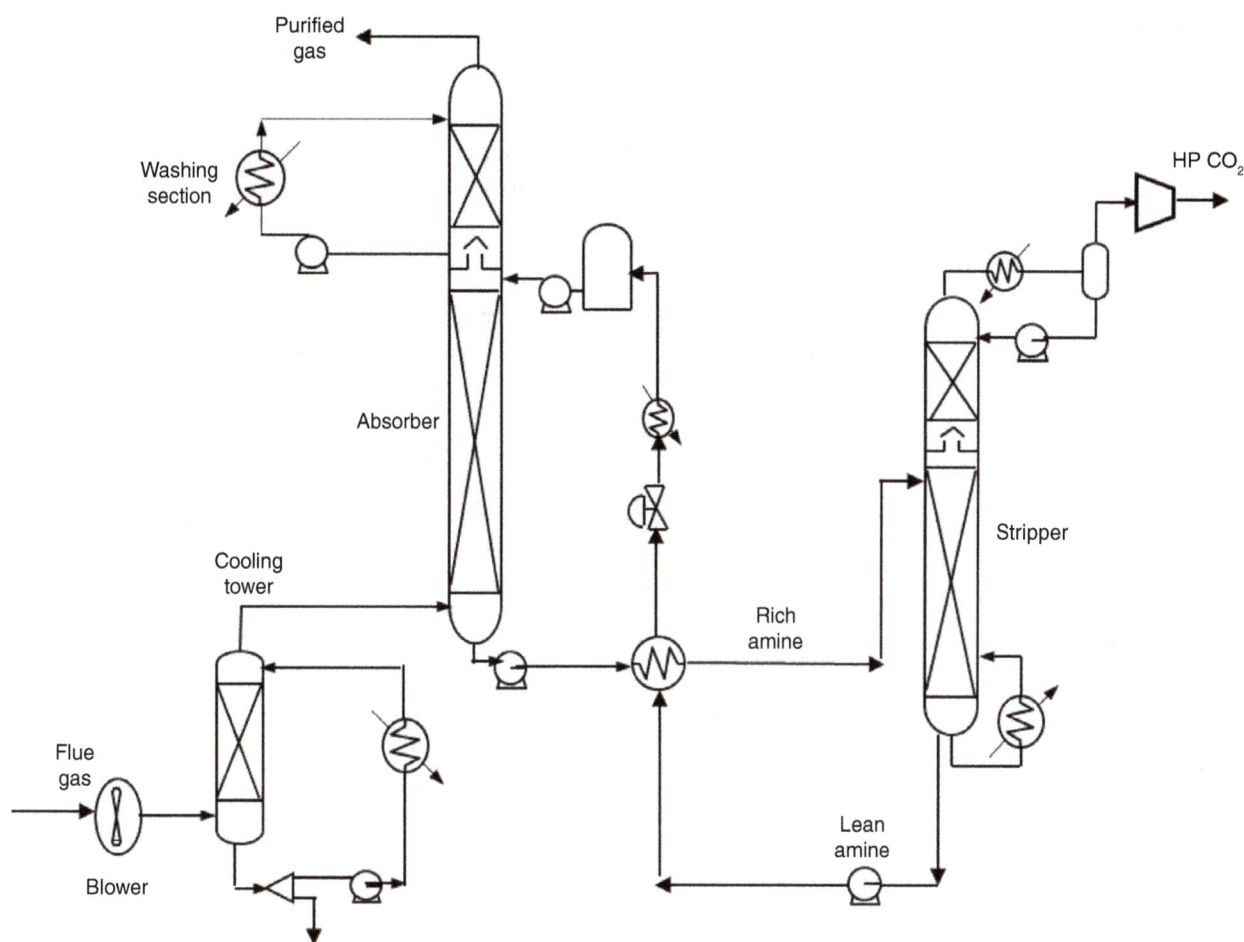

Figure 1

Simplified process flow diagram of the MEA post-combustion capture process.

pilot type being envisaged only after the end of ACACIA depending of the results obtained.

In this paper, we present the results of the techno-economical study carried out on the $DMX^{TM}$ process and the comparison with the reference process using 30 wt% MEA. For all other original routes studied (hydrates, enzymes, ionic liquid), it has not been possible to obtain sufficient data for performing process evaluation and techno-economic evaluation.

In a first part, a description of the operating conditions for the MEA (1st generation of chemical solvent) and $DMX^{TM}$ (2nd generation of chemical solvent) processes is provided. A second part is dedicated to a description of the emission case studies, the study methodology and economic assumptions. In the last part, a comparative analysis between the MEA and $DMX^{TM}$ processes is presented.

# 1 CO$_2$ CAPTURE PROCESS: 1$^{ST}$ GENERATION VERSUS 2$^{ND}$ GENERATION WITH DMX$^{TM}$ SOLVENT

## 1.1 MEA Process Description

To separate $CO_2$ from the flue gas (low pressure, low $CO_2$ content), the reference process is a chemical absorption process using 30 wt% MEA as solvent. It is widely admitted that this process is the reference for $CO_2$ capture on flue gases [3]. Within the CASTOR and CESAR FP7 European projects, this process has been demonstrated at pilot plant scale on real power plant flue gas [4-6] and some companies are able to commercialize such a process with already some large scale references existing in the food industry [7, 8]. Figure 1 shows a typical process flow diagram for a first generation process such as the reference MEA 30 wt% process.

The capture process is composed of five main sections:
- a cooling tower which purpose is first to cool down the flue gas issued at 140°C and second to perform a preconditioning of the flue gas (washing of ash, impurities, etc.);
- an absorber, operated at ambient pressure and moderate temperature, where $CO_2$ is separated from the flue gas by being contacted with the solvent;
- a washing section which ensures that the decarbonized flue gas sent to the stack does not contain any unwanted pollutants (amines, degradation products or any other volatile compounds);
- a regenerator operated at moderate pressure and high temperature, where $CO_2$ is separated from the solvent, the latter being regenerated;
- a compression section needed to deliver high-pressure pure $CO_2$ ready for storage.

All these five sections are specific to a given process and are interconnected. As an example, the cooling tower, using a first washing section, can be more or less important depending on the solvent sensitivity towards impurities contained in the flue gas such as $SO_x$ or $NO_x$. In the same idea, a process using a volatile solvent may require a large washing section downstream the absorber, while a small section may be enough for others. Similarly, the operating conditions in the regeneration section, in particular in terms of pressure may impact the compression section. It is thus mandatory to consider all the needed sections for the process at constant boundary limits, inlet flue gas and outlet $CO_2$ and treated gas as shown in Figure 1, when making comparison.

## 1.2 DMX™ Process Description

The DMX™ process has been developed and patented by *IFPEN* (see [9] or [10] for process or physical and chemical basic information respectively). It has been described with further details in [11, 12] and only a quick description is given hereafter. The main objective being to present the techno-economic comparison with the MEA process.

The DMX™ process is based on the use of very specific solvents which, for given loading and temperature conditions, can form two immiscible liquid phases. These phases have sufficient density differences that they can be separated by decantation. The light liquid phase is such that it contains almost no $CO_2$, the latter being concentrated in the heavy phase. This result is similar to what could be obtained with a high capacity. The DMX™ solvent is also characterized by an easy separation which can be performed in a standard decanter placed downstream the lean/rich heat exchanger,

downstream the absorption column, as can be seen in Figure 2. The decanter is preferably positioned after the amine/amine heat exchanger and before the regenerator in particular to make decantation easier *via* the reduction of liquid viscosity associated with the increase of temperature. Only the $CO_2$ rich loaded heavy phase is sent to the stripper, the $CO_2$ lean light phase being directly sent back to the absorber. Note that depending on the operating conditions chosen at stripper and at decanter, one may observe an important $CO_2$ gas release at decanter. The compression section is then modified turning into a possible supplementary energy reduction when the decanter is operated at a pressure higher than the pressure at stripper.

Figure 3 shows a picture of a transparent decanter that has been used on a mini-pilot at *IFPEN*. The flow goes from left to right as indicated by the plain arrow. As can be seen in the close view, the inlet flow contains $CO_2$ gas bubbles and droplets of the light phase dispersed in the heavy phase. On the right-hand-side, one observes that, very quickly, a clear separation of the phases is reached, the interface being indicated with a dashed line.

Such a process presents a significant decrease of solvent mass flow and of captured $CO_2$ sent to the regeneration column requiring less energy input. It can thus offer a significant cost reduction compared to the reference case that is the MEA 30 wt% based process.

The choice for the formulation of the demixing solvent DMX-1, was firstly based on its thermodynamic capacity which comes in addition to demixtion for reducing the solvent flow rate going to the stripper. Secondly, we paid a particular attention towards degradation performances. As shown in Raynal *et al.* [11], the DMX-1 degradation performances are much better than those of MDEA (MethylDiEthanolAmine), a commercial amine known as being much more stable than MEA. As discussed by Raynal *et al.* [12], degradation impacts many costs and not only solvent make-up; it makes possible the operation of the stripper at higher pressure/temperature operating conditions enabling $CO_2$ compression cost reduction. Last, kinetics performance and operability issues were considered.

## 2 CASE DESCRIPTION STUDY METHODOLOGY AND ECONOMIC EVALUATION: MEA *VERSUS* DMX™ PROCESS

### 2.1 Case Description

Each industry has defined the gas to be treated by the $CO_2$ capture process considered within the ACACIA project.

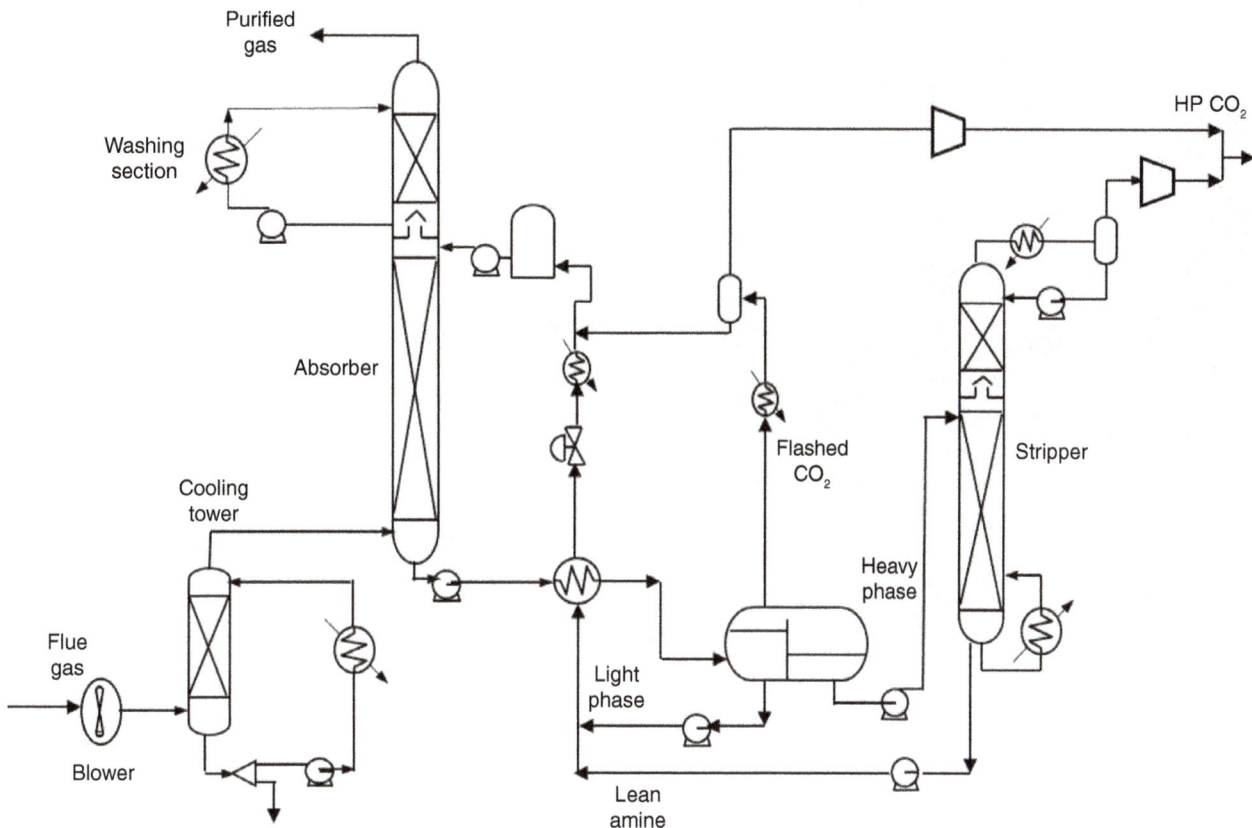

Figure 2

Simplified process flow diagram of the DMX$^{TM}$ post-combustion capture process.

The information on these gases, include:
- the flow rate, the density at standard conditions, the pressure and temperature;
- the molar composition;
- the expected impurities (dust, $SO_x$, $NO_x$).

This information is given for a nominal flowrate case of with a range of expected changes to account for the flexibility of the units.

The emitting industries concerned were the following:
- electricity production by gas-fired power plant (*GDF SUEZ*) and coal-fired power plant (Electrabel *GDF SUEZ*);
- production of cement (*Lafarge*);
- chemistry (*Rhodia Operations* and *Arkema*);
- incineration of household waste (*Veolia Environnement*).

Table 1 summarizes the characteristics of the flue gases to be decarbonized, which were given by these different industries.

For geological storage application, the $CO_2$ delivery pressure at battery limit was at 110 barg.

Compression energy to this pressure level was of course taken into account within the techno-economic evaluation.

## 2.2 Evaluation Methodology

The economic assessment methodology implemented in the ACACIA project allowed to establish a strong synergy between academic and industrial partners.

The stakeholders were as follows:
- For the MEA process:
  - *IFPEN* conducted the process studies and economic evaluation for *Veolia Environnement*, Arkema and *Rhodia* cases;
  - *GDF SUEZ* has completed the design of facilities for the collection for PC and NGCC plants;
  - *Rhodia* generated sizing and quantification of the *Lafarge* plant;
  - *Lafarge* got his own experience of the design of the system and has estimated the cost of the MEA

Figure 3

Picture of the decanter inlet of the mini-pilot of *IFPEN*. The three-phase flow, G/L/L, enters the decanter on the left-hand-side, the decantation being achieved in the large diameter section on the right-hand-side.

TABLE 1

Flue gas characteristics

| Temperature | °C | 48 to 360 |
|---|---|---|
| Pressure | bar abs | 1.01 |
| $CO_2$ flow rate | t/h | 2.5 to 581 |
| Emission flow rate | $Nm^3/h$ | 15 000 to 3 320 000 |
| Emission composition | $CO_2$ | 3 to 15 |
| | $H_2O$ | 5 to 25 |
| | $N_2$ | 57 to 77 |
| | $O_2$ | 3 to 14 |
| (vol.%) | $H_2$ | 0 |
| | CO | 0.1 max |
| | Ar | 0 to 0.9 |
| | $CH_4$ | 0 |

solution and proposed the cost estimation derived from that produced by *Rhodia*, for the purposes of *Solvay* (without compression step);

– For the DMX™ process:

Among the available demixing solvents *IFPEN*, proposed to consider the DMX-1 system which is currently the best solvent for the DMX™ process.

*IFPEN* provided to partners who have chosen the DMX™ process solution (*GDF SUEZ*, *Veolia Environnement*, *Lafarge*), the mass and energy balances as well as the sizing of the main equipments:

– *IFPEN* provided balance sheets and equipment sizing devices to *GDF SUEZ* and *Lafarge*;
– *Lafarge* and *GDF SUEZ* made their own economic evaluation based on data provided by *IFPEN* for their respective cases;
– *IFPEN* performed the entire study for the *Veolia* case.

## 2.3 Study Basis and Economic Assumptions

Emissions flows are described in the basis for studies cited above.

The economic assumptions are detailed in Table 2.

The cooling water is available on site in sufficient quantities to ensure the capture units needs. Without any previous specification, the temperature of the cooling water was taken equal to 15°C (sea water) and the maximum elevation of the cooling water was set at 10°C to reach a final temperature of 25°C. For the power plant cases (*GDF SUEZ* cases) the Low Pressure (LP) steam needed for the reboilers comes from thermal power, which reduces the production of electricity from the LP turbine. The power consumption of the various equipment is provided by the power plant.

For the other cases, steam comes from a steam generator dedicated to the $CO_2$ capture plant.

For the calculation of operating costs (OPEX) processes for MEA and DMX™, the following assumptions were made:

– labor cost (number of operator for $CO_2$ capture unit) specific for each industrial case (*Tab. 2*);
– cost of electricity (€/kWh) (*) specific for each industrial case (*Tab. 2*);
– cost of cooling water (€/m³) specific for each industrial case (*Tab. 2*);
– cost of TEG: 800 €/t and consumption of 0.094 kgTEG/tCO₂ captured (**) specific for each industrial case (*Tab. 2*);
– cost of MEA: 1 500 €/t and consumption: 1.6 kgMEA/tCO₂ captured;
– cost of the solvent DMX-1: 5 000 €/t and no consumption (no degradation);

TABLE 2

Economic assumptions for economic study

| Parameter\Case | GDF SUEZ Coal-fired power plant* | GDF SUEZ Gas-fired power plant | Veolia Environnement | Rhodia | Arkema | Lafarge |
|---|---|---|---|---|---|---|
| Depreciation (years) | 20 | 20 | 20 | 10 | 10 | 20 |
| Interest rate (%) | 10 | 10 | 10 | 7 | 7 | 10 |
| Labor (€/year) | 80 000 | 80 000 | 65 000 | 80 000 | 80 000 | 80 000 |
| Electricity (€/kWh) | 0.06 | 0.06 | 0.06 | 0.07 | 0.055 | 0.07 |
| Cooling water (€/m³) | 0.02 | 0.02 | 0.05 | 0.05 | 0.05 | 0.05 |
| LP Steam (€/t) | Included in electric production penalty | | | 15 | 20 | 15 |

* For the coal fired power plant and gas fired power plant, economic assumptions were based on past assumptions (from literature review) made for a previous project (CAPCO$_2$ project funded by the French National Agency for Research, 2006-2008).

– investment depreciation (years) specific for each industrial case (*Tab. 2*);
– interest rate (%) specific for each industrial case (*Tab. 2*).

(*) The power consumption of the capture units/compress CO$_2$ is considered a loss for the thermal power plant.

(**) TEG unit is devoted to dry CO$_2$ stream before transportation and injection.

## 3 COST EVALUATION AND ANALYSIS

The overall CO$_2$ cost, expressed in €/t CO$_2$, is obtained by combining CAPEX and OPEX costs in a complete economic analysis. The obtained value corresponds to the minimum price of CO$_2$ on market for which a CCS project is profitable. That is the minimum price for which it is more interesting to invest in a CCS project rather than buying CO$_2$ emissions rights on the market. In the present analysis, we split the different costs for the main contribution in the overall CO$_2$ cost to emphasize the advantages and weaknesses of a given process.

### 3.1 MEA Evaluation

The techno-economic studies confirmed the very high cost of CO$_2$ capture for the reference 30 wt% MEA process, whatever the considered case. The cost of capture by amine scrubbing (*Tab. 3*) ranges from 39 €/tCO$_2$ to 239 €/tCO$_2$. This is related, in the first analysis, to the scale effect. Indeed, *Arkema* case deals only

2.5 tCO$_2$/hour of CO$_2$ while the coal-fired plant emits 582 tCO$_2$/h. These two extreme costs define the minimum and maximum value for capture costs. A case as small as the *Arkema* case which could correspond to a demonstration case is associated to a very high CAPEX in particular due to building, instrumentation and control costs almost as expensive as a very large case. Otherwise, for comparable emission flow rate, costs range from 63 €/tCO$_2$ for *Veolia* to 91 €/tCO$_2$ for *Lafarge*. The explanation here comes from the fact that *Veolia* has lower operating costs related to the integration of the production of steam for regeneration.

### 3.2 DMX™ Evaluation

The *GDF SUEZ* coal-fired plant case, the *Veolia* and *Lafarge* cases were considered for the DMX™ process.

It appears that the DMX™ process could be a very interesting technology. The evaluation of this process showed significant gains on the cost of CO$_2$ capture as one can observed in Table 4. Indeed when compared to the MEA process, it appears that the DMX™ process can offer reduction of −20% and up to −50% in CO$_2$ capture cost. So, with this breakthrough technology, it is possible to meet part of the initial goal of the ACACIA project: 50% of the cost of CO$_2$ capture.

Some comments can be made on these results:
– About *Lafarge* case (cement plant):
The cost of CO$_2$ capture is halved with DMX™ compared to the reference MEA process, which was the objective of the project.

TABLE 3

MEA economic evaluation results

| Parameter\Case | GDF SUEZ Coal-fired power plant | GDF SUEZ Gas-fired power plant | Veolia Environnement | Rhodia | Arkema | Lafarge |
|---|---|---|---|---|---|---|
| Emission flow rate ($Nm^3/h$) | 2 244 000 | 3 320 000 | 245 000 | 231 000 | 17 480 | 250 000 |
| vol.% of $CO_2$ | 13.2 | 3.4 | 9 | 9.6 | 7.2 | 14.9 |
| $CO_2$ inlet (t/h) | 582 | 236 | 43.3 | 43.5 | 2.5 | 73.2 |
| $CO_2$ captured (t/h) | 524 | 212 | 39 | 39 | 2.2 | 64 |
| Depreciation (years) | 20 | 20 | 20 | 10 | 10 | 20 |
| Interest rate (%) | 10 | 10 | 10 | 7 | 7 | 10 |
| Labor (€/year) | 80 000 | 80 000 | 65 000 | 80 000 | 80 000 | 80 000 |
| Electricity (€/kWh) | 0.06 | 0.06 | 0.06 | 0.07 | 0.055 | 0.07 |
| Cooling water (€/$m^3$) | 0.02 | 0.02 | 0.05 | 0.05 | 0.05 | 0.05 |
| LP Steam (€/t) | Included in electric production penalty | | | 15 | 20 | 15 |
| TEG (€/t) | 800 | 800 | 800 | 800 | 800 | 800 |
| MEA (€/t) | 1 500 | 1 500 | 1 500 | 1 500 | 1 500 | 1 500 |
| $CO_2$ capture cost (€/t$CO_2$) | 39 | 70 | 63 | 71 | 239 | 91 |

TABLE 4

Economic comparison between DMX$^{TM}$ and MEA process

| Parameter\Case | GDF SUEZ Coal-fired power plant | | | Veolia | | | Lafarge | | |
|---|---|---|---|---|---|---|---|---|---|
| | MEA | DMX$^{TM}$ | Gain | MEA | DMX$^{TM}$ | Gain | MEA | DMX$^{TM}$ | Gain |
| $CO_2$ captured (t/h) | 524 | 524 | | 39 | 39 | | 64 | 64 | |
| Plant efficiency | 34.3 | 35.6 | + 3.8% | N/A | N/A | | N/A | N/A | |
| $CO_2$ capture cost (€/t$CO_2$) | 37.1 | 31.4 | −15.4% | 63 | 52 | −17.4% | 93 | 44.9 | −51.7% |

It is interesting to discuss how such a gain can be explained. Three main reasons can explain this result:
- a small part is due to the investment, a little lower for DMX$^{TM}$: −4%. (for the Lafarge case, it is necessary to build a boiler in order to generate the steam necessary for the solvent regeneration/size of this boiler is reduced for the DMX$^{TM}$ process);
- the most important part corresponds to variable costs and especially steam: −20%;
- the low possibility of heat integration between the cement plant and the DMX$^{TM}$ process explains the better performance of the process DMX$^{TM}$, by a significant reduction of steam consumption from utility device;

– About GDF SUEZ case (coal power plant):
The gain for the plant performance related to the use of DMX$^{TM}$ process is 1.3% for the thermal overall efficiency (see Tab. 4 above). IFPEN expects to have more than 2 points performance gain with an innovative heat integration with the power plant steam cycle.
The cost of captured $CO_2$ is estimated at 37.1 €/t$CO_2$ for the reference MEA, and 31.4 €/t$CO_2$ for DMX$^{TM}$ process, that implies a decrease of 15.4% of the capture cost.
– About Veolia Environnement case (central waste incineration):
The emissions flow rates are comparable to the cement plant and the cost of treatment with DMX$^{TM}$ is the same order of magnitude (52 €/t$CO_2$ captured).

However, the cost reduction compared to MEA is not as important, from 63 €/tCO$_2$ to 52 €/tCO$_2$.

The thermal integration is here already done for MEA case by steam extraction available on the incineration plant, the transition to DMX$^{TM}$ is less profitable even if it allows a significant reduction in cost.

## CONCLUSIONS

With the DMX$^{TM}$ process patented and developed by *IFPEN*, it is possible to have significant energy savings compared to the reference MEA. This gain was 1.3% on energy penalty for the coal power plant but studies show that it is even possible to achieve a gain of 2%. Gains on operating costs (OPEX) enable cost reduction CO$_2$ capture 15 to 50% depending on the cases.

In order to go one step further in terms of process development, it is now necessary to perform industrial demonstration of the DMX$^{TM}$ process. This is one of the goal of the European FP7 OCTAVIUS project, which started on March 1$^{st}$ 2012. Tests at large scale are scheduled to be performed in 2015-206 on the ENEL pilot plant in Brindisi which treats 10 000 Nm$^3$/h of flue gases issued from a coal fired power plant (2.5 t/h CO$_2$ captured equivalent), and for which a important revamp is planned.

## ACKNOWLEDGMENTS

This work has been carried out within the ACACIA project (*Amélioration du CAptage du CO$_2$ Industriel et Anthropique*); it was launched by the AXELERA cluster and was financially supported by the French Government through the FUI convention No. 08 2 90 6390.

## REFERENCES

1 Technology Roadmap – CCS, IEA ed., see http://www.iea.org/papers/2009/CCS_Roadmap.pdf.

2 Lecomte F., Broutin P., Lebas E. (2010) *CO$_2$ Capture Technologies to Reduce Greenhouse Gas Emissions*, Editions Technip.

3 Steeneveldt R., Berger B., Torp T.A. (2006) CO$_2$ capture and storage – closing the knowing-doing gap, *Chem. Eng. Res. Design* **84**, 739-763.

4 Abu-Zahra M., Schneiders L.H.J., Niederer J.P.M., Feron P.H.M., Versteeg G.F. (2007) CO$_2$ capture from power plants Part I. A parametric study of the technical performance based on monoethanolamine, *Int. J. of Greenhouse Gas Control* **1**, 37-46.

5 Knudsen J., Jensen J.N., Vilhelmsen P.-J., Biede O. (2009) Experience with CO$_2$ capture from coal flue gas in pilot-scale: testing of different amine solvents, *Energy Procedia* **1**, 783-790.

6 Knudsen J.N., Jensen J.N., Andersen J., Biede O. (2011) Evaluation of process upgrades and novel solvents for CO$_2$ post combustion capture in pilot-scale, *Energy Procedia* **4**, 1558-1565.

7 Chapel D.G., Mariz C.L. (1999) *Recovery of CO$_2$ from flue gases: commercial trends*, Paper Presented at the *Canadian Society of Chemical Engineers Annual Meeting*, Saskatoon, Saskatchewan, Canada, 4-6 Oct.

8 Lemaire E., Raynal L. (2009) IFP solutions for lowering the cost of post-combustion carbon capture. From HiCapt + TM to DMX$^{TM}$ and future steps, in: *12th Meeting of the IEA International Post-Combustion CO$_2$ Capture Network*, Regina, Canada, 28 Sept.-1 Oct.

9 Cadours R., Carrette P.-L., Boucot P., Mougin P. (2006) Procédé de désacidification d'un gaz avec une solution absorbante à régénération fractionnée, *French Patent* 1,656,983.

10 Aleixo M., Prigent M., Gibert A., Porcheron F., Mokbel I., Jose J., Jacquin M. (2011) Physical and Chemical Properties of DMX Solvents, *Energy Procedia* **4**, 148-155.

11 Raynal L., Alix P., Bouillon P.-A., Gomez A., le Febvre de Nailly M , Jacquin MKittel A., di Lella J., Mougin P., Trapy J. (2011) The DMXTM process: an original solution for lowering the cost of post-combustion carbon capture, *Energy Procedia* **4**, 779-786.

12 Raynal L., Bouillon P.-A., Gomez A., Broutin P. (2011) From MEA to demixing solvent and future step, a roadmap for lowering the cost of post-combustion carbon capture, *Chemical Engineering Journal* **171**, 742-752.

# 13

# CO$_2$ Absorption by Biphasic Solvents: Comparison with Lower Phase Alone

Zhicheng Xu, Shujuan Wang*, Guojie Qi, Jinzhao Liu, Bo Zhao and Changhe Chen

*Key Laboratory for Thermal Science and Power Engineering of Ministry of Education, Beijing Key Laboratory for CO$_2$ Utilization and Reduction Technology, Department of Thermal Engineering, Tsinghua University, Beijing 100084 - China*
*e-mail: wangshuj@tsinghua.edu.cn*

* Corresponding author

**Résumé** — **Absorption du CO$_2$ par des solvants biphasiques : comparaison avec la phase inférieure isolée** — Il a été montré que les mélanges de 1,4-butanediamine (BDA) 2 M et 2-(diéthylamino)-éthanol (DEEA) 4 M sont des solvants biphasiques prometteurs. Dans cette étude la composition de la phase inférieure est déterminée en utilisant un chromatographe ionique (CI) DX-1290 et un titrateur automatique *Metrohm* 909 Titrando. Les capacités cycliques, les taux de charge cycliques et les produits de réaction du solvant biphasique sont comparés à ceux d'une solution aqueuse de même concentration en amines que la phase inférieure du solvant biphasique au taux de charge riche ((2B4D)$_L$), en utilisant des équipements de screening rapide et un spectromètre à Résonance Magnétique Nucléaire (RMN) JNM ECA-600. Les vitesses d'absorption ont aussi été mesurées à différents taux de charge sur film tombant. Les résultats indiquent que la capacité cyclique et le taux de charge cyclique de (2B4D)$_L$ sont très proches de celles de 2B4D. La vitesse d'absorption de (2B4D)$_L$ est plus élevée que celle de 2B4D pour les trois taux de charge étudiés, sauf dans le cas d'une solution vierge à une pression partielle de CO$_2$ inférieure à 10 kPa. Les résultats de RMN indiquent que les produits de réaction de (2B4D)$_L$ comportent plus de bicarbamate de BDA, moins de BDA et moins de carbamate de BDA que 2B4D. Les produits de réaction de (2B4D)$_L$ avec le CO$_2$ comportent deux fois plus de carbonate/bicarbonate que dans le cas de 2B4D et moins de carbamate de BDA.

*Abstract* — *CO$_2$ Absorption by Biphasic Solvents: Comparison with Lower Phase Alone* — *The mixtures of 2 M 1,4-butanediamine (BDA) and 4 M 2-(diethylamino)-ethanol (DEEA) have been found to be promising biphasic solvents. This work identifies the composition of the lower phase using a DX-120 Ion Chromatograph (IC) and a* Metrohm *809 Titrando auto titrator. The cyclic capacities, cyclic loadings and reaction products of the biphasic solvent are compared with those of the aqueous solution with the same amine concentration as the lower phase of the biphasic solvent at the rich loading ((2B4D)$_L$) using a fast screening facility and a JNM ECA-600 Nuclear Magnetic Resonance spectrometer (NMR). Their absorption rates at different loadings are also investigated using a Wetted Wall Column (WWC). The results show that the cyclic capacity and cyclic loading of (2B4D)$_L$ are almost the same as those of 2B4D. The absorption rate of (2B4D)$_L$ is higher than 2B4D at all the 3 tested loadings, except for the fresh solutions at CO$_2$ pressure lower than 10 kPa. NMR results show that the reaction products of (2B4D)$_L$ had more BDA bicarbamate, less BDA and less BDA carbamate than 2B4D. The CO$_2$ reaction products of (2B4D)$_L$ had twice as much carbonate/bicarbonate as with 2B4D and less BDA carbamate.*

## INTRODUCTION

Over the last decades, amine gas sweetening has become a proven technology for the $CO_2$ capture from natural gas. In recent years, amine based absorption has also been widely investigated for $CO_2$ capture in power plant, due to its high flexibility and easy retrofit for existing power plant (Rochelle, 2009). Development of solvents with high efficiency is regarded as one of the most crucial issues for $CO_2$ absorption. Many solvents, such as monoethanolamine (MEA), methyldiethanolamine (MDEA), diethanolamine (DEA) and piperazine (PZ), have been applied to capture $CO_2$ (Bishnoi and Rochelle, 2002; Derks and Dijkstra, 2005; Rinker et al., 2000; Rinker and Ashour, 2000). However, this absorption-desorption process always requires lots of energy and high operation costs during solvent regeneration. Therefore, to minimize the energy penalty is of great importance for this absorption-desorption system.

Recent years, some novel concepts, such as DMX$^{TM}$ (Raynal et al., 2011) and lipophilic amine solvents (Zhang X., 2007), have been proposed for the improvement of the energy performance. Zhang J. et al. (2011) did screening tests of dipropylamine (DPA), dimethylcyclohexylamine (DMCA) and other solvents. Tan (2010) studied the kinetics and thermodynamics of DPA and DMCA blend and found that the cyclic loading of this solvent can reach 0.7 mol $CO_2$/mol amine. The precipitation of the solvent at high loading, however, is still a challenge for this system. Raynal et al. (2011) explained the DMX$^{TM}$ process, which, according to their simulations, could remarkably reduce the reboiler heat duty to 2.3 GJ/t $CO_2$. Rojey et al. (2009) found that some solvents with special structures can separate into two liquid phases after absorption, but they did not give the particular structures. Hu (2009) pointed out that biphasic solvents should consist of several compounds, including at least one activator A and one solvent B, and that the mixture composition should be 20% A plus 80% B. However, Hu (2009) did not give the exact amines in the solvents. Bruder and Svendsen (2011) found that a blend of 5 M DEEA and 2 M MAPA separated into two phases after $CO_2$ absorption, with a cyclic loading higher than that of 5 M MEA. The lower phase, however, was found to be viscous, which affected the transition of the lower phase to the stripper.

Cyclic capacity $\Delta R$, expressed as either $C_{amine}$ $(\alpha_{rich} - \alpha_{lean})$ or $R_{abs} - R_{des}$, is an important characteristic of the solvent, where $C_{amine}$ is the amine concentration, $R_{abs}$ and $R_{des}$ are the $CO_2$ concentrations in terms of moles per kilogram after absorption and desorption; $\alpha_{rich}$ and $\alpha_{lean}$ are rich and lean loading of the solvent, the difference of which is shown as cyclic loading, $\Delta\alpha$.

Cyclic efficiency, $\theta$, is the result of $\Delta\alpha$ divided by $\alpha_{rich}$. A promising solvent should have relatively higher cyclic capacity, cyclic loading and cyclic efficiency (Aronu et al., 2011).

About 30 solvents with various compositions, including solutions with Low Critical Solution Temperature (LCST) and other potential biphasic solvents, were screened in our previous research (Xu et al., 2012a,b, 2013a,b,c). Some of these biphasic solvents, such as 5 M TEA and 2 M BDA blended with 4 M DEEA, have been proven to have higher cyclic capacities, cyclic loadings or absorption rates than traditional 5 M MEA. Moreover, the aqueous solution of 2 M BDA mixed with 4 M DEEA (simplified as 2B4D hereafter) was found to have the best performance among the selected solvents, with 46% higher cyclic loading, 48% higher cyclic capacity and 11% higher cyclic efficiency than 5 M MEA, which will reduce the sensible heat requirement during regeneration (Svendsen et al., 2011).

It was found that the solution of 2B4D became two phases after $CO_2$ absorption using the fast screening facility on the absorption mode, with 97.4% of $CO_2$ existing in the lower phase and a total loading of 0.505 mol $CO_2$/mol amine. The weight percentages of the upper and lower phases after absorption on the absorption mode were 21.79% and 78.21%, respectively. The screening facility will be described in the experimental section. The amine and $CO_2$ distributions in the two phases were analyzed by IC in the previous studies. It was confirmed that the biphasic solvent separation is due to the fast reaction rate of BDA with $CO_2$ and the limited solubility of DEEA in the reaction products of BDA with $CO_2$. The reaction products of BDA in the two phases were then analyzed. The products was mainly BDA carbamate in the upper phase, while in the lower phase, at a total loading of 0.446 mol/mol amine, the mole fractions of BDA, BDA carbamate and BDA bicarbamate were 16.8%, 55.8% and 27.4% respectively (Xu et al., 2013c). Since most of the $CO_2$ absorbed existed in the lower phase after absorption, comparison of the alone lower phase and biphasic solvent is necessary to decide which one will be more efficient.

This paper will identify the composition of 2B4D lower phase after absorption. Then, the absorption and desorption properties of the alone lower phase of 2B4D (simplified as $(2B4D)_L$ hereafter), which has the same amine concentration as the 2B4D lower phase at rich loading, will be measured and compared with 2B4D. The absorption rates of 2B4D and $(2B4D)_L$ at different loadings will be compared with WWC. The reaction products of $(2B4D)_L$ with $CO_2$ during absorption will also be measured by NMR and compared with those of 2B4D.

## 1 EXPERIMENTAL

The chemicals used in this work, BDA ($\geq 98$ wt%), DEEA ($\geq 99$ wt%), dioxane ($\geq 99.5$ wt%) and $D_2O$ ($\geq 99.96$ wt%) NaOH ($\geq 96$ wt%) from *Aladdin Reagent Company*, and $CO_2$ ($\geq 99.9\%$ pure), $SO_2$ (1.51% vol%, $N_2$ balanced) and $N_2$ ($\geq 99.99\%$ pure) from *Beijing Huayuan Gas Company* were used without further purification. Distilled deionized water was used for preparing the solutions. The amine concentrations were determined by titration against 0.2 N $H_2SO_4$ using a *Metrohm* 809 Titrando auto titrator.

The BDA and DEEA structures are shown Figure 1.

The absorption and desorption capacities, and the rich and lean loadings were measured using a fast screening facility, and the loadings were confirmed by the titration method. The absorption experiments were conducted at 40°C with the desorption at 90°C at atmospheric pressure. On "absorption mode", $CO_2$ and $N_2$, controlled by mass flow controller, were used to simulate the flue gas with 12% of $CO_2$ in terms of volume. The total gas flow rate was 463 mL/min. The simulated flue gas went through the gas mixture first to mix intensively, and then to the reactor, which was made of glass and had a volume of about 150 mL. After reaction with solvent, the gas went to condenser, which was circulated by 3°C water. The condensed water went back to the reactor to avoid water losses, and then the acid washing, in case that amine vapor mixed with simulated gas and resulted in measurement error in IR $CO_2$ analyzer. After being dried by anhydrous calcium chloride, $CO_2$ concentration of the gas was measured by the IR $CO_2$ analyzer. Equilibrium was assumed to be reached when the outlet $CO_2$ concentration reached 12%. On "desorption mode", $N_2$ was used to sweep the desorbed $CO_2$ to take turns in going through condenser, acid washing, drier and analyzer with a flow rate of 874 mL/min. Lean loading of solution was assumed to be reached when the outlet $CO_2$ concentration was less than 0.1%. This assumption has been verified in our previous work (Xu *et al.*, 2012a). The volume of the fresh solution added in the reactor was 100 mL. The other details of the facility were described by Xu *et al.* (2012a).

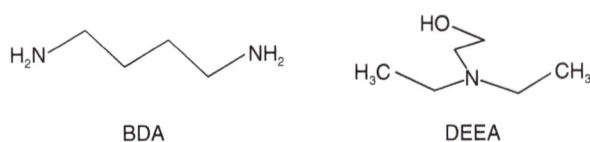

A wetted wall column was used to investigate the absorption rate of the amine solution with a contact area of about 41.45 $cm^2$. The gas-liquid contactor was constructed from a stainless steel tube, measuring 11.0 cm in height and 1.2 cm in diameter. The gas-liquid contact region was enclosed by a 31.0 cm thick-walled glass tube, separated from a water bath. More details about the original WWC system can be found in our previous research (Liu *et al.*, 2009, 2011, 2012). The experimental system used in this work shown in Figure 2 has two modifications from the original system. A saturator with the same temperature and pressure as the reactor was added before the gas entered the reactor. The gas side mass transfer coefficient was calibrated using $SO_2$ absorption into NaOH solution, to replace the previous one which was determined with $CO_2$ absorption into MEA solution.

For the analysis method, A *Dionex* DX-120 system with an IonPac Column CG17/CS17 was used to measure the individual amine concentrations in the two liquid phases. The $CO_2$ loading was determined by titration using the barium carbonate precipitation method (Hilliard, 2008). A JNM ECA-600 NMR from *JEOL Company* was used to analyze the reaction products in the solution in terms of $^{13}C$. As the natural abundance of $^{13}C$ is 1.11%, the ordinary $CO_2$ was added into the solution for further analysis. $CO_2$ was added by bubbling the gas into the solution. The total loading was determined from the weight change of the solution after bubbling with $CO_2$, and the individual loadings of the upper and lower phases were titrated. Then, about 1 mL solution was added to a WG-5000-7-50 sample tube from *Wilmad Company*. Dioxane was used as an internal standard and a small amount of $D_2O$ was added to the sample tube to get a signal lock. The magnetic field strength of the NMR was 14.096 T, with a $^{13}C$ resonance frequency of 150.91 MHz. The quantitative $^{13}C$ lasted 15 hours for each sample with a relaxation time of 20 seconds.

## 2 THEORY

### 2.1 Chemical Reactions

The chemical reactions of $CO_2$ with the amines in the solution can be addressed as follows:

Dissociation of carbon dioxide:

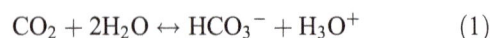

$$CO_2 + 2H_2O \leftrightarrow HCO_3^- + H_3O^+ \qquad (1)$$

Dissociation of bicarbonate ion:

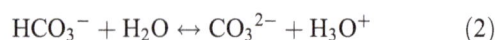

$$HCO_3^- + H_2O \leftrightarrow CO_3^{2-} + H_3O^+ \qquad (2)$$

Figure 1

BDA and DEEA structures.

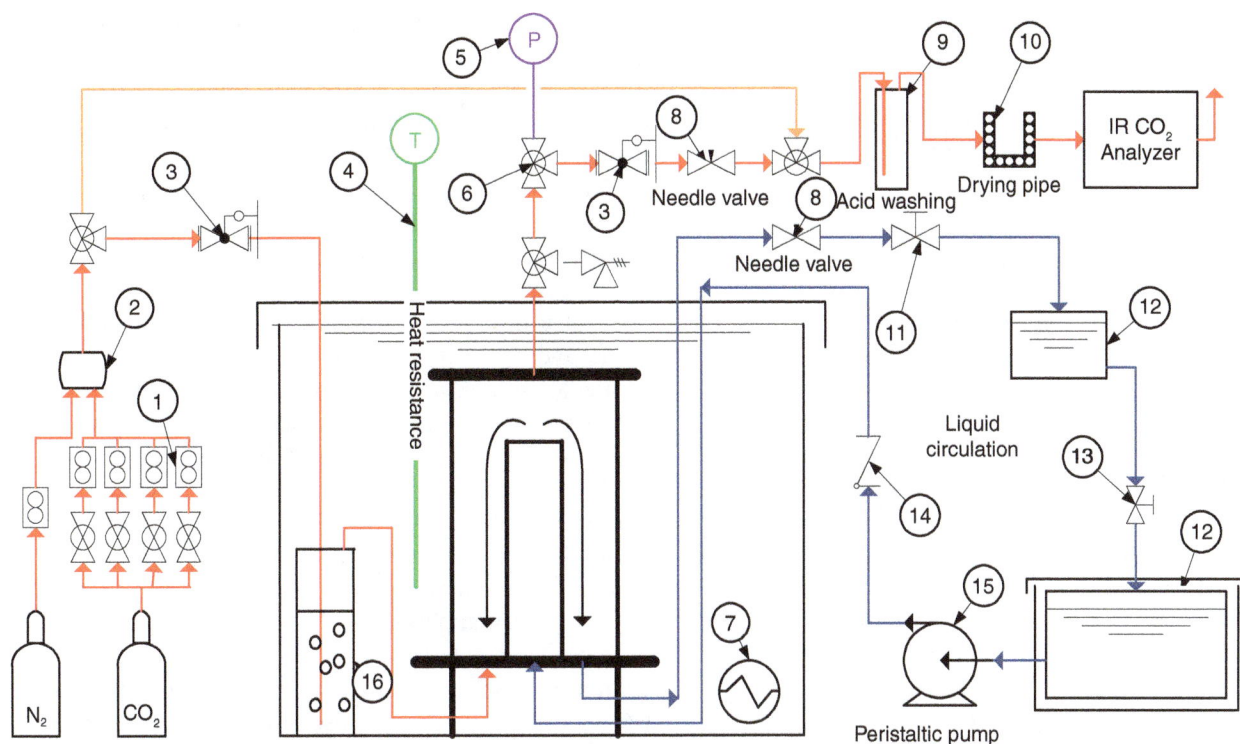

1. Mass flow controller  2. Cylinders mixed  3. Gas valves  4. Heat resistance heater  5. Pressure transmitter
6. Three-way valve  7. Electric heater  8. Precision needle valve  9. Acid wash  10. Drying pipe  11. Liquid valves
12. Liquid storage tank  13. Rotary screw valve  14. One-way valve  15. High-pressure peristaltic pump  16. Saturator

Figure 2

The experiment scheme of wetted wall column.

Dissociation of protonated amine:

$$DEEAH^+ + H_2O \leftrightarrow DEEA + H_3O^+ \qquad (3)$$

$$BDAH^+ + H_2O \leftrightarrow BDA + H_3O^+ \qquad (4)$$

Dissociation of deprotonated BDA:

$$BDAH_2^{2+} + H_2O \leftrightarrow BDAH^+ + H_3O^+ \qquad (5)$$

Formation of carbamate of BDA:

$$BDA + CO_2 + H_2O \leftrightarrow BDACOO^- + H_3O^+ \qquad (6)$$

Formation of bicarbamate of BDA:

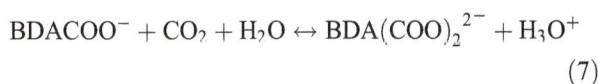

$$BDACOO^- + CO_2 + H_2O \leftrightarrow BDA(COO)_2^{2-} + H_3O^+ \qquad (7)$$

Dissociation of protonated carbamate:

$$BDAHCOO + H_2O \leftrightarrow BDACOO^- + H_3O^+ \qquad (8)$$

## 2.2 Absorption Rate Calculation of WWC

Based on the two film model, the mass flux with chemical absorption can be described as follows, where the total resistance to mass transfer is divided as the sum of the resistance from gas side and liquid side, as Equation (9) shows:

$$\frac{1}{k_g} + \frac{1}{k_G{'}} = \frac{1}{K_G} \qquad (9)$$

The overall gas transfer coefficient, $K_G$, can be calculated from the $Flux$, $P_{CO_2,in}$ and $P_{CO_2,out}$. $K_G$ is fixed for a given temperature and amine concentration and can be calculated by the $Flux$-$CO_2$ partial pressure curve:

$$K_G = \frac{Flux}{P_{CO_2,b} - P_{CO_2}^*} \qquad (10)$$

where $P_{CO_2,b}$ is the operational partial pressure of $CO_2$ in the wetted wall column, which is the log mean average, as Equation (11) shows:

$$P_{CO_2,b} = \frac{P_{CO_2,in} - P_{CO_2,out}}{Ln(P_{CO_2,in}/P_{CO_2,out})} \qquad (11)$$

The $CO_2$ pressure in the gas-liquid contact face, $P_{CO_2,i}$, can be calculated by Equation (12):

$$k_g = \frac{Flux}{P_{CO_2,b} - P_{CO_2,i}} \qquad (12)$$

The gas side mass transfer coefficient $k_g$ can be given by Equation (13):

$$Sh = \alpha(Re \times Sc \times d_h/h)^\beta \qquad (13)$$

where $Sh = RTk_g d_h D_{CO_2}$, $Re = \rho_g V_g d_h/\mu_g$, $Sc = \upsilon/D$, $d_h$ is the hydraulic diameter of the annulus and $h$ is the length of the column. As Pacheco (1998) stated, the parameters $\alpha$ and $\beta$ are fitted based on the experimental data of $SO_2$ absorption into 0.1 M NaOH solution. The fitted results will be presented in Section 3.3.

## 3 RESULTS AND DISCUSSION

### 3.1 (2B4D)$_L$ Identification and the Lean Loading Calculation

The amine and $CO_2$ concentrations in the two phases were measured by IC and auto-titrator after the $CO_2$ absorption by 2B4D reaching equilibrium using the fast screening facility on absorption mode. The measurement error has been evaluated in previous work and the results proven to be reliable (Xu et al., 2013c). At the rich loading, the BDA, DEEA and $CO_2$ concentrations in the lower phase were 2.23,

2.40 and 3.29 mol/kg solution, which can be converted to the amine concentration of (2B4D)$_L$ by subtracting the weight of $CO_2$ in the solution. The actual BDA and DEEA concentrations in fresh (2B4D)$_L$ solution were 2.61 and 2.81 mol/kg.

The absorption rates of 2B4D and (2B4D)$_L$ at lean loadings were measured to compare their performance comprehensively. Thus, the lean loadings of 2B4D and (2B4D)$_L$ were identified here. The 2B4D lean solution was got using the fast screening facility on desorption mode and the $CO_2$ loading of the lean solution was titrated. The rich and lean solutions of (2B4D)$_L$ were got using the fast screening facility on absorption and desorption mode, respectively. Then the $CO_2$ loadings of (2B4D)$_L$ rich and lean solution were titrated using the titrator. The $CO_2$ and amine concentrations in 2B4D lean solution were the mixture of upper phase and the lean solution of lower phase, the same as would occur in actual application.

Table 1 presents the BDA, DEEA and $CO_2$ concentrations in 2B4D upper and lower phases at lean and rich $CO_2$ loadings, as well as those of (2B4D)$_L$ solutions at lean and rich $CO_2$ loadings. The amine and $CO_2$ concentrations are between the concentrations for the upper phase solution and the lean solution of lower phase. Table 1 reveals that the (2B4D)$_L$ lean solution has higher BDA concentration, lower DEEA concentration, slightly higher $CO_2$ concentration and higher $CO_2$ loading than 2B4D lean solution.

### 3.2 Comparisons of the Cyclic Capacity and Cyclic Loading

Our previous works have confirmed the reliability of the screening and titration results (Xu et al., 2012a, 2013c).

TABLE 1

BDA, DEEA and $CO_2$ concentrations and loadings in 2B4D and (2B4D)$_L$ at different loadings

| Solvent | | BDA | DEEA | $CO_2$ | $CO_2$ loading |
|---|---|---|---|---|---|
| | | (mol/kg) | (mol/kg) | (mol/kg) | (mol/mol amine) |
| 2B4D | Upper phase | 0.115 | 7.265 | 0.315 | 0.043 |
| | Lower phase rich | 2.233 | 2.400 | 3.291 | 0.710 |
| | Lower phase lean | 2.477 | 2.663 | 1.050 | 0.204 |
| | 2B4D lean | 2.200 | 3.906 | 0.943 | 0.154 |
| (2B4D)$_L$ | Fresh | 2.607 | 2.806 | / | / |
| | Rich | 2.268 | 2.441 | 3.398 | 0.722 |
| | Lean | 2.490 | 2.679 | 1.076 | 0.208 |

Figure 3 shows the cyclic loadings and cyclic capacities of the 2B4D lower phase and $(2B4D)_L$, indicating that the cyclic capacity and cyclic loading of 2B4D and $(2B4D)_L$ are very similar. Therefore, $(2B4D)_L$ differs little from the lower phase of 2B4D in terms of the cyclic capacity and cyclic loading.

## 3.3 Comparison of Absorption Rate

The influence of the saturator added to the WWC system was investigated by titrating DEEA concentration against 0.2 N $H_2SO_4$ using a *Metrohm* 809 Titrando auto titrator before and after the WWC tests of aqueous DEEA solution. The results listed in Table 2 confirm that the addition of the saturator effectively balanced the water in the system.

The parameters $\alpha$ and $\beta$ in Equation (13) were determined with $SO_2$ absorption into 0.1 M NaOH solution to determine the gas side mass transfer coefficient $k_g$. The results are listed in Table 3 with a correlation in Figure 4.

Therefore, the correlated equation for the gas side mass transfer coefficient is Equation (14):

$$Sh = 6.7097(Re \times Sc \times d_h/h)^{0.5036} \qquad (14)$$

The $CO_2$ absorption rates for 2B4D and $(2B4D)_L$ were measured with WWC system at 298.15, 313.15

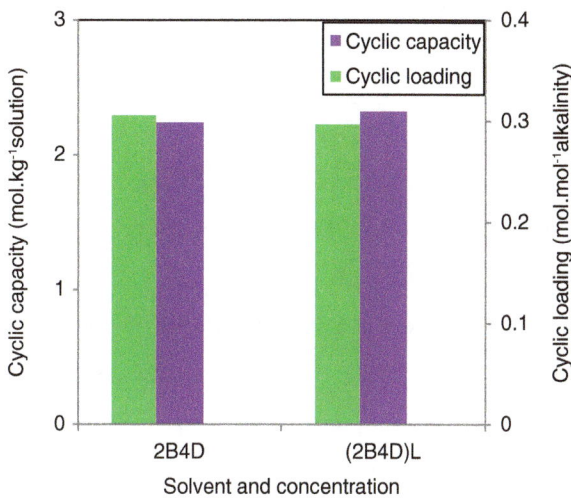

Figure 3

Comparison of cyclic capacities and cyclic loadings of 2B4D and $(2B4D)_L$.

TABLE 2

Titration results of DEEA concentrations before and after the WWC experiment

| Sample | Sample amount (g) | $H_2SO_4$ (mL) | DEEA concentration (mol·kg$^{-1}$) |
|---|---|---|---|
| Before WWC | 0.8880 | 17.8741 | 4.0538 |
| After WWC | 0.8531 | 17.1017 | 4.0374 |

TABLE 3

Correlation of gas-side mass transfer coefficient $k_g$ fitted with $SO_2$ absorption into 0.1 M NaOH solution

| Total flow rate | $P$ | $k_g \times 10^{10}$ | $Re \times Sc \times d_h/h$ | $Sh$ |
|---|---|---|---|---|
| (L/min) | (MPa) | (mol/Pa·cm$^2$·s) | | |
| 5.84 | 0.13 | 8.38 | 43.39 | 44.70 |
| 6.11 | 0.15 | 7.47 | 45.38 | 45.14 |
| 8.13 | 0.17 | 7.89 | 60.35 | 54.16 |
| 8.26 | 0.14 | 9.07 | 61.35 | 52.16 |
| 3.56 | 0.16 | 5.10 | 26.43 | 32.74 |
| 3.69 | 0.16 | 5.56 | 27.42 | 36.36 |
| 3.96 | 0.17 | 5.52 | 29.42 | 38.15 |
| 9.88 | 0.17 | 8.67 | 73.32 | 59.19 |
| 1.54 | 0.17 | 3.39 | 11.46 | 23.17 |

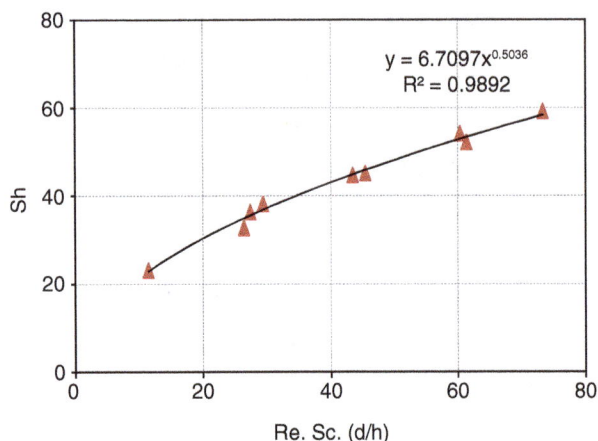

Figure 4

Correlation curve of gas-side mass transfer coefficient $k_g$ fitted by $SO_2$ absorption into 0.1 M NaOH solution.

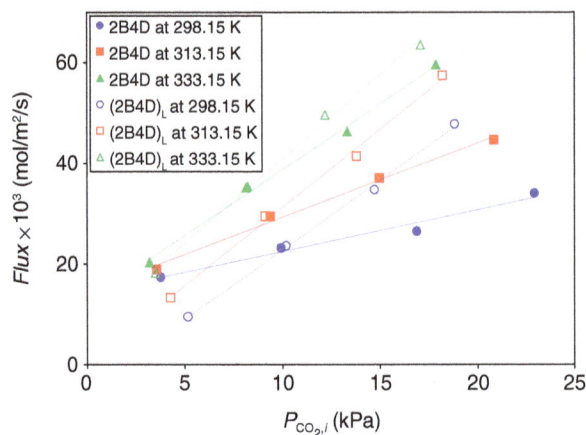

Figure 5

*Flux*-$P_{CO_2,i}$ relationship of fresh 2B4D and $(2B4D)_L$ solutions.

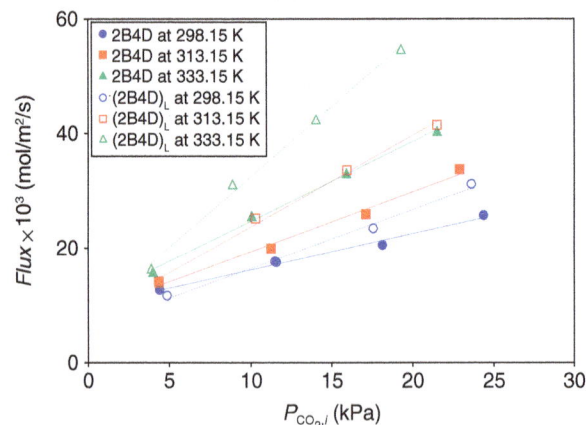

Figure 6

*Flux*-$P_{CO_2,i}$ relationship of 2B4D and $(2B4D)_L$ solution at their half lean loadings.

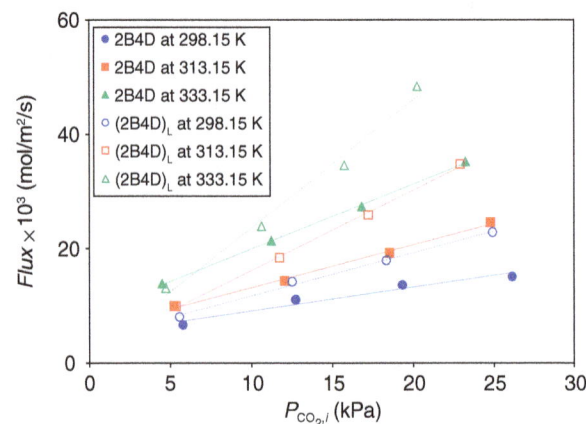

Figure 7

*Flux*-$P_{CO_2,i}$ relationship of 2B4D and $(2B4D)_L$ solution at their lean loadings.

and 333.15 K with the $CO_2$ pressure of 3-25 kPa at loadings of 0 (fresh solution), half lean loading and lean loading, as presented in Tables 4 and 5.

The Flux of 2B4D and $(2B4D)_L$ at 0 loading, half lean loading and lean loading are compared in Figures 5 to 7. Figure 5 shows that the fresh 2B4D solution reacted faster than $(2B4D)_L$ with $P_{CO_2,i}$ lower than 8 kPa, while the absorption rate of the fresh $(2B4D)_L$ solution was higher than that of fresh 2B4D solution with $P_{CO_2,i}$ higher than 10 kPa. Figure 6 reveals that at half lean loading, the absorption rate of $(2B4D)_L$ was higher than that of 2B4D for $P_{CO_2,i}$ between 5 and 25 kPa. Figure 7 shows that at the lean loading, $(2B4D)_L$ reacted faster than 2B4D. For example, the mass flux of $(2B4D)_L$ at lean loading and 313.15 K was 37% higher than that of 2B4D. Comparing the 3 figures, at 313 K and $P_{CO_2,i} = 15$ kPa, the mass flux of fresh $(2B4D)_L$ solution was twice more than that at lean loading and the mass flux of fresh 2B4D was 2.2 times more than that at lean loading.

## 3.4 Comparison of the Reaction Products

As for the reaction products, our previous research has identified the carbon numbers, as shown in Table 6 (Xu *et al.*, 2013c). The peaks positions of the fresh 2B4D solution NMR spectrum agreed well with those in the literature (Balaban *et al.*, 1985; Parker *et al.*, 2011). Comparison of the NMR data and the titration results showed that the NMR error was less than 8%. (Xu *et al.*, 2013c)

TABLE 4

Kinetic parameters of $CO_2$ absorption into 2B4D solutions at 298.15, 313.15 and 333.15 K

| Solvent | $T$ (K) | $Flux \times 10^3$ (mol/m²/s) | $P_{CO_2,b}$ (kPa) | $k_g \times 10^6$ (mol/Pa/m²/s) | $P_{CO_2,i}$ (kPa) |
|---|---|---|---|---|---|
| 2B4D fresh | 298.15 | 17.41 | 5.80 | 8.55 | 3.76 |
| | | 23.24 | 12.67 | 8.42 | 9.91 |
| | | 26.51 | 20.02 | 8.41 | 16.87 |
| | | 34.06 | 26.98 | 8.39 | 22.92 |
| | 313.15 | 18.96 | 5.85 | 8.30 | 3.56 |
| | | 29.46 | 12.93 | 8.27 | 9.37 |
| | | 37.07 | 19.46 | 8.25 | 14.97 |
| | | 44.71 | 26.25 | 8.23 | 20.83 |
| | 333.15 | 20.41 | 5.68 | 8.17 | 3.19 |
| | | 35.39 | 12.54 | 8.13 | 8.19 |
| | | 46.32 | 19.07 | 8.05 | 13.32 |
| | | 59.59 | 25.23 | 8.08 | 17.85 |
| 2B4D half lean loading | 298.15 | 12.68 | 5.90 | 8.50 | 4.41 |
| | | 17.53 | 13.65 | 8.43 | 11.57 |
| | | 20.52 | 20.58 | 8.36 | 18.13 |
| | | 25.71 | 27.43 | 8.35 | 24.35 |
| | 313.15 | 14.15 | 6.06 | 8.31 | 4.35 |
| | | 19.89 | 13.65 | 8.29 | 11.26 |
| | | 25.86 | 20.26 | 8.22 | 17.11 |
| | | 33.74 | 26.99 | 8.21 | 22.88 |
| | 333.15 | 15.80 | 5.92 | 8.24 | 4.00 |
| | | 25.55 | 13.20 | 8.10 | 10.05 |
| | | 33.01 | 19.99 | 8.09 | 15.90 |
| | | 40.34 | 26.49 | 8.07 | 21.49 |
| 2B4D lean loading | 298.15 | 6.71 | 6.57 | 8.46 | 5.78 |
| | | 11.11 | 14.02 | 8.50 | 12.71 |
| | | 13.67 | 20.93 | 8.50 | 19.32 |
| | | 15.12 | 27.92 | 8.44 | 26.12 |
| | 313.15 | 10.00 | 6.44 | 8.37 | 5.25 |
| | | 14.39 | 13.77 | 8.36 | 12.05 |
| | | 19.24 | 20.87 | 8.30 | 18.55 |
| | | 24.59 | 27.74 | 8.28 | 24.77 |
| | 333.15 | 13.90 | 6.20 | 8.19 | 4.50 |
| | | 21.36 | 13.85 | 8.11 | 11.22 |
| | | 27.40 | 20.15 | 8.21 | 16.81 |
| | | 35.21 | 27.61 | 8.03 | 23.23 |

TABLE 5

Kinetic parameters of $CO_2$ absorption into $(2B4D)_L$ solutions at 298.15, 313.15 and 333.15 K

| Solvent | $T$ (K) | $Flux \times 10^3$ (mol/m²/s) | $P_{CO_2,b}$ (kPa) | $k_g \times 10^6$ (mol/Pa/m²/s) | $P_{CO_2,i}$ (kPa) |
|---|---|---|---|---|---|
| $(2B4D)_L$ fresh | 298.15 | 9.55 | 6.26 | 8.57 | 5.14 |
| | | 23.65 | 12.92 | 8.59 | 10.16 |
| | | 34.75 | 18.73 | 8.62 | 14.70 |
| | | 47.85 | 24.34 | 8.65 | 18.81 |
| | 313.15 | 13.33 | 5.85 | 8.42 | 4.27 |
| | | 29.51 | 12.64 | 8.33 | 9.10 |
| | | 41.44 | 18.82 | 8.24 | 13.79 |
| | | 57.43 | 25.19 | 8.20 | 18.19 |
| | 333.15 | 18.30 | 5.74 | 8.12 | 3.49 |
| | | 35.16 | 12.46 | 8.08 | 8.11 |
| | | 49.63 | 18.30 | 8.10 | 12.17 |
| | | 63.44 | 24.93 | 8.07 | 17.07 |
| $(2B4D)_L$ half lean loading | 298.15 | 11.68 | 6.26 | 8.39 | 4.87 |
| | | 17.63 | 13.60 | 8.32 | 11.48 |
| | | 23.40 | 20.36 | 8.36 | 17.56 |
| | | 31.14 | 27.33 | 8.34 | 23.60 |
| | 313.15 | 13.88 | 6.05 | 8.25 | 4.36 |
| | | 25.17 | 13.36 | 8.17 | 10.28 |
| | | 33.56 | 20.02 | 8.21 | 15.93 |
| | | 41.47 | 26.55 | 8.19 | 21.48 |
| | 333.15 | 16.36 | 5.87 | 8.24 | 3.88 |
| | | 31.10 | 12.71 | 8.09 | 8.86 |
| | | 42.45 | 19.27 | 8.06 | 14.00 |
| | | 54.68 | 26.13 | 7.98 | 19.28 |
| $(2B4D)_L$ lean loading | 298.15 | 8.13 | 6.53 | 8.45 | 5.56 |
| | | 14.26 | 14.19 | 8.44 | 12.50 |
| | | 17.95 | 20.47 | 8.43 | 18.34 |
| | | 22.84 | 27.61 | 8.42 | 24.90 |
| | 313.15 | 10.00 | 6.56 | 8.32 | 5.36 |
| | | 18.45 | 13.95 | 8.30 | 11.73 |
| | | 25.90 | 20.36 | 8.28 | 17.23 |
| | | 34.79 | 27.11 | 8.26 | 22.90 |
| | 333.15 | 13.16 | 6.34 | 8.13 | 4.72 |
| | | 23.89 | 13.60 | 8.00 | 10.62 |
| | | 34.59 | 20.05 | 8.03 | 15.74 |
| | | 48.34 | 26.24 | 8.05 | 20.24 |

TABLE 6

Identification of carbon atoms in NMR spectrum (Xu et al., 2013c)

| Component | Structure and identification |
|---|---|
| DEEA/DEEAH$^+$ |  |
| BDA/BDAH$^+$/BDAH$_2^{2+}$ |  |
| BDACOO$^-$/HBDACOO |  |
| BDA(COO$^-$)$_2$ |  |
| CO$_3^{2-}$/HCO$_3^-$ |  |

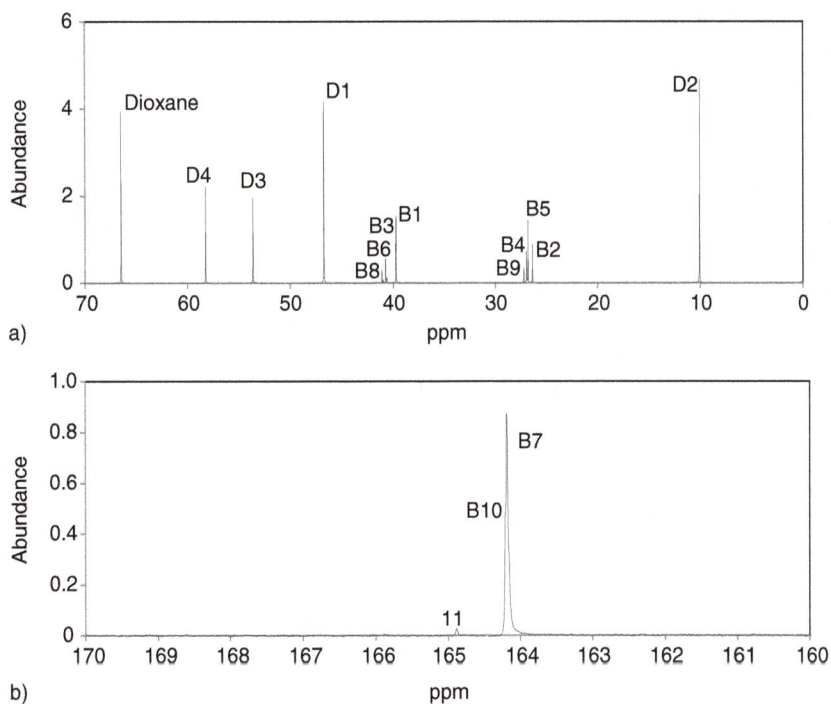

Figure 8

NMR spectrum of (2B4D)$_L$ at the loading of 0.361 mol/mol amine. a) Up field part spectrum; b) low field part spectrum.

TABLE 7

Comparison of the $CO_2$ concentrations measured by NMR and titration in $(2B4D)_L$

| Loading (mol/mol amine) | NMR results | | | | $CO_2$ by titration (mol/kg) | Error (%) |
|---|---|---|---|---|---|---|
| | BDACOO (mol/kg) | BDA(COO)$_2$ (mol/kg) | $CO_3^{2-}$/$HCO_3^-$ (mol/kg) | $CO_2$ by NMR (mol/kg) | | |
| 0.113 | 0.430 | 0.000 | 0.147 | 0.577 | 0.543 | 5.83 |
| 0.219 | 0.803 | 0.109 | 0.099 | 1.119 | 1.028 | 8.08 |
| 0.361 | 1.051 | 0.293 | 0.125 | 1.762 | 1.641 | 6.85 |
| 0.532 | 1.282 | 0.614 | 0.049 | 2.558 | 2.341 | 8.49 |
| 0.735 | 1.095 | 0.815 | 0.530 | 3.256 | 3.112 | 4.42 |

In this work, the quantitative NMR spectrum of $(2B4D)_L$ at five various loadings were acquired. The loadings of all solutions were controlled by the amount of $CO_2$ bubbled into the solution and verified by titration against 0.2 N $H_2SO_4$. The five loadings were 0.113, 0.219, 0.361, 0.532 and the rich loading 0.735 mol $CO_2$/mol amine.

The NMR results of the solution with the 0.361 mol/mol loading are shown in Figure 8 as an example. Referring to the NMR spectrum for fresh 2B4D solution (Xu et al., 2013c), the BDA and DEEA positions can be easily recognized and the positions of D1, D2, D3, D4, B1 and B2 can be identified in the figure. B3, B4, B5 and B6 represent BDA carbamate, the peak areas of which must be equal and close to those of B1 and B2, so they can also be identified in the spectrum. Then, the remaining peaks are B8 and B9 in Figure 8a. The peaks in Figure 8b are B7, B10 and 11. As previous researchers have pointed out, due to the fast proton exchange with water, it is not possible to distinguish signals of amine from protonated amine, or carbonate from bicarbonate. Thus, the position of these peaks in the spectra are the average of the chemical shift of amine and proton amine, or carbonate and bicarbonate (Jakobsen et al., 2005, 2008; Hartono et al., 2007). Thus in the following tables and figure legends of all species represent species and their protonated forms. For example, BDA represents the sum of BDA, $BDAH^+$ and $BDAH_2^{2+}$.

The NMR results in this paper are compared with the titration results to verify its accuracy, as shown in Table 7. The $CO_2$ concentration measured by the NMR result is the sum of BDA carbamate, bicarbamate and carbonate/bicarbonate. The difference between the NMR and titration results indicates the accuracy of NMR results of this work.

The concentrations of the various $(2B4D)_L$ species at different loadings are shown in Figure 9. It is obvious

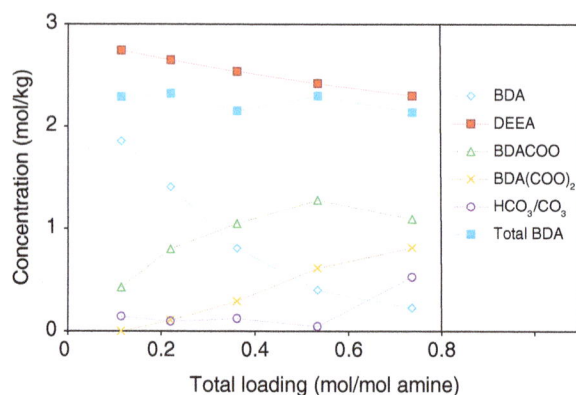

Figure 9

Concentrations of $(2B4D)_L$ species at various loadings.

that BDA concentration decreased from 1.859 mol/kg solution at the loading of 0.113 mol/mol to 0.206 mol/kg at the rich loading of 0.735 mol/mol. The BDA carbamate concentration first increased and then decreased slightly after the loading of 0.532 mol/mol, possibly due to the formation of bicarbamate. BDA bicarbamate kept increasing with increased loading. BDA bicarbamate concentration was higher than the BDA concentration at high loadings. The carbonate/bicarbonate concentration increased sharply after the loading of 0.532 mol/mol as a result of DEEA reaction with $CO_2$.

Figure 10 presents the species mole fraction variations of BDA with the increased loading. Since the total BDA concentration did not vary much during the absorption, the tendency of mole fractions is similar to the variations in Figure 9. Figure 10 also demonstrates that at higher loadings, the mole fraction of BDA bicarbamate was relatively high. At rich loading, the mole fractions of BDA, $BDACOO^-$ and $BDA(COO)_2^{2-}$ were 10.62%, 51.23% and 38.15%, respectively.

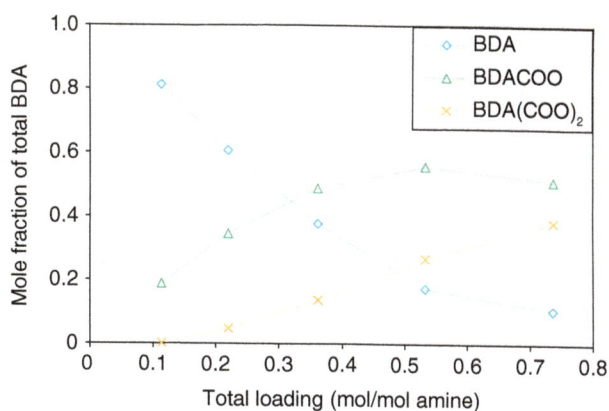

Figure 10

Mole fractions of BDA species in $(2B4D)_L$ at various loadings.

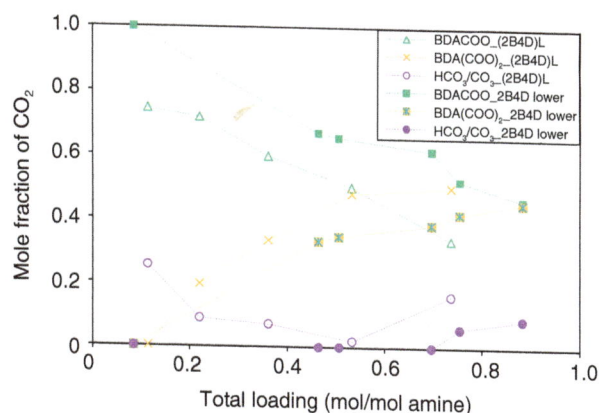

Figure 11

Comparison of $CO_2$ distribution in terms of BDACOO, $BDA(COO)_2$ and $CO_3^{2-}/HCO_3^-$ between 2B4D and $(2B4D)_L$ at different loadings.

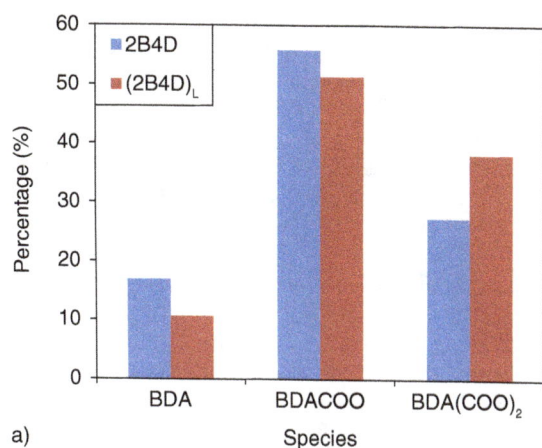

a)

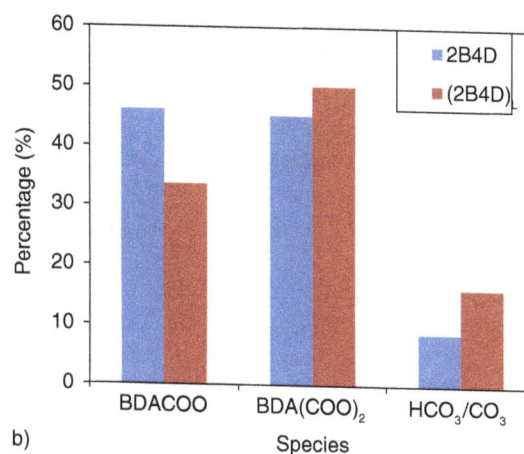

b)

Figure 12

BDA and $CO_2$ reaction products distributions in 2B4D and $(2B4D)_L$ at the rich loading. a) BDA distribution; b) $CO_2$ distribution.

Figure 11 illustrates the distribution of $CO_2$ in terms of $BDACOO^-$, $BDA(COO)_2^{2-}$ and $CO_3^{2-}/HCO_3^-$. The current NMR spectrum, unfortunately, had no peaks around 125 ppm, which would indicate free $CO_2$. As the results of Jokobsen *et al.* (2005) and Suda *et al.* (1996) proved that the free $CO_2$ was less than 5% of its total amount in the solution, the free $CO_2$ in the solution was neglected in this work. Given that BDA bicarbamate has two carbon atoms, its mole fraction in the figure is twice as its actual fraction. Figure 11 shows that at the beginning, the $CO_2$ formed BDA carbamate, while the BDA carbamate fraction decreased and bicarbamate increased with increasing $CO_2$ loading. At the rich loading, $CO_2$ in the solution was distributed as 33.64%

carbamate, 50.10% bicarbamate and 16.26% carbonate/bicarbonate.

Figure 11 also compares the $CO_2$ distribution between 2B4D lower phase and $(2B4D)_L$ at different loadings. The $CO_2$ distribution in 2B4D lower phase can be referred to Xu *et al.* (2013c). It shows that in $(2B4D)_L$ solution, the mole fraction of $CO_2$ existed in terms of carbamate was more than that in 2B4D lower phase, while the mole fractions of $CO_2$ in terms of bicarbamate and carbonate/bicarbonate were slightly higher than those in 2B4D lower phase.

The reaction products and the BDA and $CO_2$ distributions in 2B4D and $(2B4D)_L$ at the rich loading are compared in Figure 12. Figure 12a shows that the $(2B4D)_L$

reaction products had more BDA bicarbamate than the 2B4D products, with less BDA and less BDA carbamate. Figure 12b demonstrates that the $CO_2$ reaction products of $(2B4D)_L$ had twice as much carbonate/bicarbonate as 2B4D and less BDA carbamate. The reason for more BDA bicarbamate of $(2B4D)_L$ is that the BDA concentration in $(2B4D)_L$ was almost constant during absorption, while the equivalent BDA concentration (including BDA carbamate and bicarbamate) in the 2B4D lower phase kept decreasing while approaching the rich loading. The BDA concentration of 2B4D lower phase at different loadings can be referred to Xu et al. (2013c).

## CONCLUSIONS

The present study identifies the $(2B4D)_L$ composition with 2.607 mol/kg BDA and 2.806 mol/kg DEEA. The cyclic capacity and cyclic loading of $(2B4D)_L$ are almost the same as those of 2B4D. For reaction conditions of 298.15, 313.15 and 333.15 K and $CO_2$ pressure of 3-25 kPa, the absorption rates of fresh $(2B4D)_L$ solution, $(2B4D)_L$ solutions at half lean loading and lean loading are higher than those of fresh 2B4D solution, and 2B4D solutions at half lean loading and lean loading, respectively, except for the fresh solutions at $CO_2$ pressure lower than 10 kPa.

NMR measurements of the reaction products of $(2B4D)_L$ show that in $(2B4D)_L$ solution, the BDA bicarbamate concentration increases with increasing $CO_2$ loading, while the BDA carbamate concentration decreases a little at loadings greater than 0.532 mol/mol. The reaction products of $(2B4D)_L$ have more BDA bicarbamate, less BDA and less BDA carbamate than 2B4D. The $CO_2$ reaction products of $(2B4D)_L$ have twice as much carbonate/bicarbonate as with 2B4D and less BDA carbamate. Further study on the performance of $(2B4D)_L$ is necessary.

## ACKNOWLEDGMENTS

Financial support from the Chinese MOST project "Key Technology Research and Development on Advanced Coal Conversion and Power Generation" (2010DFA72730) is greatly appreciated.

## REFERENCES

Aronu U.E., Hoff K.A., Svendsen H.F. (2011) $CO_2$ capture solvent selection by combined absorption-desorption analysis, Chem. Eng. Res. Des. **89**, 1197-1203.

Balaban A.T., Dinculescu A., Elguero J., Faure R. (1985) Carbon-13 NMR studies of primary amines and their corresponding 2,4,6-trimethyl-pyridinium salts, Magn. Reson. Chem. **23**, 553-558.

Bishnoi S., Rochelle G.T. (2002) Absorption of carbon dioxide in aqueous piperazine/methyldiethanolamine, AIChE J. **48**, 2788-2799.

Bruder P., Svendsen H.F. (2011) Solvent comparison for post combustion $CO_2$ capture, 1st Post Combustion Capture Conference, 17-19 May, Abu Dhabi, Kingdom of Saudi Arabia.

Derks P.W.J., Dijkstra H.B.S. (2005) Solubility of carbon dioxide in aqueous piperazine solutions, AIChE J. **51**, 2311-2327.

Hartono A., da Silva E.F., Grasdalen H., Svendsen H.F. (2007) Qualitative determination of species in DETA-$H_2O$-$CO_2$ system using $^{13}C$ NMR spectra, Ind. Eng. Chem. Res. **46**, 249-254.

Hilliard M.D. (2008) A predictive thermodynamic model for an aqueous blends of potassium carbonate, piperazine, and monoethanolamine for carbon dioxide capture from flue gas, PhD Thesis, University of Texas, Austin.

Hu L. (2009) Phase transitional absorption method, United States Patent, 7541001.

Jakobsen J.P., Krane J., Svendsen H.F. (2005) Liquid-phase composition determination in $CO_2$-$H_2O$-alkanolamine system: an NMR study, Ind. Eng. Chem. Res. **44**, 9894-9903.

Jakobsen J.P., da Silva E.F., Krane J., Svendsen H.F. (2008) NMR study and quantum mechanical calculations on the 2-[(2-aminoethyl) amino]- ethianol- $H_2O$- $CO_2$ system, J. Magn. Reson. **191**, 304-314.

Liu J., Wang S., Zhao B., Tong H., Chen C. (2009) Absorption of carbon dioxide in aqueous ammonia, Energy Procedia **1**, 933-940.

Liu J., Wang S., Qi G., Zhao B., Chen C. (2011) Kinetics and mass transfer of carbon dioxide absorption into aqueous ammonia, Energy Procedia **4**, 525-532.

Liu J., Wang S., Zhao B., Chen C. (2012) Study on mass transfer and kinetics of $CO_2$ absorption into aqueous ammonia and piperazine blended solutions, Chem. Eng. Sci. **75**, 298-308.

Pacheco M.A. (1998) Mass transfer, kinetics and rate-based modeling of reactive absorption, PhD Thesis, University of Texas, Austin.

Parker E., Leconte N., Godet T., Belmont P. (2011) Solver-catalyzed furoquinolines synthesis: from nitrogen effects to the use of silver imidazolate polymer as a new and robust silver catalyst, Chem. Commun. **47**, 343-345.

Raynal L., Alix P., Bouillon P.A., Gomez A., de Nailly M.F., Jacquin M., Kittel J., di Lella A., Mougin P., Trapy J. (2011) The DMX$^{TM}$ process: An original solution for lowering the cost of post-combustion carbon capture, Energy Procedia **4**, 779-786.

Rinker E.B., Ashour S.S. (2000) Absorption of carbon dioxide into aqueous blends of diethanolamine and methyldiethanolamine, Ind. Eng. Chem. Res. **39**, 4346-4356.

Rochelle G.T. (2009) Amine scrubbing for $CO_2$ capture, Science **325**, 1652-1654.

Rojey A., Cadours R., Carrette P.L., Boucot P. (2009) Method of deacidizing a gas by means of an absorbent solution with fractionated regeneration through heating, United States Patent. Pub. No.: US 2009/0199709 A1.

Suda T., Iwaki T., Mimura T. (1996) Facile determination of dissolved species in $CO_2$-amine-$H_2O$ system by NMR spectroscopy, *Chem. Lett.* **9**, 777-778.

Svendsen H.F., Hessen E.T., Mejdell T. (2011) Carbon dioxide capture by absorption, challenges and possibilities, *Chem. Eng. J.* **171**, 718-724.

Tan Y.H. (2010) Study of $CO_2$-absorption into thermomorphic lipophilic amine solvents, *PhD Thesis*, University of Dortmund, Germany.

Xu Z., Wang S., Liu J., Chen C. (2012a) Solvents with low critical solution temperature for $CO_2$ capture, *Energy Procedia* **23**, 64-71.

Xu Z., Wang S., Zhao B., Chen C. (2012b) Study on potential biphasic solvents: absorption capacity, $CO_2$ loading and reaction rate, *11th International Conference on Greenhouse Gas Technologies*, 18-22 Nov., Kyoto, Japan.

Xu Z., Wang S., Chen C. (2013a) Experimental study of $CO_2$ absorption by biphasic solvents, *J. Tsinghua University (Science and Technology)* **53**, 3, 336-341.

Xu Z.C., Wang S.J., Chen C.H. (2013b) Experimental Study of $CO_2$ absorption by MAPA, DEEA, BDA and BDA/DEEA mixtures, *J. Combust. Sci. Technol.* **19**, 2, 103-108.

Xu Z., Wang S., Chen C. (2013c) $CO_2$ absorption by biphasic solvents: mixtures of 1,4-Butanediamine and 2-(Diethylamino)-ethanol, *Int. J. Greenhouse Gas Control* **16**, 107-115.

Zhang J., Agar D.W., Zhang X., Geuzebroek F. (2011) $CO_2$ absorption in biphasic solvents with enhanced low temperature solvent regeneration, *Energy Procedia* **4**, 67-74.

Zhang X. (2007) Studies on multiphase $CO_2$ capture system, *PhD Thesis*, University of Dortmund, Germany.

# 14

# Amine Solvent Regeneration for $CO_2$ Capture Using Geothermal Energy with Advanced Stripper Configurations

D.H. Van Wagener[1], A. Gupta[2], G.T. Rochelle[1*] and S.L. Bryant[2]

[1] Department of Chemical Engineering – C0400, The University of Texas at Austin, Austin, TX 78712 - USA
[2] Department of Petroleum and Geochemical Engineering – C0304, The University of Texas at Austin, Austin, TX 78712 - USA
e-mail: dvanwage@utexas.edu - abhishek.k.gupta@mail.utexas.edu - gtr@che.utexas.edu - steven_bryant@mail.utexas.edu

* Corresponding author

**Résumé — Régénération d'un solvant de captage du $CO_2$ utilisant l'énergie géothermique et des configurations améliorées pour le régénérateur** — Le captage du $CO_2$ dans les fumées des centrales électriques au charbon utilisant des alcanolamines nécessite un apport d'énergie substantiel. Les designs typiques prévoient l'utilisation de vapeur de pression faible à intermédiaire prélevée sur les turbines pour fournir la chaleur au rebouilleur. L'énergie géothermique issue de saumures chaudes offre une alternative à cette perte importante dans le cycle de génération d'électricité. Nous avons évalué les besoins (nombre et espacement des puits d'extraction et d'injection) pour produire de la chaleur à 150 °C à l'échelle pilote (60 $MW_e$) et pour une unité de taille réelle (900 $MW_e$) pendant trente ans. Les calculs sont basés sur les propriétés d'un aquifère géopressurisé/géothermique situé près de la côte du Golf du Texas. Cet aquifère est situé entre 3 050 et 3 350 m (10 000 et 11 000 pieds) de la surface à proximité d'une importante centrale électrique au charbon. Nous présentons un nouveau design pour le régénérateur du procédé basé sur un échange de chaleur avec la saumure, délivrant une saumure à 100 °C. Les résultats montrent que le procédé complet est faisable et que les coûts sont du même ordre de grandeur que pour le design standard.

*Abstract — Amine Solvent Regeneration for $CO_2$ Capture Using Geothermal Energy with Advanced Stripper Configurations — Absorption/stripping using alkanolamine solvents for removing $CO_2$ from the flue gas of coal-fired power plants requires a substantial amount of energy. Typical designs anticipate the use of steam extraction between the Intermediate Pressure (IP) and Low Pressure (LP) turbines to provide heat for the reboiler. Geothermal energy in the form of hot brine offers an alternative to this large parasitic load on the power generation cycle. We investigate the requirements (number and spacing of extraction/injection well pairs) to provide heat at 150°C for a pilot scale (60 $MW_e$) and a full scale (900 $MW_e$) capture process for thirty years. The calculations are based on properties of a geopressured/geothermal aquifer near the Texas Gulf Coast. In the vicinity of a large coal-fired power plant in South Texas, this aquifer lies between 3 050 and 3 350 m (10 000 and 11 000 ft) below the surface. We present a novel design of the stripper/regenerator process based on heat exchange with the brine, discharging the brine at 100°C. The results indicate that the overall process is feasible and that costs are of similar magnitude to standard designs.*

## NOMENCLATURE

| | |
|---|---|
| $A$ | Well drainage area (m$^2$) |
| $C_{p,\,brine}$ | Volumetric heat capacity of brine (J/g/K) |
| $D$ | Well spacing between injector and producer of an isolated doublet (cm) |
| $E_T$ | Thermal power generated (J/s) |
| $H$ | Reservoir thickness (cm) |
| $h$ | Water level drawdown (m) |
| $h_t$ | Threshold water level drawdown for the specific formation (m) |
| $K_R$ | Thermal conductivity of cap rock and bed rock (cal/cm/s/°C) |
| $m_{brine}$ | Mass flow rate of brine (g/s) |
| $P_{in}$ | Pressure of overhead vapor entering compressor (bar) |
| $Q$ | Constant production rate (m$^3$/s) |
| $Q_i$ | Heat duty in reboiler/cross exchanger (kJ/mol CO$_2$) |
| $Q_{max}$ | Maximum production rate for the specific pressure drawdown (m$^3$/s) |
| $Q_s$ | Brine recycling rate at surface condition (cm$^3$/s) |
| $r_e$ | Aquifer drainage radius (m) |
| $r_w$ | Wellbore radius (m) |
| $S$ | Aquifer storage coefficient |
| $T_{brine,in}$ | Temperature of incoming brine to stripper (K) |
| $T_{brine,out}$ | Temperature outlet brine from stripper (K) |
| $T_i$ | Temperature of reboiler (K) |
| $T_{i,f}$ | Exiting temperature of brine from heat exchanger (K) |
| $T_{i,o}$ | Entering temperature of brine from heat exchanger (K) |
| $T_{inj}$ | Surface temperature of injecting brine (K) |
| $T_{prod}$ | Surface temperature of producing brine (K) |
| $T_r$ | Aquifer transmissivity (m$^2$/s) |
| $T_{sink}$ | Sink temperature for turbine calculations 313 K |
| $W_{comp}$ | Compression work (kJ/mol CO$_2$) |
| $W_{eq}$ | Equivalent work (kJ/mol CO$_2$) |
| $W_{heat}$ | Electric penalty from heat usage (kJ/mol CO$_2$) |
| $W_{pump}$ | Pump work (kJ/mol CO$_2$) |
| $\Delta t$ | Time (s) |
| $\rho_R C_R$ | Volumetric heat capacity of reservoir (cal/cm$^3$/°C) |
| $\rho_w C_w$ | Volumetric heat capacity of formation brine (cal/cm$^3$/°C) |
| $\varphi$ | Reservoir average porosity (fraction) |

## INTRODUCTION

Absorption/stripping using alkanolamine solvents is the state-of-the-art technology for removing CO$_2$ from the flue gas of coal-fired power plants (Rochelle, 2009). Lean solvent fed to the top of the absorber column counter-currently contacts the flue gas to absorb the CO$_2$ by chemical reaction. A typical coal-fired power plant emits flue gas with 12% CO$_2$ and the absorber captures 90% to leave 1.3%. The rich solvent is heated in a cross exchanger and then fed to the stripper. Heat applied to the rich solvent reverses the reaction and liberates the CO$_2$ to regenerate the lean solvent. The gaseous CO$_2$ is compressed for sequestration and the lean amine solvent is recycled to the absorber. The base case absorption/stripping flowsheet using 7 m Monoethanolamine (MEA) with a simple stripper column has been estimated to reduce the output of a coal-fired power plant by 25 to 30% due to the energy requirement of the reboiler, pumps and multi-stage compressor (Van Wagener, 2011). The pumps and multi-stage compressor would directly draw generated electricity but the reboiler would reroute steam from between the Intermediate Pressure (IP) and Low Pressure (LP) turbines to provide heat.

Because this re-routed steam accounts for much of the power plant output penalty, a heat source such as geothermal energy is potentially attractive alternative. The heat in the Earth's interior originated from its consolidation from dust and gases over 4 billion years ago and is continually regenerated from the decay of radioactive elements that occur in the rock. High temperature geothermal reservoirs can be utilized to generate electricity from hot water/steam. Moderate to low temperature reservoirs are generally best suited for direct use in space heating/cooling and industrial or agricultural processes. A number of geopressured/geothermal reservoirs are present along the Texas Gulf coast which can produce hot brine (150°C) at rates high enough to operate the reboiler in a CO$_2$ stripper. However, using geothermal brine would be very different from using steam in several respects. The heat supplied from the brine will arrive at different temperatures, depending on the reservoir, and over time, the brine produced from the same reservoir will cool if re-injected brine makes its way to the extraction wells. A typical steam reboiler would not make efficient use of geothermal brine because there would be a large approach temperature on the hot side of the exchanger and a pinch on the cold side. Alternatively, using a cross exchanger with the hot brine and cool rich solvent could more effectively take advantage of the high temperature brine by balancing the temperature approach throughout the exchanger.

Recent work has investigated the use of multi-stage flash flowsheets to strip $CO_2$ as opposed to the typical simple stripper column. An isothermal 2-stage flash in place of the stripper column was demonstrated to be beneficial (Van Wagener and Rochelle, 2010). The heat supplied for the previous multi-stage flash configurations was delivered in a reboiler with steam. A form of the multi-stage flash configuration is proposed here that incorporates cool rich solvent cross exchangers to heat the solvent with brine with flashing at two different pressures.

## 1 STRIPPER MODELS

Concentrated aqueous piperazine (8 m PZ) was selected as the primary solvent for this study. It is thermally stable to at least 150°C. Monoethanolamine (MEA) is less useful with geothermal heat because it degrades above 120°C. As discussed in Section 3, brine is available at approximately 150°C. At a concentration of 8 m, PZ has twice the $CO_2$ carrying capacity of 7 m MEA, which improves the energy performance of PZ compared to MEA (Freeman *et al.*, 2010). 5deMayo is an available thermodynamic model for aqueous PZ in *Aspen Plus*® (Van Wagener, 2011). This model was regressed in the electrolyte nonrandom two liquid (e-NRTL) framework to represent $CO_2$ solubility (Ermatchkov *et al.*, 2006; Hilliard, 2008; Dugas, 2009; Xu and Rochelle, 2011), heat capacity, and amine volatility (Nguyen *et al.*, 2011).

8 m PZ was simulated in an advanced 2-stage, 2-pressure flash (2T2PFlash) (*Fig. 1*). The configuration utilized an arrangement of five heat exchangers to remove heat from brine and the returning lean solvent. The heating in this configuration was different from traditional arrangements in that the rich solvent was fully heated before entering the two adiabatic flash vessels in series. The first flash had the higher temperature and pressure, and the second flash dropped in both temperature and pressure.

All unit operations were modeled with chemical equilibrium within and between the gas and liquid phases. Several conditions were specified to be constant while others were optimized. A constant rich loading of 0.4 mol $CO_2$/mol alkalinity was specified. The input temperature of the rich solvent coming from the absorber was specified to be a constant 50°C. The lean solvent would be fed to the absorber at 40°C and a rise in solvent temperature of 10°C can be expected due to the exothermic heat of absorption. The Log Mean Temperature Difference (LMTD) was 5°C for all exchangers and a minimum approach of 1°C was specified for either side. In the base case the brine was supplied at 150°C but in all cases the return temperature for the brine was set to be 50°C cooler than the supply temperature. The temperature difference between flash vessels was varied to ensure that equal moles of vapor were generated in each flash. This specification was made to improve the reversibility of the process.

The work required for the multi-stage compressor was calculated using a correlation derived from a separate *Aspen Plus*® simulation (*Eq. 1*). The work calculation assumed a pre-condenser and intercoolers cooling to 40°C with water knockout. The number of compression stages was set to the minimum required to keep a compression ratio between stages less than 2. The final compressed $CO_2$ pressure was 150 bar. The split of solvent between exchangers 2 and 3 was set to 80% toward exchanger 3.

$$W_{comp}\left(\frac{kJ}{mol\,CO_2}\right)$$
$$= \begin{cases} 4.572\,ln(150P_{in}) - 4.096, P_{in} \le 4.56\,bar \\ 4.023\,ln(150P_{in}) - 2.181, P_{in} > 4.56\,bar \end{cases} \quad (1)$$

The temperature of the first flash vessel was allowed to float to match the desired outlet brine temperature. The lean loading was manipulated by varying the brine flow rate. The overall work requirement for the pumps, multi-stage compressor, and heat duty was determined using an equivalent work expression, which calculated the total electricity usage normalized by the $CO_2$ removal rate (*Eq. 2*). The work requirement of the heat duty was calculated by Equation (3). This has been used in previous work to calculate the amount of electricity that would be generated by the power plant turbines if the steam had not been used to heat the reboilers.

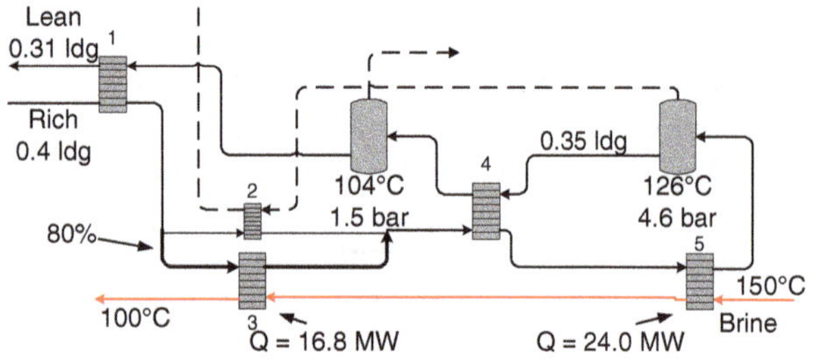

Figure 1

Advanced 2-Stage, 2-Pressure Flash (2T2Pflash) for amine solvent regeneration with geothermal brine heating.

This equation required the temperature of the steam, $T_i$. For this work, Equation (3) was integrated between the inlet and outlet temperatures to account for the changing value of heat at different temperatures, assuming that each unit of heat flow resulted in the same change in temperature along the entire temperature range. The inlet and outlet temperatures in each heater $i$ were $T_{i,o}$ and $T_{i,f}$, respectively. This integration gave Equation (4).

$$W_{eq} = W_{heat} + W_{pump} + W_{comp} \qquad (2)$$

$$W_{heat} = \sum_{i=1}^{n_{heaters}} 0.75 Q_i \left( \frac{T_i + 5K - T_{sink}}{T_i + 5K} \right) \qquad (3)$$

$$W_{heat} = \sum_{i=1}^{n_{heaters}} 0.75 Q_i \left( \frac{T_{i,f} - T_{i,o} - T_{sink} \ln\left(\frac{T_{i,f}}{T_{i,o}}\right)}{T_{i,f} - T_{i,o}} \right) \qquad (4)$$

For a comparison study for application at a demonstration being planned by NRG Energy (Stopek *et al.*, 2011), 9 m MEA was used in an alternative stripper configuration. MEA was represented by another e-NRTL model developed in *Aspen Plus*® (Hilliard, 2008). This model was developed similarly to the PZ model; model parameters were regressed to fit laboratory measurements. The configuration that was analyzed with MEA was a simple stripper with an adiabatic flash on the lean solvent (*Fig. 2*). This configuration has been patented by Fluor (Reddy *et al.*, 2007). The demonstration is designed for MEA, so the same solvent was selected for this modeling with geothermal heating. The brine heated a reboiler and a rich feed preheater. Unfortunately, the re-

boiler had a large hot side approach temperature since the solvent temperature was constant but this case represented a reconfiguration that adapted the Fluor configuration to use brine if it was already constructed to use steam from the power plant. The only additional process unit would be the cross exchanger to preheat the rich feed. The same constants were specified as for the 2-stage flash. The rich loading was specified to be 0.5 mol $CO_2$/mol alkalinity. This loading had an equilibrium partial pressure of $CO_2$ ($P^*_{CO_2}$) of 5 kPa at 40°C, a typical absorber temperature. The rich loading of 0.4 for PZ also had a $P^*_{CO_2}$ of 5 kPa at 40°C.

## 2 GEOTHERMAL ENERGY EXTRACTION

Geothermal energy can be extracted by producing hot water resident in the geothermal reservoir. The life of a geothermal reservoir depends mainly on the type of extraction method. Extraction methods for geothermal reservoirs are essentially the same as for conventional oil and gas reservoirs, except that reservoir life may be shortened because of lack of heat in addition to lack of pressure. There are mainly two development schemes considered to extract the energy from geothermal reservoirs, which we briefly review.

### 2.1 Production without Reinjection

The simplest development scheme to extract energy from geothermal reservoirs involves production of hot water from the geothermal reservoir with no reinjection of heat-depleted cold water back to the reservoir. In this case, the water will be produced at a constant temperature equal to the reservoir temperature minus the heat loss to surrounding earth formations from the bottom to the top of the well. The reservoir life in this development scheme depends on the volume of water present in the reservoir and the reservoir pressure. The water can be produced as long as the reservoir pressure is sufficient for the production rate to be sustained. The water production rate from the single well present in the center of a reservoir depends on the reservoir boundary conditions. Single well behavior can be approximated by simple analytical models that can be used to determine maximum production rate and reservoir life time.

### 2.1.1 Constant Pressure Boundary

For a uniform, homogeneous, isotropic, circular reservoir with constant pressure at the boundary, the

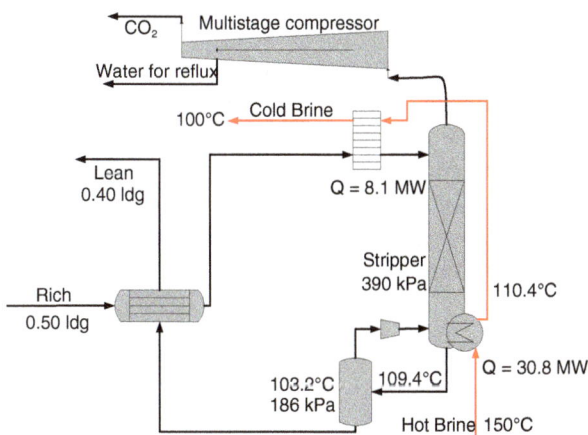

Figure 2

Fluor configuration modified for geothermal heating.

maximum pressure drawdown at the wellbore can be obtained under steady state conditions as (Dietz, 1965):

$$h = \frac{Q}{2\pi T_r} \ln \left( \frac{r_e}{r_w} \right) \qquad (5)$$

Thus the maximum production rate that can be achieved for the threshold pressure drawdown is:

$$Q_{max} = \frac{2\pi T_r h_t}{\ln \left( \frac{r_e}{r_w} \right)} \qquad (6)$$

### 2.1.2 No-Flow Boundary

For a uniform, homogeneous, isotropic, square-shaped reservoir with no-flow boundary, the maximum pressure drawdown for a well present at the center of the reservoir can be determined as (Matthews and Russell, 1967):

$$h = \frac{Q\Delta t}{SA} + \frac{Q}{4\pi T_r} \ln \left( \frac{A}{3.7 r_w^2} \right) \qquad (7)$$

Thus the maximum production rate that can be achieved for the threshold pressure drawdown is:

$$Q_{max} = \frac{h_t}{\frac{\Delta t}{SA} + \frac{1}{4\pi T_r} \ln \left( \frac{A}{3.7 r_w^2} \right)} \qquad (8)$$

### 2.2 Production with Reinjection

Production with reinjection involves a doublet type of development in which all produced water is reinjected into the reservoir after the heat is extracted. This helps in maintaining the reservoir pressure, and therefore in maintaining production rates, and also permits the recovery of the much larger amount of heat contained in the reservoir rock. This procedure creates a zone of injected water around the injector at a different temperature which grows with time and will eventually reach the producer. This arrival decreases temperature of produced hot water and reduces drastically the efficiency of the operation. Thus it is important to design a system of doublets with sufficient well spacing to prevent injected water breakthrough during the operation of the capture facility.

### 2.2.1 Single Producer and Injector Doublet

Gringarten and Sauty (1975) developed an analytical model for temperature front movement in a geothermal reservoir with single producer and injector doublet. They developed the following analytical model to determine well spacing $D$ between injector and producer of an isolated doublet to prevent injected water breakthrough before a specified time interval $\Delta t$. See Equation (9).

### 2.2.2 Multi Producer and Injector Doublets

Multiple doublets are required to produce the hot water from the geothermal reservoir to meet the capture process requirements for a large power plant ($> 500$ MW). A study was performed by Gringarten (1977) to evaluate the effect of different doublet patterns in reservoir life and ultimate heat recovery from the reservoir. They found that the ratio of the reservoir lifetime with various two doublet patterns to that of a single doublet depends on the ratio of distance between two doublets and the well spacing between injector and producer in an individual doublet. The ratio of reservoir lifetime increases with the ratio of distance between two doublets and the well spacing between injector and producer of an individual doublet. They also performed a similar study for different multi-doublet (more than two) patterns and found that the reservoir lifetime for a multi-doublet development scenario is always less than that of single doublet in an infinite system.

## 3 RESERVOIR MODELING

Similar to conventional hydrocarbon reservoirs, geothermal reservoirs are complex hydrothermal systems involving single-or multi-phase fluid flow. Numerical simulation of such reservoirs is helpful in estimating the recoverable energy, in determining the optimum management techniques and improving the understanding of the reservoir geometry, and in estimating boundary conditions and rock properties. CMG-STARS (Steam, Thermal and Advanced processes Reservoir Simulator) reservoir simulator was used to model fluid and temperature front movement in a geothermal brine reservoir.

$$D = \left[ \frac{2Q_s \Delta t}{\left( \varphi + (1-\varphi) \frac{\rho_R C_R}{\rho_w C_w} \right) H + \left\{ \left( \varphi + (1-\varphi) \frac{\rho_R C_R}{\rho_w C_w} \right)^2 H^2 + 2 \frac{K_R \rho_R C_R}{(\rho_w C_w)^2} \Delta t \right\}^{0.5}} \right]^{0.5} \qquad (9)$$

## 4 CASE STUDY: WILCOX RESERVOIR

The Wilcox group is one of the oldest thick sand/shale sequences within the Gulf coast tertiary system. It crops out in a 16 to 32 km wide band and dips coastward into subsurface. The presence of growth faults along the shoreline of large delta lobes in the neighborhood of the Wilcox group of formations restricts the fluid flow across the growth faults (Bebout et al., 1979). This caused an increase in pressure gradient at depths greater than 10 000 ft from a normal hydrostatic pressure gradient of 0.44 psi/ft to between 0.7 and 1.0 psi/ft and temperature gradient from a normal 1.8°C/100 m to

between 2.7 and 3.6°C/100 m (Bebout et al., 1979). This faulted downdip section of the Wilcox group, which exhibits a large pressure gradient and temperature exceeding 149°C (300°F), comprises the Wilcox geothermal corridor as given in Figure 3. The Wilcox group of reservoirs is located beneath large coal-fired power plants in Texas where $CO_2$ capture is being considered. Thus geothermal fairways of the Wilcox group of reservoirs can be utilized to provide geothermal energy to capture $CO_2$ produced by coal-fired electric power plants.

Due to its good porosity and permeability characteristics the Dewitt fairway in the Wilcox group is used to

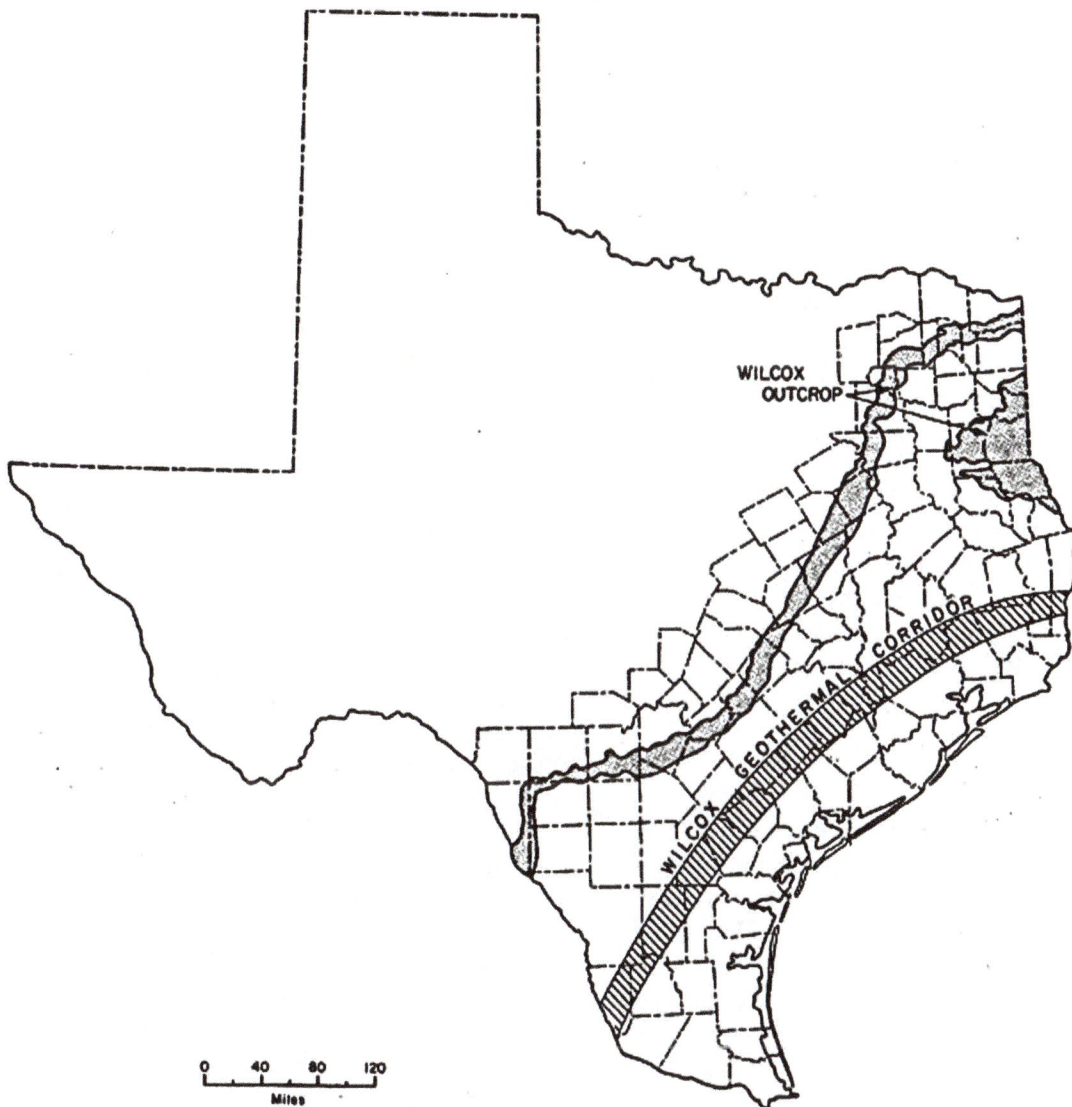

Figure 3

Texas geographical map showing the location of Wilcox outcrop and Wilcox geothermal corridor (Bebout et al., 1979).

TABLE 1

Reservoir rock and brine properties used in all the reservoir simulation studies

| Property | Unit | Value |
|---|---|---|
| Average porosity, $\varphi$ | % | 20 |
| Average matrix permeability, $k_h$ | mD | 100 |
| $k_v/k_h$ | Fraction | 1.0 |
| Reservoir depth | Ft | 10 500 |
| Reservoir thickness | Ft | 550 |
| Initial reservoir temperature | °F | 302 |
| Initial reservoir pressure | Psi | 4 900 @10 500 ft |
| Reservoir rock thermal conductivity | Btu/ft-day-°F | 44 |
| Reservoir brine thermal conductivity | Btu/ft-day-°F | 8.6 |
| Volumetric heat capacity of cap rock and bed rock | Btu/ft$^3$-°F | 35 |
| Thermal conductivity of cap rock and bed rock | Btu/ft-day-°F | 44 |
| Injection water temperature | °F | 208.4 |
| Well bore tubing outside diameter | Inch | 6.625 |
| Down hole pump power for producing well | kW | 1 000 |
| Extraction period | Years | 30 |

study the feasibility of geothermal energy extraction in this paper. At the site of the W. A. Parish plant in South Texas, the Dewitt is 10 500 ft beneath the surface. Here, we report on a strategy to generate geothermal power sufficient to drive a pilot-scale capture process for 60 MWe power generation. We then scale up 15 times to examine the well spacing requirements for a full-scale capture process for a 900 MWe coal-fired power plant.

## 4.1 Geothermal Modeling

A homogeneous reservoir of 640 km$^2$ (64 km × 10 km) surface area with no-flow at the reservoir boundary is used for the study. The reservoir is located 10 500 ft below the Earth surface. The whole reservoir is divided in 100 grid cells in $x$-direction (larger dimension) and 60 grid cells in $y$-direction (smaller dimension). Wells are drilled mainly in the center of the reservoir, thus fine grid sizing is used at the center with coarser grid cells towards the boundary of the reservoir. The reservoir thickness of 550 ft is divided in 11 equally sized layers. Reservoir and brine properties used in the simulation study are given in Table 1. Injection and production

wells are drilled with well spacing required to maintain constant temperature brine production at the surface.

Overburden and underburden heat gain from the cap rock and bed rock respectively is considered in all the simulations. CMG-STARS utilizes a semi-analytical model to determine heat gain due to overburden and underburden of cap rock and bed rock (Vinsome and Westerveld, 1980). The semi-analytical model calculates heat transfer rate and its derivative with respect to temperature on the basis of thermal conductivity and heat capacity of the cap and bed rock. A separate simulation study is performed to evaluate the effect of overburden and underburden heat gain on temperature front movement in the reservoir. In this simulation study, a similar set of grid cells with zero porosity is included above and below the actual reservoir to allow gradual temperature variation from cap and bed rock to the brine reservoir. The results from this simulation study indicate that the effective heat transfer coefficient in CMG-STARS approximates the more accurate solution of heat transfer from cap and bed rock to the brine reservoir. The temperature of cap rock and bed rock is considered the same as the reservoir temperature at initial conditions.

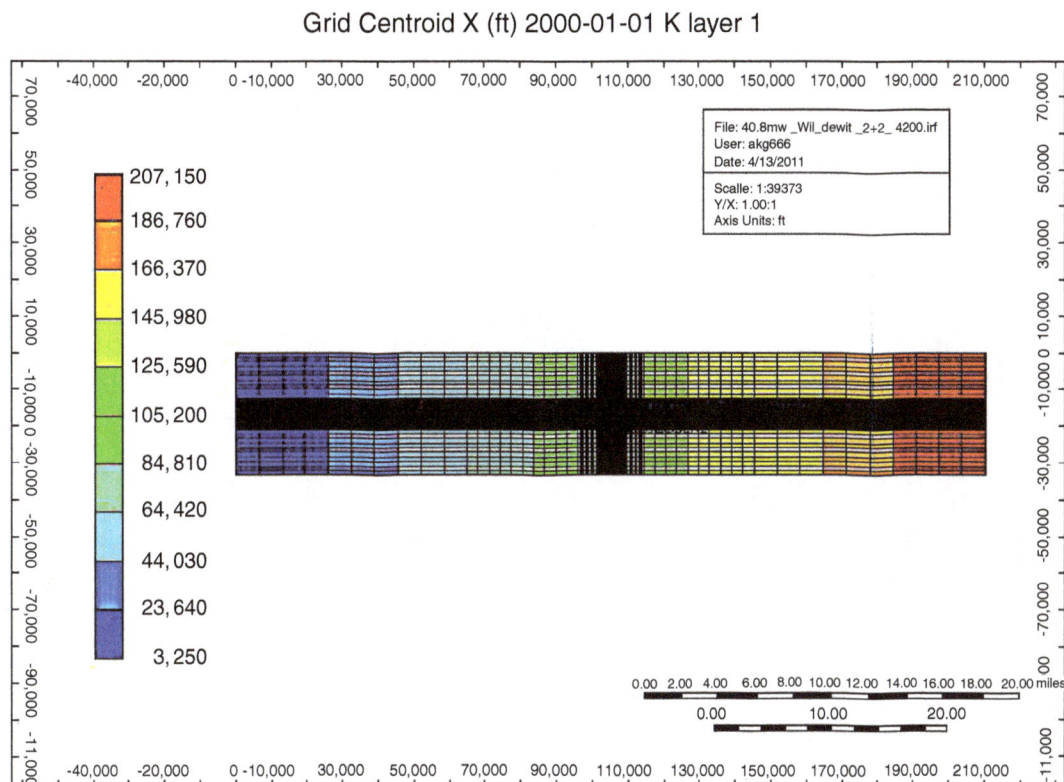

Figure 4

Topmost layer of the grid structure for pilot study showing reservoir size (64 km by 10 km). The black portion in the plot is fine grid cells with grid size of 210 ft in $x$ and $y$-directions. The injection/extraction well doublets are located within the central portion of the fine grid (cf. zoomed-in view in *Fig. 5*). The color bar indicates the distance from the leftmost boundary of the reservoir with blue color showing reservoir portion close to left boundary and red color showing reservoir portion close to the rightmost boundary of the reservoir.

### 4.1.1 Pilot Study

A reservoir simulation study for a pilot power generation of 60 MW$_e$ is performed in the CMG-STARS reservoir simulator. The target operation period is 30 years. Brine extraction/injection rate required for obtaining 41.2 MW of thermal energy for the given temperatures of produced and injected brine is determined by the simple relationship between brine mass and energy generated:

$$E_T = m_{brine} C_{p,brine} (T_{prod} - T_{inj}) \qquad (10)$$

The brine rate required for the given set of conditions is 108 000 bbl/day. Two injector/producer doublets are used to achieve the required recycling rate. Each well is produced/injected at a rate of 54 000 bbl/day of brine. Two downhole pumps, each with 1 000 kW capacity, are required to pump the hot brine to the surface at the required rate. Thus a part of the extracted geothermal energy will be utilized to the run the project.

Historically, the cost of drilling and completing an onshore well such as is required for the pilot study is approximately 5 million US dollars per well, making an estimated total of 20 million US dollars for drilling and completion of the 4 wells for the pilot study. The well spacing between an injector and a producer of a doublet required to maintain the constant temperature brine production at surface for the duration of the project is determined by using the analytical model given in Equation (9). The well spacing ($D$) calculated using the above model for the given reservoir and fluid parameters is 130 332 cm (4 276 ft). The well spacing between two doublets is considered the same as the well spacing between the injector and producer of a doublet.

The topmost layer of the grid structure used in the simulation is given in Figure 4. The grid structure shows the reservoir size in $x$-direction (larger dimension). From the grid structure, it is evident that the wells are drilled at the center of the reservoir which has fine gridding at the

Temperature (F) 2030-01-01 K layer 1

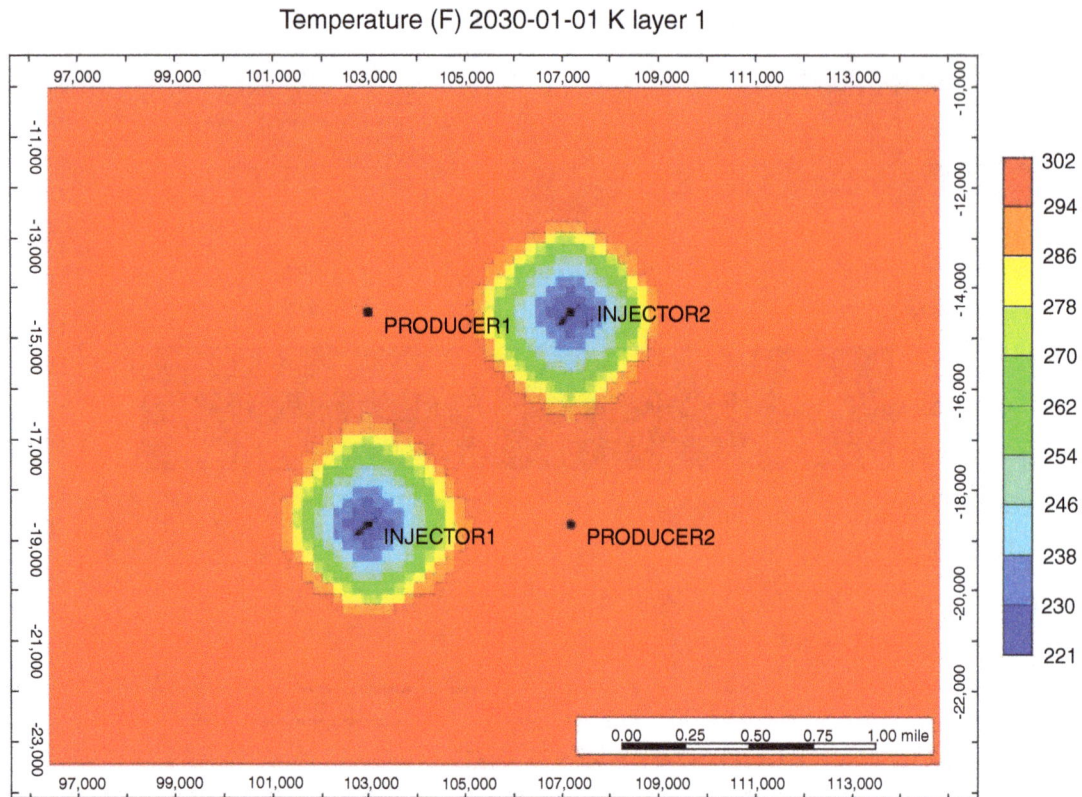

Figure 5

Topmost layer of the reservoir showing temperature front movement from injector to producer at the end of extraction period (30 years from the beginning). Wells flow at 54 000 B/D and are spaced about 4 300 ft apart. Cool regions near the injection wells show the advance of injected brine temperature; the red color corresponds to the initial reservoir temperature.

center and coarse gridding towards the boundary of the reservoir.

Figure 5 is a zoomed-in map view of the top most layer of the reservoir showing temperature front movement at the end of the extraction period (30 years from the beginning). Blue color close to injector representing low temperature values close to injection brine temperature, shows the cooling of reservoir rock due to cold brine injection. The temperature front is moving from injectors towards the producers. The red color close to producer representing high temperature value close to initial reservoir temperature shows that temperature front has not yet reached the producer.

From Figure 5, it is evident that the well separation of 4 300 ft permits constant temperature brine production for the target operating period of thirty years.

Figure 6 is the topmost layer of the reservoir showing injected brine front movement from injectors towards the producers at the end of extraction period (30 years from the beginning). The well pattern permits very

effective sweep of native formation brine from injectors to the producers. From Figures 5 and 6, it is evident that the brine front travels faster than the temperature front in the reservoir. The lag of the temperature front occurs because the reservoir rock releases heat to the cooler injected brine. A smaller contribution to the lag is the additional heat gain from the cap rock and the bed rock.

### 4.1.2 Scaled-Up Study

The pilot project to capture $CO_2$ from 60 $MW_e$ electric power is scaled up to 900 $MW_e$ of electric power. The geothermal power requirement scales proportionately. Thus 618 MW thermal power will be required. The number of wells is also increased by the same proportion to 30 doublets of injector and producer pairs, each having a brine recycle rate of 54 000 bbl/day. Thirty downhole pumps for each producer with 1 000 kW capacity are required to pump geothermal brine to the

Global mole fraction (trac_wat) 2030-01-01 K layer 1

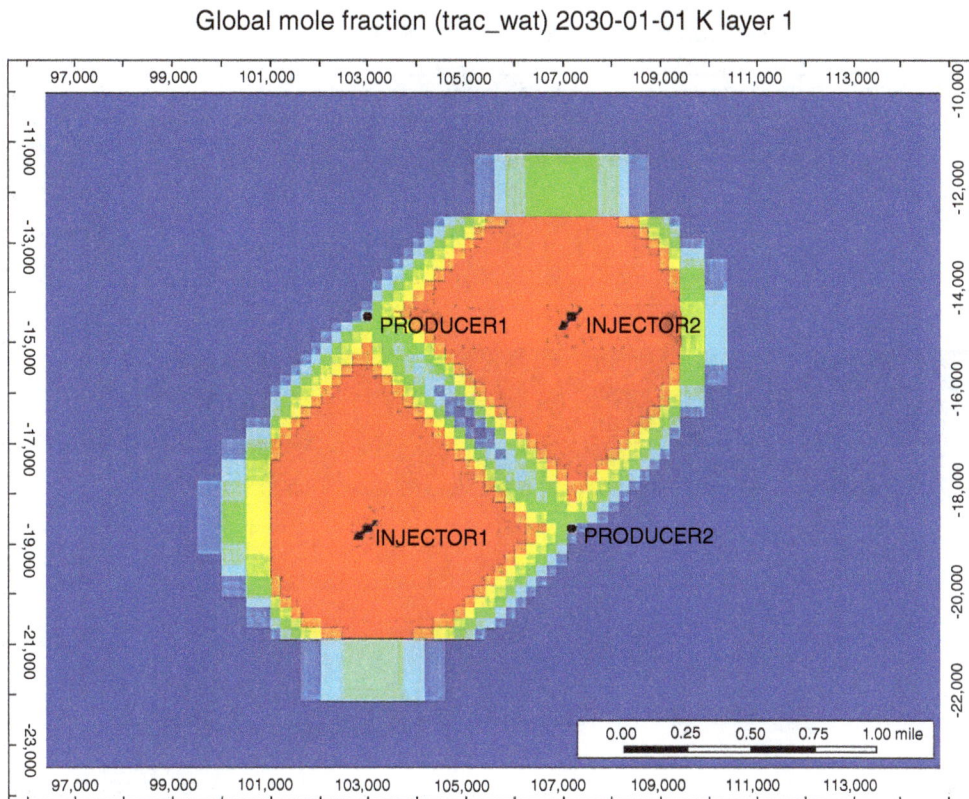

Figure 6

Same view as Figure 5 showing injection brine front movement (indicated by concentration of a tracer that moves at fluid velocity) at the end of extraction period (30 years from the beginning). Blue color shows 100% native brine mole fraction and red color shows maximum injection brine mole fraction.

surface at the required rate. Thus 30 $MW_e$ of the total extracted geothermal power will be utilized to run down-hole pumps. The well spacing between two doublets and an injector and a producer of an individual doublet are kept constant at 4 276 ft. The total well footprint area required to extract 618 MW of thermal power for 30 years is 33 $mile^2$. In this scenario, 30 producers and 30 injectors are placed in 33 $mile^2$ area and a surface transportation system is required to transport the produced brine from well heads to the stripping plant and injection brine from stripping plant to various injectors. The topmost layer of the grid structure used in the scaled-up reservoir simulation study is shown in Figure 7.

Figure 8 is the topmost layer of the reservoir showing temperature front at the end of the extraction period (30 years from the beginning). As in Figure 5, high temperature value close to initial reservoir temperature shows that temperature front has not yet reached the producers.

Figure 9 is the topmost layer of the reservoir showing injected brine front movement from injectors towards

the producers at the end of extraction period (30 years from the beginning). From the plot it is evident that the well pattern permits the maximum sweep of native formation brine from the injectors to the producers.

## 4.2 Surface Utilization

The 2T2PFlash configuration was scaled to regenerate enough solvent to treat the flue gas of a 60 $MW_e$ power plant. The flue gas from this size power plant was estimated by scaling an industrial estimate. Approximately 1 195 ton $CO_2$/day would be removed for 60 $MW_e$ (Fisher et al., 2005). The lean loading was optimized to minimize the overall work requirement. Relevant operating conditions for the optimized case are detailed in Figure 1.

Figure 10 shows the behavior of both equivalent work and total heat duty as a function of lean loading. The optimum equivalent work was at a lean loading of approximately 0.33 but the heat duty was minimized at

Grid centroid X (ft) 2045-01-01 K layer 1

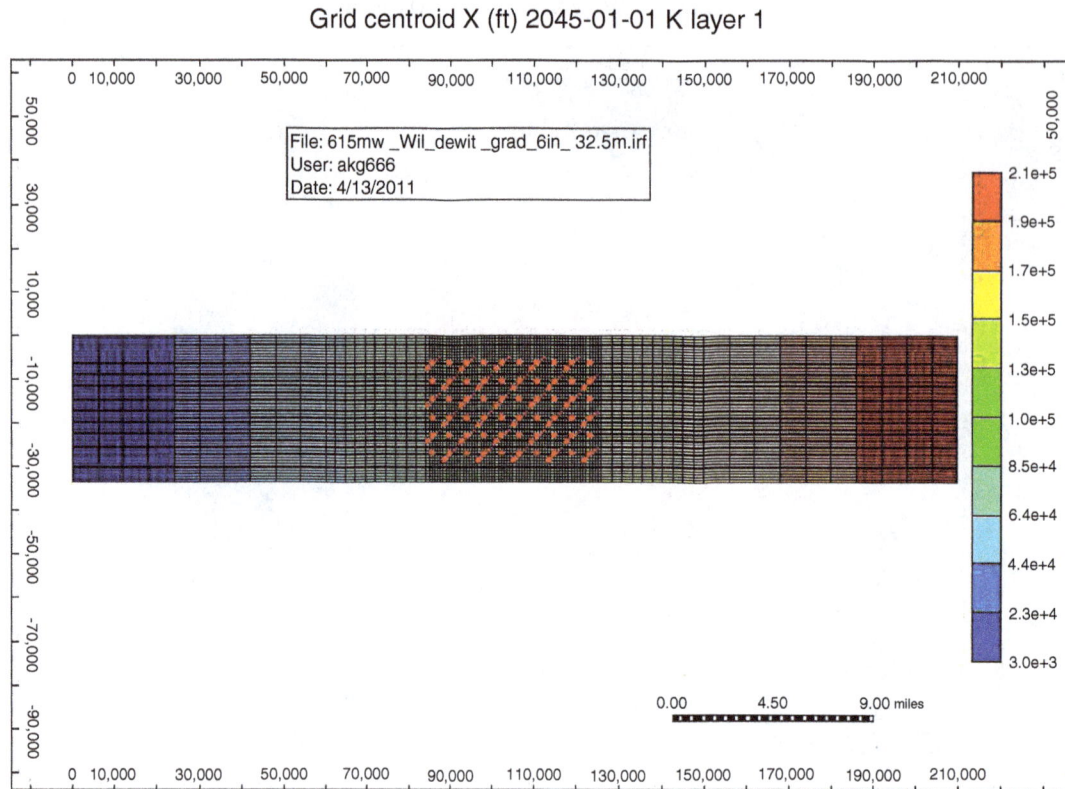

Figure 7

Topmost layer of the grid structure for scaled up study showing a 33 square mile well field of 30 injector/producer pairs in a 64 km by 10 km reservoir. Color bar as in Figure 4.

a slightly higher lean loading of 0.335. These results were calculated using a rich loading of 0.4, corresponding to a $P^*_{CO_2}$ of 5 kPa at 40°C. At the lean loading of 0.33, the equivalent work was 35.1 kJ/mole $CO_2$.

The $P^*_{CO_2}$ at 40°C for the optimum lean loading of 0.335 was approximately 0.85 kPa. Since the absorber would operate with 90% removal from $CO_2$ partial pressures of 12 kPa to 1.2 kPa, solvent concentrations representing a gas side removal of less than 90% might not provide adequate absorber performance. An over-stripped lean solvent would perform well in the absorber because it would have a significant driving force to achieve the desired clean gas purity. Additionally, the lower lean loading would reduce the solvent circulation rate. Conversely, an understripped lean solvent would have trouble attaining the desired purity of 1.2% without using chilled water for cooling or excessive packing. For this reason, the operation point was chosen to have a lean loading of 0.31, where the $P^*_{CO_2}$ at 40°C was

0.5 kPa. At this lower lean loading, the equivalent work was 35.5 kJ/mole $CO_2$.

Since the temperature of the extracted brine was expected to decline over the length of the project, the sensitivity of the stripper performance with brine temperature was investigated. The change in temperature of the brine across the process was held constant at 50°C for all extraction temperatures. The base case temperature of 150°C required 40.8 MW of heat. The expected decrease in brine temperature over a 30-year period was 2°C. A reduction in brine temperature from 150°C to 148°C would change the heat duty to 41.2 MW, only a 2.4% increase from the design case. An extreme scenario where the brine temperature dropped to 145°C required 42.4 MW of heat, only 3.7% greater than the design case. If a brine formation that could supply heat at 160°C was found, the heat duty would decrease to 38.8 MW, a 4.5% drop from the design case. Figure 11 displays the increase in heat duty and the equivalent

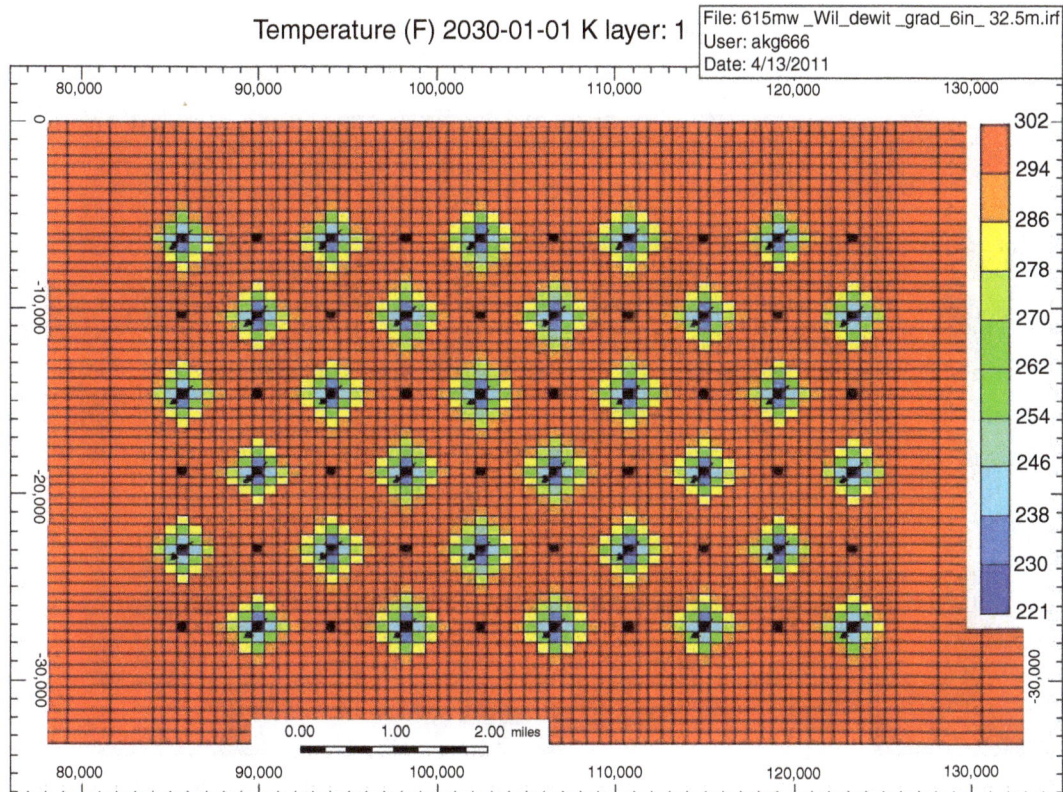

Figure 8

Topmost layer of the reservoir for 30 well pairs producing 618 MW geothermal energy shows temperature front is still far from producers at the end of extraction period (30 years from the beginning). The blue color shows low temperature close to injection brine temperature and red color shows the high temperature close to initial reservoir temperature.

work with decreasing brine temperature. Each simulation converged multiple heat exchange recycle loops at once, and the tolerance set on each recycle loop resulted in a small variability of each point. However, a general negative linear trend was observed.

The Fluor configuration was also optimized for lean loading with 9 m MEA. As has been found in previous work with MEA, the optimum lean loading was in the overstripping region (Van Wagener and Rochelle, 2010). The minimum equivalent work was 36.3 kJ/mole $CO_2$ at a lean loading of 0.39, seen in Figure 12. The overall heating requirement for a 60 MW plant was 38.6 MW, a lower heat duty than the 40.8 MW required in the PZ calculation. Previous work demonstrated a similar outcome, where a 2-stage flash with 8 m PZ had a higher heat duty than a simple stripper with 9 m MEA. Even though the heat duty was less for MEA, the PZ solvent made up in overall performance by operating at a higher pressure, so the 2-stage flash had a sig-

nificantly smaller compression work. Overall, 9 m MEA had a higher equivalent work requirement than 8 m PZ. These calculations with MEA used a rich loading of 0.5 with a $P^*_{CO_2}$ at 40°C of 5 kPa, and the optimum lean loading of 0.39 had a $P^*_{CO_2}$ at 40°C of 0.13 kPa. Therefore, the optimum lean loading was an acceptable range to be coupled with an absorber and expect adequate performance. Relevant operating conditions for this optimal case are detailed in Figure 12.

The difference in proportions of the three work contributions demonstrated that each configuration/solvent combination could have its own application. Using the 2T2Pflash with 8 m PZ would be advantageous when aiming to minimize the overall energy usage. However, the Fluor configuration with 9 m MEA would be advantageous if electricity was cheap and the goal was to minimize the heat usage as much as possible. The Fluor configuration with 9 m MEA reduced the heat duty from the 2-stage flash design case by 5.3%.

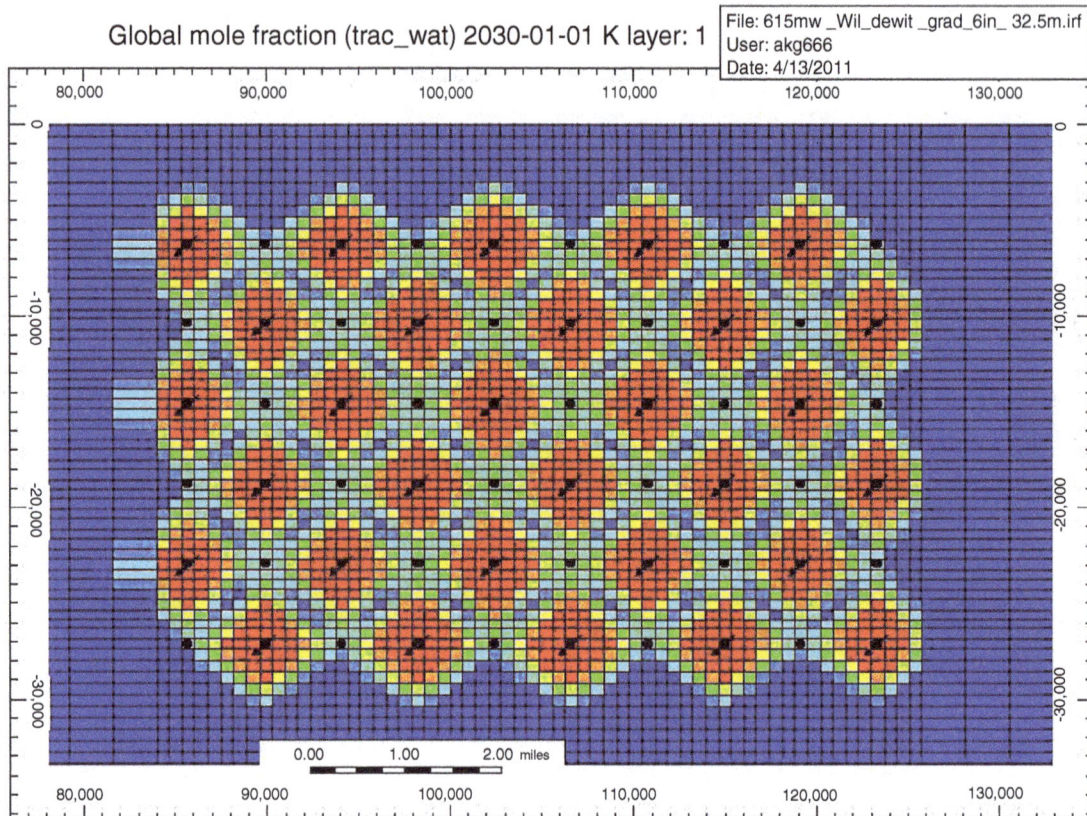

Figure 9

Topmost layer of the reservoir in scaled-up study showing injection brine front movement at the end of extraction period (30 years from the beginning). Blue color shows 100% native brine mole fraction and red color shows maximum injection brine mole fraction.

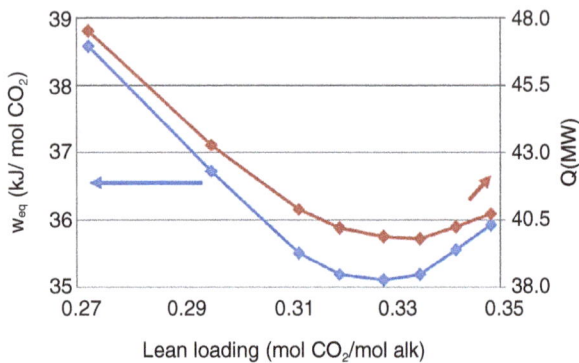

Figure 10

Lean loading optimization for 2T2Pflash with 8 m PZ. 0.40 rich loading, $T_{brine,in}= 150$°C, $T_{brine,out} = 100$°C, 5°C LMTD on heat exchangers, $CO_2$ compression to 150 bar.

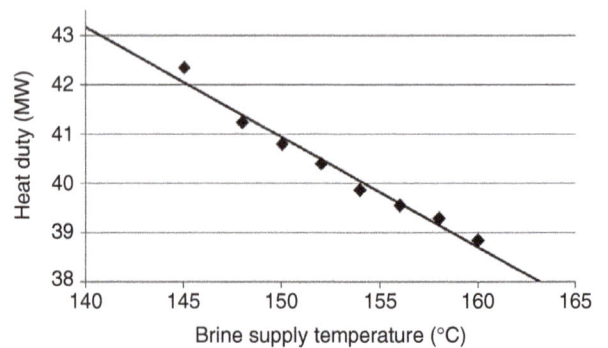

Figure 11

Reduction in total heat duty with increasing brine temperature for 2T2Pflash with 8 m PZ. 0.40 rich loading, $\Delta T_{brine} = 50$°C, 5°C LMTD on heat exchangers, $CO_2$ compression to 150 bar. Points = simulation results, line = approximate linear representation.

Previous work calculated a potential minimum energy requirement for an optimized interheated stripper column with a maximum temperature of 150°C to use 30.5 kJ/mol $CO_2$ (Van Wagener, 2011). The best performance from this work using PZ and the 2T2P Flash configuration was 35.5 kJ/mol $CO_2$, an increase of 16%.

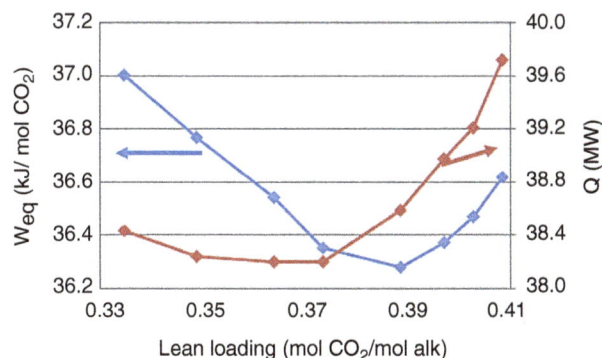

Figure 12

Lean loading optimization for Fluor configuration with 9 m MEA. 0.5 rich loading, $T_{brine,in} = 150°C$, $T_{brine,out} = 100°C$, $CO_2$ compression to 150 bar.

## CONCLUSIONS

Well spacing of 4 276 ft in 30 years of operation is satisfactory to provide a constant brine temperature over the lifetime of the wells.

The reservoir simulation studies demonstrated that enough thermal energy can be extracted to regenerate the amine solvent for $CO_2$ capture. The reservoir models followed the proposed analytical method by Gringarten and Sauty (1975) to ensure that the simulations capture the full physics of the temperature front movement in a geothermal reservoir.

The advanced 2-stage flash with 8 m PZ demonstrated the best performance. The design case was selected to be the advanced 2-stage flash using 8 m PZ, treating flue gas generated by the production of 60 $MW_e$. A conservative estimate of the brine extraction temperature was 148°C, allowing for heat loss during the transportation from underground reservoir to well head along the wellbore.

Assuming a rich loading of 0.40, the heating requirement was 41.2 MW, and the overall equivalent work was 11.1 $MW_e$ or 35.5 kJ/mol $CO_2$. Of the overall equivalent work, the total contribution from heating was 6.5 $MW_e$ or 20.6 kJ/mol $CO_2$. The balance of the total work, 4.6 $MW_e$, was electricity directly drawn for pump work and $CO_2$ compression to 150 bar. This electricity would be drawn directly from the generation of the turbines, as in any proposed post-combustion carbon capture with amines. The energy requirement is 16% greater than optimized flowsheets using higher temperature steam from the coal-fired power plant. However, this flowsheet would avoid disrupting the steam cycle to draw heat for the solvent regeneration. The process

integration and control of this heating option, therefore, could be very beneficial.

In the pilot-scale study, four wells would be required to provide the geothermal heat to regenerate the solvent and save 11.1 $MW_e$ that would otherwise be required as steam heat from the power cycle. At $5 million/well, the estimated capital cost of saving 11.1 $MW_e$ would be about $1 800/kW. This is comparable to the capital cost of replacing coal-fired power capacity with new plants. The extracted thermal energy could be increased to full-scale with a similarly-scaled increase in the number of wells and land area used by wells.

## ACKNOWLEDGMENTS

This work was supported by the Luminant Carbon Management Program and by the Geologic $CO_2$ Storage Industrial Affiliates Program at the Center for Petroleum and Geosystems Engineering, both at the University of Texas at Austin.

## REFERENCES

Bebout D.G., Weise B.R., Gregory A.R., Edwards M.B. (1979) *Wilcox sandstone reservoirs in the deep surface along the Texas Gulf coast, their potential for production of geopressured geothermal energy*, Bureau of Economic Geology, The University of Texas at Austin, October.

Dietz D.N. (1965) Determination of average reservoir pressure from build-up surveys, *J. Petrol. Technol.* **17**, 8, 955-959, SPE-1156-PA.

Dugas R.E. (2009) Carbon Dioxide Absorption, Desorption, and Diffusion in Aqueous Piperazine and Monoethanolamine, *PhD Thesis*, University of Texas at Austin.

Ermatchkov V., Kamps A.P.-S., Speyer D., Maurer G. (2006) Solubility of Carbon Dioxide in Aqueous Solutions of Piperazine in the Low Gas Loading Region, *J. Chem. Eng. Data* **51**, 5, 1788-1796.

Fisher K.S., Beutkerm C., Rueter C., Searcy K. (2005) *Integrating MEA Regeneration with CO₂ Compression and Peaking to Reduce CO₂ Capture Costs*, Final DOE Report.

Freeman S.A., Dugas R.E., Van Wagener D.H., Nguyen T., Rochelle G.T. (2010) Carbon dioxide capture with concentrated, aqueous piperazine, *Int. J. Greenhouse Gas Control* **4**, 2, 119-124.

Gringarten A.C., Sauty J.P. (1975) A Theoretical Study of Heat Extraction from Aquifers with Uniform Regional Flow, *J. Geophys. Res.* **80**, 35, 49-56.

Gringarten A.C. (1977) Reservoir Lifetime and Heat Recovery Factor in Geothermal Aquifers used for Urban Heating, *Pure Appl. Geophys.* **117**, 1-2, 297-308.

Hilliard M.D. (2008) A Predictive Thermodynamic Model for an Aqueous Blend of Potassium Carbonate, Piperazine and Monoethanolamine for Carbon Dioxide Capture from Flue Gas, *PhD Thesis*, The University of Texas at Austin.

Matthews C.S., Russell D.G. (1967) Pressure buildup and flow tests in wells, SPE Monograph Series **1**, ISBN: 978-0-89520-200-0.

Nguyen T., Hilliard M.D., Rochelle G.T. (2011) Volatility of aqueous amines in $CO_2$ capture, *Energy Procedia* **4**, 1624-1630.

Reddy S., Gilmartin J., Francuz V. (2007) *Integrated Compressor/Stripper Configurations and Methods*, International Patent WO 2007075466.

Rochelle G.T. (2009) Amine Scrubbing for $CO_2$ Capture, *Science* **325**, 5948, 1652-1654.

Stopek D., Scherffius J., Klumpyan J., Smith R., Reddy S. (2011) *Carbon Capture Demonstration Project at WA Parish Station*, Electric Power 2011, Rosemont, Illinois.

Van Wagener D.H. (2011) Stripper Modeling for $CO_2$ Removal Using Monoethanolamine and Piperazine Solvents, *PhD Thesis*, The University of Texas at Austin.

Van Wagener D.H., Rochelle G.T. (2010) Stripper Configurations for $CO_2$ Capture by Aqueous Monoethanolamine, *Chem. Eng. Res. Des.* **89**, 9, 1639-1646, doi: 10.1016/j.cherd.2010.11.011.

Vinsome P.K.W., Westerveld J. (1980) A simple method for predicting cap and base rock heat losses in thermal reservoir simulators, *J. Can Petrol Technol.* **19**, 87-90.

Xu Q., Rochelle G.T. (2011) Total Pressure and $CO_2$ Solubility at High Temperature in Aqueous Amines, *Energy Procedia* **4**, 117-124.

# Permissions

The contributors of this book come from diverse backgrounds, making this book a truly international effort. This book will bring forth new frontiers with its revolutionizing research information and detailed analysis of the nascent developments around the world.

We would like to thank all the contributing authors for lending their expertise to make the book truly unique. They have played a crucial role in the development of this book. Without their invaluable contributions this book wouldn't have been possible. They have made vital efforts to compile up to date information on the varied aspects of this subject to make this book a valuable addition to the collection of many professionals and students.

This book was conceptualized with the vision of imparting up-to-date information and advanced data in this field. To ensure the same, a matchless editorial board was set up. Every individual on the board went through rigorous rounds of assessment to prove their worth. After which they invested a large part of their time researching and compiling the most relevant data for our readers.

The editorial board has been involved in producing this book since its inception. They have spent rigorous hours researching and exploring the diverse topics which have resulted in the successful publishing of this book. They have passed on their knowledge of decades through this book. To expedite this challenging task, the publisher supported the team at every step. A small team of assistant editors was also appointed to further simplify the editing procedure and attain best results for the readers.

Apart from the editorial board, the designing team has also invested a significant amount of their time in understanding the subject and creating the most relevant covers. They scrutinized every image to scout for the most suitable representation of the subject and create an appropriate cover for the book.

The publishing team has been an ardent support to the editorial, designing and production team. Their endless efforts to recruit the best for this project, has resulted in the accomplishment of this book. They are a veteran in the field of academics and their pool of knowledge is as vast as their experience in printing. Their expertise and guidance has proved useful at every step. Their uncompromising quality standards have made this book an exceptional effort. Their encouragement from time to time has been an inspiration for everyone.

The publisher and the editorial board hope that this book will prove to be a valuable piece of knowledge for researchers, students, practitioners and scholars across the globe.

# List of Contributors

**Stephanie A. Freeman**
Department of Chemical Engineering, The University of Texas at Austin, 1 University Station C0400, Austin, TX 78712 - USA

**Xi Chen**
Department of Chemical Engineering, The University of Texas at Austin, 1 University Station C0400, Austin, TX 78712 - USA

**Thu Nguyen**
Department of Chemical Engineering, The University of Texas at Austin, 1 University Station C0400, Austin, TX 78712 - USA

**Humera Rafique**
Department of Chemical Engineering, The University of Texas at Austin, 1 University Station C0400, Austin, TX 78712 - USA

**Qing Xu**
Department of Chemical Engineering, The University of Texas at Austin, 1 University Station C0400, Austin, TX 78712 - USA

**Gary T. Rochelle**
Department of Chemical Engineering, The University of Texas at Austin, 1 University Station C0400, Austin, TX 78712 - USA

**S. M. Ghaderi**
Department of Chemical and Petroleum Engineering, University of Calgary, 2500 University Drive N.W., Calgary, Alberta - Canada

**C. R. Clarkson**
Department of Chemical and Petroleum Engineering, University of Calgary, 2500 University Drive N.W., Calgary, Alberta - Canada

**D. Kaviani**
Department of Chemical and Petroleum Engineering, University of Calgary, 2500 University Drive N.W., Calgary, Alberta - Canada

**Ismaël Aouini**
Normandie Université, INSA de Rouen, LSPC EA4704, Avenue de l'Université, BP 8, 76801 Saint-Étienne-du-Rouvray Cedex - France

**Alain Ledoux**
Normandie Université, INSA de Rouen, LSPC EA4704, Avenue de l'Université, BP 8, 76801 Saint-Étienne-du-Rouvray Cedex - France

**Lionel Estel**
Normandie Université, INSA de Rouen, LSPC EA4704, Avenue de l'Université, BP 8, 76801 Saint-Étienne-du-Rouvray Cedex - France

**Soazic Mary**
Veolia Environnement Recherche et Innovation, 10 rue Jacques Daguerre, 92500 Rueil-Malmaison - France

**E. Chabanon**
Laboratoire Réactions et Génie des Procédés, LRGP (UPR CNRS 3349), 1 rue Grandville, 54000 Nancy - France

**E. Kimball**
TNO, Leeghwaterstraat 46, 2628 CA Delft - The Netherlands

**E. Favre**
Laboratoire Réactions et Génie des Procédés, LRGP (UPR CNRS 3349), 1 rue Grandville, 54000 Nancy - France

**O. Lorain**
Polymem, impasse du Palayré, 31100 Toulouse - France

**E. Goetheer**
TNO, Leeghwaterstraat 46, 2628 CA Delft - The Netherlands

**D. Ferre**
IFP Energies nouvelles-Lyon, Rond-point de l'échangeur de Solaize, BP 3, 69360 Solaize France

**A. Gomez**
IFP Energies nouvelles-Lyon, Rond-point de l'échangeur de Solaize, BP 3, 69360 Solaize France

**P. Broutin**
IFP Energies nouvelles-Lyon, Rond-point de l'échangeur de Solaize, BP 3, 69360 Solaize France

**Nan Yang**
School of Chemical & Environmental Engineering, China University of Mining & Technology (Beijing), Beijing 100083 - P.R. China
CSIRO Energy Technology, 10 Murray Dwyer Circuit, Mayfield West, NSW 2304 - Australia

**Hai Yu**
CSIRO Energy Technology, 10 Murray Dwyer Circuit, Mayfield West, NSW 2304 - Australia

**Lichun Li**
CSIRO Energy Technology, 10 Murray Dwyer Circuit, Mayfield West, NSW 2304 - Australia
Institute of Catalysis, Zhejiang University of Technology, Hangzhou, Zhejiang 310014 - P.R. China

**Dongyao Xu**
School of Chemical & Environmental Engineering, China University of Mining & Technology (Beijing), Beijing 100083 - P.R. China

**Wenfeng Han**
Institute of Catalysis, Zhejiang University of Technology, Hangzhou, Zhejiang 310014 - P.R. China

**Paul Feron**
CSIRO Energy Technology, 10 Murray Dwyer Circuit, Mayfield West, NSW 2304 - Australia

**Eric Lemaire**
IFP Energies nouvelles, Rond-point de l'échangeur, BP 3, 69360 Solaize, France

**Pierre Antoine Bouillon**
IFP Energies nouvelles, Rond-point de l'échangeur, BP 3, 69360 Solaize, France

**Kader Lettat**
IFP Energies nouvelles, Rond-point de l'échangeur, BP 3, 69360 Solaize, France

**H. Djizanne**
LMS, École Polytechnique ParisTech, École des Mines de Paris, École des Ponts et Chaussées,
UMR 7649 CNRS, route de Saclay, 91120 Palaiseau - France

**P. Bérest**
LMS, École Polytechnique ParisTech, École des Mines de Paris, École des Ponts et Chaussées,
UMR 7649 CNRS, route de Saclay, 91120 Palaiseau - France

**B. Brouard**
Brouard Consulting, 101 rue du temple, 75003 Paris - France

**A. Frangi**
Politecnico di Milano, Piazza Leonardo da Vinci, 32, 20133 Milano - Italy

**Erin Kimball**
TNO, Leeghwaterstraat 46, 2628 CA Delft - The Netherlands

**Adam Al-Azki**
E-ON New Build & Technology Ltd, Ratcliffe on Soar Nottingham, NG11 0EE - United Kingdom

**Adrien Gomez**
IFP Energies nouvelles, Rond-point de l'échangeur de Solaize, BP 3, 69360 Solaize - France

**Earl Goetheer**
TNO, Leeghwaterstraat 46, 2628 CA Delft - The Netherlands

**Nick Booth**
E-ON New Build & Technology Ltd, Ratcliffe on Soar Nottingham, NG11 0EE - United Kingdom

**Dick Adams**
E-ON New Build & Technology Ltd, Ratcliffe on Soar Nottingham, NG11 0EE - United Kingdom

**Daniel Ferre**
IFP Energies nouvelles, Rond-point de l'échangeur de Solaize, BP 3, 69360 Solaize - France

**J.-M. Herri**
Centre SPIN, Département PROPICE, UMR CNRS 5307, École Nationale Supérieure des Mines de Saint-Étienne, 158 Cours Fauriel, 42023 Saint-Étienne - France

**A. Bouchemoua**
Centre SPIN, Département PROPICE, UMR CNRS 5307, École Nationale Supérieure des Mines de Saint-Étienne, 158 Cours Fauriel, 42023 Saint-Étienne - France

**M. Kwaterski**
Centre SPIN, Département PROPICE, UMR CNRS 5307, École Nationale Supérieure des Mines de Saint-Étienne, 158 Cours Fauriel, 42023 Saint-Étienne - France

**P. Brântuas**
Centre SPIN, Département PROPICE, UMR CNRS 5307, École Nationale Supérieure des Mines de Saint-Étienne, 158 Cours Fauriel, 42023 Saint-Étienne - France

**A. Galfré**
Centre SPIN, Département PROPICE, UMR CNRS 5307, École Nationale Supérieure des Mines de Saint-Étienne, 158 Cours Fauriel, 42023 Saint-Étienne - France

**B. Bouillot**
Centre SPIN, Département PROPICE, UMR CNRS 5307, École Nationale Supérieure des Mines de Saint-Étienne, 158 Cours Fauriel, 42023 Saint-Étienne - France

**J. Douzet**
Centre SPIN, Département PROPICE, UMR CNRS 5307, École Nationale Supérieure des Mines de Saint-Étienne, 158 Cours Fauriel, 42023 Saint-Étienne - France

**Y. Ouabbas**
Centre SPIN, Département PROPICE, UMR CNRS 5307, École Nationale Supérieure des Mines de Saint-Étienne, 158 Cours Fauriel, 42023 Saint-Étienne - France

**A. Cameirao**
Centre SPIN, Département PROPICE, UMR CNRS 5307, École Nationale Supérieure des Mines de Saint-Étienne, 158 Cours Fauriel, 42023 Saint-Étienne - France

**Hanna Knuutila**
Norwegian University of Science and Technology (NTNU), Department of Chemical Engineering, Sem Saelands vei 4, Trondheim NO 7491 - Norway

**Naveed Asif**
Norwegian University of Science and Technology (NTNU), Department of Chemical Engineering, Sem Saelands vei 4, Trondheim NO 7491 - Norway

**Solrun Johanne Vevelstad**
Norwegian University of Science and Technology (NTNU), Department of Chemical Engineering, Sem Saelands vei 4, Trondheim NO 7491 - Norway

**Hallvard F. Svendsen**
Norwegian University of Science and Technology (NTNU), Department of Chemical Engineering, Sem Saelands vei 4, Trondheim NO 7491 - Norway

**G. D. Pirngruber**
IFP Energies nouvelles, Rond-point de l'échangeur de Solaize, BP 3, 69360 Solaize - France

**V. Carlier**
IFP Energies nouvelles, Rond-point de l'échangeur de Solaize, BP 3, 69360 Solaize - France

**D. Leinekugel-le-Cocq**
IFP Energies nouvelles, Rond-point de l'échangeur de Solaize, BP 3, 69360 Solaize - France

**A. Gomez**
IFP Energies nouvelles, Rond-point de l'échangeur de Solaize, BP 3, 69360 Solaize - France

**P. Briot**
IFP Energies nouvelles, Rond-point de l'échangeur de Solaize, BP 3, 69360 Solaize - France

**L. Raynal**
IFP Energies nouvelles, Rond-point de l'échangeur de Solaize, BP 3, 69360 Solaize - France

**P. Broutin**
IFP Energies nouvelles, Rond-point de l'échangeur de Solaize, BP 3, 69360 Solaize - France

**M. Gimenez**
Lafarge, IPC, 95 rue du Montmurier, BP 70, 38291 Saint-Quentin-Fallavier Cedex - France

**M. Soazic**
VEOLIA Environnement Recherche et Innovation, 291 avenue Dreyfous Ducas, Zone Portuaire de Limay, 78520 Limay - France

**P. Cessat**
VEOLIA Environnement Recherche et Innovation, 291 avenue Dreyfous Ducas, Zone Portuaire de Limay, 78520 Limay - France

**S. Saysset**
GDF SUEZ, CRIGEN, 361 avenue du Président Wilson, BP 33, 93211 Saint-Denis-La-Plaine Cedex - France

**Zhicheng Xu**
Key Laboratory for Thermal Science and Power Engineering of Ministry of Education, Beijing Key Laboratory for CO2 Utilization and Reduction Technology, Department of Thermal Engineering, Tsinghua University, Beijing 100084 - China

**Shujuan Wang**
Key Laboratory for Thermal Science and Power Engineering of Ministry of Education, Beijing Key Laboratory for CO2 Utilization and Reduction Technology, Department of Thermal Engineering, Tsinghua University, Beijing 100084 - China

**Guojie Qi**
Key Laboratory for Thermal Science and Power Engineering of Ministry of Education, Beijing Key Laboratory for CO2 Utilization and Reduction Technology, Department of Thermal Engineering, Tsinghua University, Beijing 100084 - China

**Jinzhao Liu**
Key Laboratory for Thermal Science and Power Engineering of Ministry of Education, Beijing Key Laboratory for CO2 Utilization and Reduction Technology, Department of Thermal Engineering, Tsinghua University, Beijing 100084 - China

**Bo Zhao**
Key Laboratory for Thermal Science and Power Engineering of Ministry of Education, Beijing Key Laboratory for CO2 Utilization and Reduction Technology, Department of Thermal Engineering, Tsinghua University, Beijing 100084 - China

**Changhe Chen**
Key Laboratory for Thermal Science and Power Engineering of Ministry of Education, Beijing Key Laboratory for CO2 Utilization and Reduction Technology, Department of Thermal Engineering, Tsinghua University, Beijing 100084 - China

**D. H. Van Wagener**
Department of Chemical Engineering – C0400, The University of Texas at Austin, Austin, TX 78712 - USA

**A. Gupta**
Department of Petroleum and Geochemical Engineering – C0304, The University of Texas at Austin, Austin, TX 78712 - USA

**G.T. Rochelle**
Department of Chemical Engineering – C0400, The University of Texas at Austin, Austin, TX 78712 - USA

**S. L. Bryant**
Department of Petroleum and Geochemical Engineering – C0304, The University of Texas at Austin, Austin, TX 78712 - USA